Discrete Mathematics By Example

Discrete Mathematics By Example

Andrew Simpson
Oxford Brookes University

The McGRAW-HILL COMPANIES

London · Burr Ridge IL · New York · St Louis · San Francisco · Auckland
Bogotá · Caracas · Lisbon · Madrid · Mexico · Milan
Montreal · New Delhi · Panama · Paris · San Juan · São Paulo
Singapore · Sydney · Tokyo · Toronto

Discrete Mathematics By Example
Andrew Simpson
ISBN 0 07 709840 4

 Published by McGraw-Hill Education
Shoppenhangers Road
Maidenhead
SL6 2QL
Telephone: +44(0)1628 502500
Fax: 44(0)1628 770224
Web site: http://www.mcgraw-hill.co.uk

British Library Cataloguing in Publication Data
A catalogue record for this book is available from the British Library

Library of Congress Cataloging in Publication Data
Library of Congress data for this book is available from the Library of Congress,
Washington, D.C.

Web site address: http://www.mcgraw-hill.co.uk/textbooks/simpson

Senior Production Manager: Max Elvey
New Media Developer: Doug Greenwood

Produced for McGraw-Hill by Steven Gardiner Ltd
Cover Design by: Design Delux
Printed and bound in Great Britain by Bell and Bain Ltd, Glasgow

Contents

1	**Introduction**	**1**
	1.1　Motivation	1
	1.2　Material	2
	1.3　Organisation	3
2	**Numbers**	**5**
	2.1　The natural numbers	5
	2.2　Peano arithmetic	9
	2.3　Other classes of numbers	11
	2.4　Further exercises	12
	2.5　Solutions	12
3	**Propositional logic**	**17**
	3.1　Atomic propositions	17
	3.2　Truth values	18
	3.3　Negation	19
	3.4　Conjunction	21
	3.5　Disjunction	24
	3.6　Implication	28
	3.7　Equivalence	30
	3.8　Precedence	32
	3.9　Tautologies, contradictions, and contingencies	33
	3.10　Truth tables	34
	3.11　Equational reasoning	35
	3.12　Natural deduction	37
	3.13　Further exercises	43
	3.14　Solutions	46
4	**Set theory**	**61**
	4.1　Sets	61
	4.2　Singleton sets	63
	4.3　The empty set	64

4.4	Set membership	64
4.5	Subsets	65
4.6	Supersets	70
4.7	Set union	71
4.8	Set intersection	73
4.9	Set difference	76
4.10	Reasoning about sets	80
4.11	Cardinality	81
4.12	Finite and infinite sets	82
4.13	Power sets	83
4.14	Generalised operations	85
4.15	Further exercises	87
4.16	Solutions	92
5	**Boolean algebra**	**103**
5.1	Introduction	103
5.2	Propositional logic revisited	104
5.3	Set theory revisited	104
5.4	Fundamentals	105
5.5	Conventions	109
5.6	Precedence	109
5.7	The Boolean algebra of sets	109
5.8	The Boolean algebra of propositions	110
5.9	Isomorphic Boolean algebras	110
5.10	Duality	111
5.11	Further exercises	112
5.12	Solutions	114
6	**Typed set theory**	**123**
6.1	The need for types	124
6.2	The empty set revisited	127
6.3	Set comprehension	127
6.4	Characteristic tuples	131
6.5	Abbreviations	132
6.6	Cartesian products	133
6.7	Axiomatic definitions	139
6.8	Further exercises	141
6.9	Solutions	144
7	**Predicate Logic**	**155**
7.1	The need for quantification	155
7.2	Universal quantification	155
7.3	Existential quantification	158
7.4	Satisfaction and validity	161
7.5	The negation of quantifiers	161
7.6	Free and bound variables	163

7.7	Substitution	165
7.8	Restriction	169
7.9	Uniqueness	172
7.10	Equational reasoning	175
7.11	Natural deduction	178
7.12	The one-point rule	182
7.13	Further exercises	183
7.14	Solutions	186

8 Relations **199**

8.1	Binary relations	199
8.2	Reasoning about relations	201
8.3	Domain and range	201
8.4	Relational inverse	203
8.5	Operations on relations	205
8.6	Relational composition	210
8.7	Homogeneous and heterogeneous relations	214
8.8	Properties of relations	215
8.9	Orderings and equivalences	223
8.10	Closures	224
8.11	n-ary relations	229
8.12	Further exercises	230
8.13	Solutions	234

9 Functions **251**

9.1	A special kind of relation	251
9.2	Total functions	253
9.3	Function application	255
9.4	Overriding	257
9.5	Properties of functions	259
9.6	Recursively defined functions	263
9.7	Further exercises	266
9.8	Solutions	269

10 Sequences **277**

10.1	Bags	277
10.2	A need for order	279
10.3	Modelling sequences	280
10.4	The empty sequence	282
10.5	Length	282
10.6	Concatenation	283
10.7	Head and tail	285
10.8	Restrictions	287
10.9	Reversing	290
10.10	Injective sequences	292
10.11	Recursively defined functions revisited	293

10.12	Further exercises	294
10.13	Solutions	298

11 Induction **307**
11.1	Mathematical induction	307
11.2	Structural induction	311
11.3	Further exercises	313
11.4	Solutions	314

12 Graph theory **321**
12.1	Graphs	321
12.2	Representing graphs as sets and bags	329
12.3	Representing graphs as matrices	331
12.4	Isomorphic graphs	335
12.5	Paths	336
12.6	Cycles	341
12.7	Trees	342
12.8	Weighted graphs	344
12.9	Directed graphs	348
12.10	Binary trees	350
12.11	Further exercises	352
12.12	Solutions	354

13 Combinatorics **369**
13.1	The factorial function	369
13.2	Binomial coefficients	371
13.3	Counting	375
13.4	Permutations	377
13.5	Combinations	379
13.6	Tree diagrams	381
13.7	Sampling	382
13.8	Further exercises	385
13.9	Solutions	386

14 Examples **401**
14.1	Modelling program variables	401
14.2	Meta search engines	403
14.3	Sequences for stacks and queues	410
14.4	Digital circuits	414
14.5	A school database	417
14.6	Knowledge-based systems	422
14.7	Solutions	424

Bibliography **443**

Index **445**

For Becky, Mam, and Ian

Chapter 1

Introduction

1.1 Motivation

Discrete mathematics, for many students of computer science and related disciplines, can be a very difficult subject to learn. Furthermore, it can often be a very difficult subject to teach. One of the main reasons for this difficulty is that, for many students—even those with a reasonable mathematical background—learning discrete mathematics often involves having to think in a completely new way. A second, and somewhat more contentious, reason for this difficulty is that some of today's undergraduate computing students do not appear to have been provided with the mathematical training prior to embarking upon their university careers that their predecessors enjoyed. A third, and more fundamental, reason for this difficulty is that some of the techniques taught as part of discrete mathematics courses are relatively complex.

I have taught discrete mathematics at four very different institutions (the University of Oxford, Oxford Brookes University, the University of North London, and Ruskin College, Oxford), to undergraduate, post-graduate and pre-university students, some of whom were full-time students and some of whom were part-time students. No matter the level or type of student, two familiar cries are often emitted from those studying discrete mathematics.

The first of these is

"I think I understand this technique, but I need some more examples and exercises to make sure."

In the author's opinion, discrete mathematics is a subject in which many students learn best by example. Discrete mathematics is, in many ways, a 'contact sport', and, as such, to be proficient at it, students need a great deal of practice. There is by no means a shortage of texts on the subject of discrete mathematics. However, with some notable exceptions, most discrete mathematics texts fail to provide an extensive collection of exercises or examples to support those students who find the material difficult to come to terms with.

The second of these familiar cries is

"What has this got to do with computing?"

Many discrete mathematics texts aimed at computing students fail to relate what is being taught to the world of computing. Despite being academically excellent, many textbooks in this area often fail to motivate their readers by relating the techniques described to computing applications in a sufficient manner.

Thus, the aims of this book are two-fold.

- First, to provide an accessible, example- and exercise-led introduction to the subject of discrete mathematics (Chapters 2 to 13).

- Second, to relate the theoretical material introduced in Chapters 2 to 13 to a number of appropriate examples (Chapter 14).

I have decided to include solutions to all exercises at the end of each chapter. This decision has been taken to ensure that this text can be used as a study aid to support students as much as possible. I sincerely hope that readers do not abuse this privilege by consulting the solutions prematurely!

1.2   Material

The choice of material featured in this book has been influenced by the fact that—in the author's opinion—discrete mathematics fulfils two purposes in computer science. The first is to underpin concepts associated with theoretical computer science. Most 'traditional' discrete mathematics courses teach the subject for this purpose. The second role that discrete mathematics plays within computer science is to form the mathematical basis for formal description techniques (mathematical notations for describing and verifying properties of computer systems). Some discrete mathematics courses—especially those taught at some of the United Kingdom's newer universities—introduce discrete mathematics specifically for this purpose. This influence can be detected in Chapter 3, in which the natural deduction system presented is that of the Z formal description technique (see, for example, [WD96]). It is the author's intention that this book should be appropriate for students of both types of discrete mathematics course. As such, those readers expecting a traditional treatment of discrete mathematics may be surprised to find a chapter on sequences. This may be skipped by such readers, as may Chapter 14, in which a number of case studies are presented. Furthermore, those readers expecting an introduction to typed set theory and predicate logic for formal methods may be surprised to come across chapters on subjects such as Boolean algebra, graph theory and combinatorics. Again, such readers may feel free to avoid these chapters.

The choice of references associated with this text are influenced by two factors. The first is that the very nature of this text means that some topics are not covered in sufficient depth for some courses or readers; in such cases, pointers are given to appropriate texts for further study or for further exercises. This is most apparent in Chapters 12 and 13, in which a lack of space has meant that two complex topics—graph theory and combinatorics—have had to be covered at a pace and at a depth that might leave some readers thirsting for more detailed material. In both cases, suitable complementary material is suggested. Second, in recent years there has been a rise in popularity of 'popular science' texts. For example, biographies of famous scientists, philosophers and mathematicians, and histories of mathematical theorems (and even numbers) have all made the popular best seller lists recently. In this vein, interested readers are pointed—at appropriate times in the text—to sources that describe the lives and work of some of the 'key players' in the evolution of the various topics of discrete mathematics described in this book.[1]

---

[1] In this respect, [Gri99] is an excellent source of such references.

1.3 Organisation

This book is organised as follows.

In Chapter 2 we provide a brief introduction to the fundamental notion of numbers. In particular, we concern ourselves with the natural numbers and Peano arithmetic. The representation of natural numbers via Peano arithmetic is used to introduce the central notions of proof and recursively defined functions in a relatively familiar setting before we return to these topics in later chapters.

In Chapter 3 we concern ourselves with propositional logic: atomic propositions, and the propositional logic operators are presented. A number of methods for establishing the truth or falsity of a given proposition are introduced, namely, substitution, truth tables, proof trees, and equational reasoning. Although proof via proof trees (or natural deduction) is a generic method of establishing theorems of propositional logic, the rules which may be applied are often specific to a particular system. The system which we have chosen is that associated with the formal description technique, Z.

In Chapter 4 we look at the topic of (untyped) set theory, and introduce Venn diagrams, together with the basic set operators.

Chapter 5 is concerned with Boolean algebra, of which propositional logic and set theory are both examples. We start by presenting the abstract notion of a Boolean algebra, before showing how this relates to the structures of propositional logic and set theory.

Chapter 6 returns to the subject of set theory, explains the drawbacks of untyped set theory, and introduces the notion of typed set theory. Having established what typed set theory is, we go on to consider further set theoretic notions such as set comprehension and Cartesian products, which together provide the foundations for Chapters 8, 9 and 10.

Chapter 7 builds on Chapters 2 and 6, and introduces the topic of predicate logic. The existential and universal quantifiers are introduced, and then we go on to discuss free and bound variables and substitution.

Having introduced Cartesian products in Chapter 6, we are in a position—in Chapter 8—to consider relations. We introduce the basic concept of a relation, before considering operations on, and properties of, relations.

Chapter 9 describes the concept of a function, which is, essentially, a special type of relation. As was the case with relations, we consider operations on, and properties of, functions. In addition, Chapter 9 introduces the more advanced topic of recursively defined functions.

Chapter 10 is concerned with sequences. We describe the need for sequences, how they can be represented in terms of functions, and then introduce a number of operators on sequences.

Chapter 11 describes how we might prove properties of formally defined structures using the powerful techniques of mathematical induction and structural induction. We use the former to prove properties of the natural numbers, whereas we use the latter to prove properties of structures such as sequences.

Chapter 12 introduces the topic of graph theory. The basic terminology of graph theory is presented, together with a description of how graphs may be represented using both matrices and set theory. We then describe some special types of graph which play an important role in computing, namely, trees, weighted graphs, and binary trees.

Chapter 13 discusses the subject of combinatorics, which is, essentially, concerned with determining the number of possible outcomes for a series of events. We start by revisiting mathematical concepts such as the factorial function, before moving on to consider the topics of permutations and combinations. Finally, we show how a set of possible outcomes may be represented visually using tree diagrams.

Finally, Chapter 14 illustrates how the techniques described in the preceding chapters can be used to model and analyse properties of different systems, and also indicates some of the ways in which the topics presented relate to computer science. The examples presented are of various sizes and complexity, and attempt to motivate the preceding material by placing it in a relevant computing context. We start by considering how functions may be used to model a small programming language. Then we concern ourselves with World Wide Web search engines and how we might model such entities. Next, we look at how the computing concepts of stacks and queues relate to sequences. The fourth section looks at how the concept of Boolean algebra and the techniques introduced in Chapter 5 are applicable to digital circuit design. Next, we give an example of a formal specification by showing how predicate logic and typed set theory can be used to specify a simple database. Finally, we see how predicate logic can be used in the context of knowledge based systems.

Acknowledgments

Thanks are due to a number of people, without whom this book would never have been written. First and foremost, I would like to thank my publishers at McGraw-Hill for guiding this project to a relatively smooth conclusion. In particular, I would like to thank David Hatter for his initial enthusiasm for the project, and Sarah Douglas and Conor Graham for their patience and professionalism in the final stages. Second, I owe a debt of gratitude to all of the students to whom I have taught discrete mathematics over the past few years, and especially those who have indicated a need for a text such as this.

Invaluable discussions with some of my colleagues at Oxford Brookes have helped shape this book. Many constructive comments were received at appropriate times from David Duce. In addition, more general comments from Ken Brownsey, Nigel Crook, Barry Holmes and David Lightfoot have also helped. Furthermore, discussions over a long period of time with Jim Davies, Andrew Martin and Jim Woodcock of the University of Oxford, and Steve King of the University of York have also helped shape both the structure, content, and—in Jim Davies' case—the appearance of this book. Paul Goddard made some insightful comments, as did Jon Whiteley, who was incredibly thorough in his proof-reading. Duncan Brydon, Jon Hill, Rick McPhee, and Andrew Spencer have also contributed to the cause.

I would like to thank the reviewers for their positive response to the proposal for this book, as well as for their constructive and, in some cases, very kind, comments. I am grateful for the text reviewers for being so thorough. In particular, I am appreciative of Clive Mingham's efforts, which were far greater than any author should reasonably expect.

It should go without saying that all errors, inaccuracies and inconsistencies remain totally and solely my responsibility.

Finally, I would like to thank Becky for putting up with me, and providing a great deal of moral support during the writing of this book.

Andrew Simpson
Oxford
October, 2001

Chapter 2

Numbers

In this chapter we introduce some basic mathematical concepts that we shall rely on throughout the remainder of this book. We start by introducing the natural numbers. We then consider a formal representation of the natural numbers, called Peano arithmetic, which we use to introduce the notion of proof. Finally, we introduce two further classes of numbers: the integers and the real numbers.

2.1 The natural numbers

To many, the building blocks of discrete mathematics are set theory and propositional logic, both of which we shall meet in due course. Before studying these topics, however, there is an even more fundamental concept which we should discuss: that of the *natural numbers*.

The natural numbers are the non-negative whole numbers starting from 0, that is, the numbers $0, 1, 2, 3$, etc. We denote this collection of numbers by $\mathbb{N}$.[1]

Example 2.1 Numbers such as 57 appear in $\mathbb{N}$, but numbers such as -57, 5.7 and -5.7 do not. □

Exercise 2.1 Which of the following are natural numbers?

1. 0

2. -1

3. 1.2

□

Two operations on the natural numbers which might be unfamiliar to some readers, but will be essential in the following, are **div** and **mod**. These operations are, essentially, the division and remainder operations taught at secondary (or high) school: m **div** n returns the whole number result of dividing one natural number, m, by another, n (provided that n is not 0), while m **mod** n returns the remainder of that division.

---

[1] Note that some discrete mathematics textbooks do not include 0 in the natural numbers; we choose to do so in this book.

Example 2.2

$$7 \text{ div } 3 = 2$$
$$7 \text{ mod } 3 = 1$$

$\square$

Exercise 2.2 Calculate the following.

1. 10 div 10

2. 10 div 1

3. 10 div 3

4. 10 mod 10

5. 10 mod 1

6. 10 mod 3

$\square$

Two further operations on the natural numbers that we shall make use of in the following chapters are the *sum* and *product* operators, written $\sum$ and $\prod$ respectively. The former acts as a generalised addition operator, whereas the latter acts as a generalised multiplication operator.

Example 2.3 The sum operator can be used to denote the summation of the numbers one to five as follows.

$$\sum_{i=1}^{5} i$$

Here, the subscript $i = 1$ indicates that the variable i starts with the value 1: the overall statement then enumerates all possible values of i up to 5—the value provided by the superscript—and sums all enumerated values. As such,

$$\sum_{i=1}^{5} i = 1 + 2 + 3 + 4 + 5$$
$$= 15$$

$\square$

Example 2.4 The statement $\sum_{i=5}^{10} (i^2)$ denotes the summation of all squares of those natural numbers between 5 and 10 (inclusive). Here,

$$\sum_{i=5}^{10} (i^2) = 5^2 + 6^2 + 7^2 + 8^2 + 9^2 + 10^2$$
$$= 25 + 36 + 49 + 64 + 81 + 100$$
$$= 355$$

$\square$

Exercise 2.3 Calculate the following.

1. $\sum_{i=1}^{5} (i^3)$

2. $\sum_{i=1}^{5} (i - 1)$

3. $\sum_{i=1}^{5} (i \bmod 3)$

□

The term $\sum_{i=1}^{5} (i^3)$ is concerned with summing the possible instances of i^3, with the value of i ranging from 1 to 5. However, if we were to write

$$\sum_{i=1}^{0} (i^3)$$

then this expression would be concerned with summing the possible instances of i^3, with the value of i ranging from 1 to 0. There are, of course, no such possibilities. As such, the result is 0.

The following law represents the general case.

Law 2.1   Given any natural numbers m and n and any mathematical term, t, if $m < n$ then

$$\sum_{i=n}^{m} t = 0$$

□

Example 2.5

$$\sum_{i=1}^{0} (i^3) = 0$$

□

Example 2.6

$$\sum_{i=10}^{3} (i - 1) = 0$$

□

The syntax of the product operator, $\prod$, is the same as that of the sum operator.

Example 2.7 The statement $\prod_{i=1}^{5} i$ denotes the product—or multiplication—of all natural numbers between 1 and 5. As such,

$$\prod_{i=1}^{5} i = 1 \times 2 \times 3 \times 4 \times 5$$
$$= 120$$

□

Example 2.8

$$\prod_{i=5}^{10} (i \times 2) = (5 \times 2) \times (6 \times 2) \times (7 \times 2) \times (8 \times 2) \times (9 \times 2) \times (10 \times 2)$$
$$= 10 \times 12 \times 14 \times 16 \times 18 \times 20$$
$$= 9,676,800$$

□

Exercise 2.4 Calculate the following.

1. $\prod_{i=1}^{5} i^3$

2. $\prod_{i=1}^{5} (i \text{ div } 2)$

3. $\prod_{i=1}^{5} (\sum_{j=i}^{i+2} j)$

□

As was the case with summing, if the upper bound for the variable of a product is lower than the lower bound for the variable, then the result of the product is 0.

Law 2.2 Given any natural numbers m and n and any mathematical term, t, if $m < n$ then

$$\prod_{i=n}^{m} t = 0$$

□

Example 2.9

$$\prod_{i=1}^{0} (i^3) = 0$$

□

Example 2.10

$$\prod_{i=10}^{3} (i - 1) = 0$$

□

A further mathematical concept—one that we shall be concerned with primarily in Chapter 12—is that of the *logarithm*. In particular, we are concerned with *base 2 logarithms*.

Given any number n, we write $log_2 \, n$ to denote the *base 2 logarithm* of n. This is the number x that satisfies the equation

$$2^x = n$$

Example 2.11 We find that $log_2\, 8 = 3$ as $2^3 = 8$. In addition, $log_2\, 32 = 5$ as $2^5 = 32$ and $log_2\, 64 = 6$ as $2^6 = 64$. □

More often than not, the value of $log_2\, n$ for some natural number n is not a natural number. For example, we may conclude that $log_2\, 3$ is some number between 1 and 2, as $log_2\, 2 = 1$ and $log_2\, 4 = 2$. However, it is clear that $log_2\, 3$ is not a natural number. In such circumstances, we may 'round up' or 'round down' the result. As such, if we write $\lfloor log_2\, 3 \rfloor$, we mean the nearest whole number which is less than $log_2\, 3$. Similarly, if we write $\lceil log_2\, 3 \rceil$, we mean the nearest whole number which is greater than $log_2\, 3$. As such, $\lfloor log_2\, 3 \rfloor = 1$ and $\lceil log_2\, 3 \rceil = 2$.

Exercise 2.5 Calculate the following.

1. $log_2\, 1$
2. $log_2\, 2$
3. $log_2\, 128$
4. $\lceil log_2\, 50 \rceil$
5. $\lfloor log_2\, 50 \rfloor$
6. $\lceil log_2\, 70 \rceil$
7. $\lfloor log_2\, 70 \rfloor$

□

Because most readers will be familiar with the notion of natural numbers, it is through these that we motivate the notion of *proof*—which we do in the following section.

2.2 Peano arithmetic

Peano arithmetic[2] consists of a small collection of rules that capture the fundamental attributes of the natural numbers. Four of the rules, or axioms, of the system are as follows (we shall meet the fifth in Chapter 11).

1. 0 is a natural number.
2. If x is a natural number, then $x + 1$ is also a natural number.
3. There is no natural number, z, such that $z + 1 = 0$.
4. Given natural numbers x and y, if $x + 1 = y + 1$ then $x = y$.

The first two axioms help us to 'construct' the natural numbers. First, 0 is a natural number. As 0 is a natural number, axiom 2 allows us to conclude that 0+1 is also a natural number. As 0+1 is a natural number, axiom 2 allows us to conclude that 0+1+1 is also a natural number. As 0+1+1 is a natural number, axiom 2 allows us to conclude that 0+1+1+1 is also a natural number. Of course, this process can continue forever to construct every natural number.

---

[2]Named after Giuseppe Peano (1858 - 1932). The interested reader should consult [Ken80] for an account of the life and work of Peano.

The third axiom dictates that the process for constructing natural numbers really does begin at 0—there are no natural numbers less than 0.

Finally, the fourth axiom allows us to determine exactly when two natural numbers are equal to each other.

One important thing to note is that Peano arithmetic shows that the natural numbers can be defined entirely from two concepts: a unique element, 0, and an operation, $+1$.

Exercise 2.6 Using Peano's axioms, prove that $0 + 1$ is a natural number. □

Establishing the theorem of Exercise 2.6 allows us to demonstrate, using the rules of Peano arithmetic, the fact that $0 + 1$ is a natural number. Peano arithmetic is a formal treatment of the natural numbers which allows us to prove such properties; everything that follows in this book is concerned with presenting formal systems which allow us to model real-world entities, and—just as importantly—to prove properties of these formal representations.

Of course, the natural numbers would be of limited use to us if all we could do with them was to say when two natural numbers are equal. In Chapter 9 we shall meet the notion of a *function*; functions on the natural numbers include such concepts as subtraction, multiplication, and division. We can, of course, define such functions in terms of Peano arithmetic.

Example 2.12 We can define the function *subtract* on natural numbers as follows.

$$subtract\,(x + 1, 0) = x + 1$$
$$subtract\,(x + 1, y + 1) = subtract(x, y)$$

As an illustration,

$$subtract\,(0 + 1 + 1, 0) = 0 + 1 + 1$$

Note that $subtract\,(m, n)$ is only defined if m is greater than or equal to n, as there is no notion of negative numbers here. □

Exercise 2.7 Given the above definition of *subtract*, calculate

$$subtract\,(0 + 1 + 1 + 1, 0 + 1)$$

□

Example 2.13 We can define the function *add_two* on natural numbers as follows.

$$add\_two\,(0) = 0 + 1 + 1$$
$$add\_two\,(x + 1) = (x + 1) + 1 + 1$$

□

Both *subtract* and *add_two* are examples of *recursively defined functions*. We shall meet such functions formally in Chapter 9.

Exercise 2.8 Using our definition of the function *add_two*, prove that

$$add\_two\,(0 + 1) = 0 + 1 + 1 + 1$$

□

In general, our proofs will require a series of steps to establish a truth, rather than just one step (as was the case for the solution to the previous exercise). Under such circumstances, we format our proofs using a sensible, formal layout. The following is a simple example (consisting of only two steps) of such a proof.[3]

Example 2.14 The following proof establishes that

$$add\_two\,(x+1+1) = add\_two\,(x+1) + 1$$

holds for all natural numbers x.

$$
\begin{aligned}
add\_two\,(x+1+1) &= (x+1+1)+1+1 \quad &&[\text{Definition of } add\_two]\\
&= (x+1+1+1)+1 \quad &&[\text{Property of } +]\\
&= add\_two\,(x+1)+1 \quad &&[\text{Definition of } add\_two]
\end{aligned}
$$

□

Example 2.15 The following proof establishes that

$$subtract\,(add\_two\,(x+1+1), 0+1+1) = x$$

holds for all natural numbers x.

$$
\begin{aligned}
&subtract\,(add\_two\,(x+1+1), 0+1+1)\\
={}&subtract\,((x+1+1)+1+1, 0+1+1) \quad &&[\text{Definition of } add\_two]\\
={}&subtract\,((x+1+1)+1, 0+1) \quad &&[\text{Definition of } subtract]\\
={}&subtract\,(x+1+1, 0) \quad &&[\text{Definition of } subtract]\\
={}&x+1+1 \quad &&[\text{Definition of } subtract]
\end{aligned}
$$

□

2.3   Other classes of numbers

As we have seen, $\mathbb{N}$ denotes the collection of non-negative whole numbers. There are two further collections of numbers that we shall be concerned with in this book. The first is the integers, denoted $\mathbb{Z}$. As well as the non-negative whole numbers, this collection also contains the negative whole numbers. That is to say that $\mathbb{Z}$ contains $0, 1, -1, 2, -2, 3, -3$, and so on. Furthermore, the real numbers, denoted $\mathbb{R}$, include all decimal numbers such as $1.2, -1.3, 1.333$, etc., numbers such as $\sqrt{2}$, in addition to the whole numbers.

From the above, we can see that *every* number that is contained in $\mathbb{N}$ is also contained in $\mathbb{Z}$, and every number contained in $\mathbb{Z}$ is also contained in $\mathbb{R}$.

Exercise 2.9 State in which of $\mathbb{N}$, $\mathbb{Z}$, or $\mathbb{R}$ the following numbers are contained.

1. 23

2. -23

---

[3]Those readers requiring further guidance in writing mathematical proofs are referred to [Sol90].

3. 2.3

4. -2.3

□

Exercise 2.10 Is it the case that every number contained in $\mathbb{N}$ is contained in $\mathbb{R}$? □

2.4   Further exercises

Exercise 2.11 Define the 'normal' addition operator, $+$, in terms of $+1$. □

Exercise 2.12 Define the operation *minus_one* on $\mathbb{N}$ using Peano arithmetic. Assume that $minus\_one\,(0) = 0$. □

Exercise 2.13 Define the multiplication operator, *mult*, on $\mathbb{N}$ using Peano arithmetic. □

Exercise 2.14 Define the square operator, *square*, on $\mathbb{N}$ using Peano arithmetic and the function *mult*. □

Exercise 2.15 Prove that, for any natural numbers m and n,

$$mult\,(add\_two\,(n),\,m) = mult\,(n,\,m) + mult\,(add\_two\,(0),\,m)$$

□

2.5   Solutions

Solution 2.1 Of these numbers, only 0 appears in $\mathbb{N}$. □

Solution 2.2

1. 1

2. 10

3. 3

4. 0

5. 0

6. 1

□

Solution 2.3

1. $\displaystyle\sum_{i=1}^{5} i^3 = 1^3 + 2^3 + 3^3 + 4^3 + 5^3$
   $= 1 + 8 + 27 + 64 + 125$
   $= 225$

2. $\displaystyle\sum_{i=1}^{5}(i-1) = (1-1)+(2-1)+(3-1)+(4-1)+(5-1)$
$$= 0+1+2+3+4$$
$$= 10$$

3. $\displaystyle\sum_{i=1}^{5}(i \bmod 3) = (1 \bmod 3)+(2 \bmod 3)+(3 \bmod 3)+$
$$(4 \bmod 3)+(5 \bmod 3)$$
$$= 1+2+0+1+2$$
$$= 6$$

$\square$

Solution 2.4

1. $\displaystyle\prod_{i=1}^{5} i^3 = 1^3 \times 2^3 \times 3^3 \times 4^3 \times 5^3$
$$= 1 \times 8 \times 27 \times 64 \times 125$$
$$= 1,728,000$$

2. $\displaystyle\prod_{i=1}^{5}(i \text{ div } 2) = (1 \text{ div } 2) \times (2 \text{ div } 2) \times (3 \text{ div } 2) \times (4 \text{ div } 2) \times (5 \text{ div } 2)$
$$= 0 \times 1 \times 1 \times 2 \times 2$$
$$= 0$$

3. $\displaystyle\prod_{i=1}^{5}\left(\sum_{j=i}^{i+2} j\right) = \sum_{j=1}^{3} j \times \sum_{j=2}^{4} j \times \sum_{j=3}^{5} j \times \sum_{j=4}^{6} j \times \sum_{j=5}^{7} j$
$$= (1+2+3) \times (2+3+4) \times (3+4+5) \times (4+5+6) \times (5+6+7)$$
$$= 6 \times 9 \times 12 \times 15 \times 18$$
$$= 174,960$$

$\square$

Solution 2.5

1. As $2^0 = 1$, it follows that $log_2\, 1 = 0$.
2. As $2^1 = 2$, it follows that $log_2\, 2 = 1$.
3. As $2^7 = 128$, it follows that $log_2\, 128 = 7$.
4. As $2^5 = 32$ and $2^6 = 64$, it follows that $\lceil log_2\, 50 \rceil = 6$.
5. As $2^5 = 32$ and $2^6 = 64$, it follows that $\lfloor log_2\, 50 \rfloor = 5$.
6. As $2^6 = 64$ and $2^7 = 128$, it follows that $\lceil log_2\, 70 \rceil = 7$.
7. As $2^6 = 64$ and $2^7 = 128$, it follows that $\lfloor log_2\, 70 \rfloor = 6$.

$\square$

Solution 2.6 Axiom 1 states that 0 is a natural number. Axiom 2 indicates that if 0 is a natural number, then $0+1$ is also a natural number. Therefore, $0+1$ is a natural number. $\square$

Solution 2.7

$$subtract\,(0+1+1+1, 0+1) = subtract\,(0+1+1, 0)$$
$$= 0+1+1$$

□

Solution 2.8 By our definition of *add_two*,

$$add\_two\,(0+1) = 0+1+1+1$$

□

Solution 2.9

1. $\mathbb{N}$, $\mathbb{Z}$, and $\mathbb{R}$

2. $\mathbb{Z}$ and $\mathbb{R}$

3. $\mathbb{R}$

4. $\mathbb{R}$

□

Solution 2.10 Yes. Assume that a number n is contained in $\mathbb{N}$. It was stated above that if n is contained in $\mathbb{N}$ then it is contained in $\mathbb{Z}$. Therefore, n is contained in $\mathbb{Z}$. Furthermore, it was also stated above that if n is contained in $\mathbb{Z}$ then it is contained in $\mathbb{R}$. Therefore, n is contained in $\mathbb{R}$.
□

Solution 2.11

$$0 + n = n$$
$$(m+1) + n = (m+n) + 1$$

□

Solution 2.12

$$minus\_one\,(0) = 0$$
$$minus\_one\,(n+1) = n$$

□

Solution 2.13

$$mult\,(0, n) = 0$$
$$mult\,(0+1, n) = n$$
$$mult\,(m+1, n) = mult\,(m, n) + n$$

□

Solution 2.14

$$square\,(n) = mult\,(n, n)$$

□

Solution 2.15

$$
\begin{aligned}
mult\,(add\_two\,(n), m) &= mult\,(n + 1 + 1, m) & & \text{[Definition of } add\_two\text{]} \\
&= mult\,(n + 1, m) + m & & \text{[Definition of } mult\text{]} \\
&= mult\,(n, m) + m + m & & \text{[Definition of } mult\text{]} \\
&= mult\,(n, m) + mult\,(0 + 1, m) + m & & \text{[Definition of } mult\text{]} \\
&= mult\,(n, m) + mult\,(0 + 1 + 1, m) & & \text{[Definition of } mult\text{]} \\
&= mult\,(n, m) + mult\,(add\_two\,(0), m) & & \text{[Definition of } add\_two\text{]}
\end{aligned}
$$

□

Chapter 3

Propositional logic

In this chapter we provide an introduction to the subject of propositional logic. We start by considering the notion of a proposition, and then introduce the propositional logic operators. Then we show a number of techniques that can be used to establish the validity (or otherwise) of propositions. Those readers interested in studying this topic further after reading this chapter are advised to consult [BMN97] or [CLP00] as a first step. Those requiring a more advanced text might consider [Men87].

3.1 Atomic propositions

Often in life we are faced with statements that may be either true or false. In discrete mathematics, we refer to such statements as *propositions*. The *truth value* associated with a proposition may always be true or false (see, for example, Example 3.1), or it may change according to certain circumstances (see, for example, Example 3.2). The fact remains, however, that we can attempt to associate truth values with such statements.

Example 3.1 The truth value of the proposition "Tuesday is the day before Wednesday" is always true. □

Example 3.2 The truth value of the proposition "Today is Tuesday" may be true or false, depending, of course, upon when the statement is made. □

The fact that we may attempt to determine the truth or falsity of the above statements without first having to determine the truth or falsity of other propositions means that we may refer to them as *atomic propositions*: each one is an atomic entity, the truth or falsity of which is independent of any other propositions.

Examples of other atomic propositions include the following.

- "Jim is a vegetarian."

- "It is raining."

- "Becky likes biscuits."

- "There is nothing on the television."

Of course, without further information (i.e., which Jim or Becky it is that we are talking about), we cannot determine whether some of these statements are true or false. Furthermore, the ambiguity of the final statement exemplifies why the formality of mathematics is often preferable to the vagueness of natural language—does this statement mean "there is nothing showing on the television", or "there is nothing on top of the television", or "there is nothing worth watching on the television"?

3.2 Truth values

Every atomic proposition may be associated with a truth value. In our treatment of propositional logic, there are two possible truth values which may be associated with atomic propositions: *true* and *false*. Furthermore, every proposition may be associated with only one of these values at any given time. For example, the statement "Wednesday is the day immediately after Tuesday" has the associated truth value *true*, whereas "April is the month immediately before June" has the associated value *false*.

Although we might not be able to determine the truth or falsity of a given proposition, we know it must be true or false. Furthermore, although the truth value of a given proposition might change according to when it is stated, a proposition can never be both true and false at the same time.

Example 3.3 We do not know whether the proposition "Jack likes Jill" is true or false, but we know it must be one or the other. □

Example 3.4 The proposition "Today is Tuesday" is sometimes true and sometimes false, but is always either true or false. □

Exercise 3.1 Determine which of the following statements are equivalent to *true* and which are equivalent to *false*.

1. $0 < 1$
2. $1 + 1 = 2$
3. $1 \times 1 = 2$

□

Some treatments of propositional logic concern themselves with a third truth value: *undefined*. Such *three-valued logics* are able to associate a truth value with propositions, the truth or falsity of which are uncertain. For example, without further information, it is difficult to attach a truth value to the statement "Jeremy is happy". In a three-valued logic, we would be quite content to attach the value *undefined* to this statement. However, in this book, we concern ourselves only with two truth values: *true* and *false*. As such, in our approach, we may conclude that this statement is equivalent to either *true* or *false*, but we have no means of determining precisely which.

Exercise 3.2 Which of the following atomic propositions can we attach a truth value to without the need for further information?

1. "Jon is older than 35 years old."

2. "The earth revolves around the moon."

3. "3 is greater than 2."

4. "Jon's name contains three letters."

□

Atomic propositions, together with the truth values *true* and *false* are the building blocks of propositional logic. Just as natural numbers can be combined via addition or multiplication to form arithmetical terms, so atomic propositions can be combined via a variety of *propositional operators* to form *compound propositions*. The first of these propositional operators is called *negation*.

3.3 Negation

Given any proposition p, we may talk about its *negation*, which is denoted $\neg\, p$. Here, $\neg\, p$ is also a proposition, with $\neg\, p$ being read "not p".

Example 3.5 Given the proposition t, which represents the statement "Today is Tuesday", the proposition $\neg\, t$ represents the statement "Today is not Tuesday". □

Example 3.6 Given the proposition r, which represents the statement "Rick is a vegetarian", the proposition $\neg\, r$ represents the statement "Rick is not a vegetarian". □

Exercise 3.3 State the negation of each of the following propositions.

1. "It is snowing."

2. "Jon likes Ali."

3. "x is greater than y."

□

If the truth value associated with a proposition p is *true*, then the truth value associated with its negation, $\neg\, p$, is *false*. Furthermore, if the truth value associated with a proposition q is *false*, then the truth value associated with its negation, $\neg\, q$, is *true*.

Example 3.7 If the proposition "Today is Tuesday" is true, then the proposition "Today is not Tuesday" must be false. □

At this point, we encounter our first law of propositional logic.

Law 3.1

$$(\neg\ true) \Leftrightarrow false$$
$$(\neg\ false) \Leftrightarrow true$$

□

In the above, the statement $(\neg\ true) \Leftrightarrow false$ means that the propositions $\neg\ true$ and $false$ are *logically equivalent*; we shall present a formal definition of this notion later in this chapter. For now, it is sufficient to regard the statement $p \Leftrightarrow q$ simply as meaning that "p has the same truth value as q".[1]

Our second law below states that if we negate the negation of a proposition, we arrive at the value of the original proposition.

Law 3.2 For any proposition p,

$$(\neg\ \neg\ p) \Leftrightarrow p$$

□

Example 3.8

$$(\neg\ \neg\ true) \Leftrightarrow true$$
$$(\neg\ \neg\ false) \Leftrightarrow false$$

□

Truth tables provide us with a means of representing the truth or falsity of logical statements. The possible values of all atomic propositions contained in the proposition are listed, and the truth value of the overall proposition can thus be calculated for each possible combination. The truth table for negation is given below.

p	$\neg\ \mathbf{p}$
true	**false**
false	**true**

Note that the possible values of p are listed in the left-hand column, and the results are displayed in the right-hand column. The result in the first row is *false* due to the fact that $\neg\ true$ is logically equivalent to *false*. The result in the second row is *true* due to the fact that $\neg\ false$ is logically equivalent to *true*. Throughout this chapter we shall denote the result column of each truth table by presenting it in a bold font.

Exercise 3.4 Calculate the truth values of the following propositions.

1. $\neg\ (0 < 1)$

2. $\neg\ (1 + 1 = 2)$

3. $\neg\ (\text{"The earth revolves around the moon"})$

□

---

[1] Note that some texts use $p \equiv q$ to denote the fact that propositions p and q are logically equivalent.

3.4   Conjunction

Negation was the first of our propositional logic operators; it is a *unary* operator, so-called because it takes one argument, i.e., it operates on one proposition—the proposition that is to be negated. All of the other propositional logic operators that we shall encounter are *binary* operators: they all take two arguments.

The first of these binary operators that we shall consider is *conjunction*, or 'and'. This operator works by taking two propositions, and returning the result *true* if both propositions are equivalent to *true*, and returning the result *false* otherwise. That is, the overall proposition is equivalent to *true* if, and only if, both parts of the proposition are equivalent to *true*.

Given two propositions p and q, the conjunction of p and q is written $p \wedge q$. This proposition is read "p and q". It is also possible to say that $p \wedge q$ is "p conjoined with q".

Example 3.9 The conjunction of "Rick is a vegetarian", which is denoted by v, and "Rick eats chocolate", which is denoted by c, is written $v \wedge c$. $\square$

Example 3.10 The statement $0 < 1 \wedge 1 < 0$ is false, as it is not the case that both $0 < 1$ and $1 < 0$ are true. $\square$

Exercise 3.5 Calculate the truth values of the following propositions.

1. $(1 < 0) \wedge (2 < 1)$
2. $(0 < 1) \wedge (2 < 1)$
3. $(0 < 1) \wedge (1 < 2)$

$\square$

The truth table for conjunction is given below.

p	q	$\mathbf{p \wedge q}$
true	*true*	**true**
true	*false*	**false**
false	*true*	**false**
false	*false*	**false**

Note that a proposition $p \wedge q$ is equivalent to *true* if, and only if, both p and q are themselves equivalent to *true*. Note also the enumeration of the possible values of p and q. There are four possible combinations: p being *true* and q being *true*; p being *true* and q being *false*; p being *false* and q being *true*; and p being *false* and q being *false*.

As one might expect, we can use truth tables to determine the truth values of propositions involving both of the operators we have met thus far.

Example 3.11 The truth table for $p \wedge (\neg q)$ is given below.

p	q	$\neg q$	$\mathbf{p} \wedge (\neg \mathbf{q})$
true	true	false	**false**
true	false	true	**true**
false	true	false	**false**
false	false	true	**false**

From the above, we can see that $p \wedge (\neg q)$ is equivalent to *true* if, and only if, p is equivalent to *true* and q is equivalent to *false*. Note that this truth table is calculated in three stages: first, the possible values of p and q are listed, then the possible values of $\neg q$ are calculated, and, finally, p and $\neg q$ are combined via conjunction in the 'result' column.

We may also reason about propositions containing the truth values *true* and *false* using truth tables: of course *true* always has the value *true*, and *false* always has the value *false*.[2]

Example 3.12 The truth table for $p \wedge true$ is given below.

p	$true$	$\mathbf{p} \wedge \mathbf{true}$
true	true	**true**
false	true	**false**

Exercise 3.6 Write out the truth table for $(\neg p) \wedge (\neg q)$. □

Exercise 3.7 Write out the truth table for $\neg (p \wedge q)$. □

Exercise 3.8

1. The truth tables of Exercises 3.6 and 3.7 both have four rows. How many rows will the truth table for the proposition $(p \wedge q) \wedge r$ have?

2. In general, if a proposition contains n atomic propositions, how many rows will the truth table for that proposition have?

3. If a proposition consists of only the truth values *true* and *false*, how many rows will the truth table for that proposition have?

□

There are a number of laws which hold for conjunction. The first of these states that the truth value of any proposition p conjoined with itself is simply the truth value of p. This is referred to as the *idempotence* of conjunction.

---

[2]Some texts use different names or symbols to distinguish between the *syntactic* embodiment of truth, i.e., the proposition *true*, and the *semantic* embodiment of truth, i.e., the truth value *true*. For example, some texts use T for the former, and tt for the latter. We choose not to make this distinction in this text. In this book, *true* represents both the proposition that is always true and also plays the role of one of the two truth values. This is also the case for *false*.

Law 3.3 Conjunction is idempotent. That is, for any proposition p,

$$(p \wedge p) \Leftrightarrow p$$

☐

Second, the truth value of *true* conjoined with any proposition p is equivalent to the truth value of p.

Law 3.4 For any proposition p,

$$(p \wedge true) \Leftrightarrow p$$

☐

Third, the truth value of *false* conjoined with any proposition p is equivalent to *false*.

Law 3.5 For any proposition p,

$$(p \wedge false) \Leftrightarrow false$$

☐

Referring to the truth table for $\wedge$, we find that the following law also holds.

Law 3.6 For any proposition p,

$$(p \wedge (\neg\, p)) \Leftrightarrow false$$

☐

Exercise 3.9 Using the laws of propositional logic seen so far, calculate the truth values of the following propositions.

1. $\neg\,\neg\,(true \wedge false)$
2. $true \wedge (p \wedge false)$
3. $(\neg\,(p \wedge false)) \wedge true$

☐

There are two further laws for conjunction. The first states that $\wedge$ is *commutative* (i.e., the order in which the propositions are written makes no difference to the overall result), while the second states that $\wedge$ is *associative* (i.e., parentheses are not necessary when considering a series of conjuncts).

Law 3.7 Conjunction is commutative. That is, for any propositions p and q,

$$(p \wedge q) \Leftrightarrow (q \wedge p)$$

☐

Law 3.8 Conjunction is associative. That is, for any propositions p, q, and r,

$$p \wedge (q \wedge r) \Leftrightarrow (p \wedge q) \wedge r$$

☐

Exercise 3.10 Using the laws of propositional logic seen so far, calculate the truth values of the following propositions.

1. $(false \wedge p) \wedge q$
2. $p \wedge (false \wedge q)$

☐

3.5 Disjunction

Our second binary operator for propositional logic is *disjunction*, or 'or'. This operator works by taking two propositions, and returning *true* if at least one of the propositions is equivalent to *true*, and returning *false* otherwise. That is, the overall proposition is equivalent to *true* if, and only if, at least one part of the proposition is equivalent to *true*.

Given two propositions p and q, the disjunction of p and q is written $p \vee q$. This proposition is read "p or q". It is also possible to say that $p \vee q$ is "p disjoined with q".

Example 3.13 The disjunction of "Rick is a vegetarian", which is denoted by v, and "Rick eats chocolate", which is denoted by c, is written $v \vee c$. ☐

Example 3.14 As a further example, the statement $(0 < 1) \vee (1 < 0)$ is true, as at least one part of the disjunction—specifically, $0 < 1$—is true. ☐

Exercise 3.11 Calculate the truth values of the following propositions.

1. $(1 < 0) \vee (2 < 1)$
2. $(0 < 1) \vee (2 < 1)$
3. $(0 < 1) \vee (1 < 2)$

☐

The truth table for disjunction is given below.

p	q	$\mathbf{p \vee q}$
true	*true*	**true**
true	*false*	**true**
false	*true*	**true**
false	*false*	**false**

Exercise 3.12 Write out the truth table for $\neg (p \vee q)$. ☐

Exercise 3.13 Write out the truth table for $(\neg\, p) \vee (\neg\, q)$. □

Note the similarity between the truth table for Exercise 3.6 and the truth table for Exercise 3.12, and also the similarity between the truth table for Exercise 3.7 and the truth table for Exercise 3.13, all of which are given below.

p	q	$\neg\, p$	$\neg\, q$	$(\neg\, \mathbf{p}) \wedge (\neg\, \mathbf{q})$
true	*true*	*false*	*false*	**false**
true	*false*	*false*	*true*	**false**
false	*true*	*true*	*false*	**false**
false	*false*	*true*	*true*	**true**

p	q	$p \vee q$	$\neg\, (\mathbf{p} \vee \mathbf{q})$
true	*true*	*true*	**false**
true	*false*	*true*	**false**
false	*true*	*true*	**false**
false	*false*	*false*	**true**

p	q	$p \wedge q$	$\neg\, (\mathbf{p} \wedge \mathbf{q})$
true	*true*	*true*	**false**
true	*false*	*false*	**true**
false	*true*	*false*	**true**
false	*false*	*false*	**true**

p	q	$\neg\, p$	$\neg\, q$	$(\neg\, \mathbf{p}) \vee (\neg\, \mathbf{q})$
true	*true*	*false*	*false*	**false**
true	*false*	*false*	*true*	**true**
false	*true*	*true*	*false*	**true**
false	*false*	*true*	*true*	**true**

The result columns of the first two truth tables are identical, as are the result columns of the second two truth tables. The fact that the result columns are identical indicates that the proposition $\neg\, (p \wedge q)$ is logically equivalent to $(\neg\, p) \vee (\neg\, q)$ and that the proposition $\neg\, (p \vee q)$ is logically equivalent to $(\neg\, p) \wedge (\neg\, q)$.

These equivalences are captured by *de Morgan's laws*,[3] which are given below.

Law 3.9 (de Morgan's laws) For any propositions p and q,

$$\neg\, (p \wedge q) \Leftrightarrow ((\neg\, p) \vee (\neg\, q))$$
$$\neg\, (p \vee q) \Leftrightarrow ((\neg\, p) \wedge (\neg\, q))$$

□

---

[3]Named after Augustus de Morgan (1806 - 1871). The interested reader is referred to [Mer90] for a description of de Morgan's work in this area.

As was the case with $\land$, $\lor$ is idempotent.

Law 3.10 Disjunction is idempotent. For any proposition p,

$$(p \lor p) \Leftrightarrow p$$

□

The combining, via disjunction, of any proposition with *false* gives a result that is logically equivalent to the original proposition.

Law 3.11 For any proposition p,

$$(p \lor false) \Leftrightarrow p$$

□

The result of combining any proposition with *true* via disjunction is *true*.

Law 3.12 For any proposition p,

$$(p \lor true) \Leftrightarrow true$$

□

Exercise 3.14 Using the laws of propositional logic seen so far, calculate the truth values of the following propositions.

1. $(p \lor false) \lor true$
2. $\lnot \, ((p \land false) \lor true)$

□

As was the case with conjunction, disjunction is both associative and commutative.

Law 3.13 Disjunction is associative. For any propositions p, q, and r,

$$p \lor (q \lor r) \Leftrightarrow (p \lor q) \lor r$$

□

Law 3.14 Disjunction is commutative. For any propositions p and q,

$$p \lor q \Leftrightarrow q \lor p$$

□

Exercise 3.15 Using the laws of propositional logic seen so far, calculate the truth values of the following propositions.

1. $(\lnot \, (true \lor p)) \land (p \lor false)$

  2. $p \vee (true \vee q)$

□

Referring to the truth table for disjunction, we can see that, given a proposition p, it is always the case that $p \vee (\neg p)$ is equivalent to *true*: if p is true, then $p \vee (\neg p)$ is true because the left-hand disjunct is true; on the other hand, if p is false, then $p \vee (\neg p)$ is true because the right-hand disjunct is true. This is known as *the law of the excluded middle*, and is stated formally below.

Law 3.15 For any proposition p,

  $((\neg p) \vee p) \Leftrightarrow true$

□

Our final two laws for disjunction and conjunction state that disjunction *distributes* through conjunction, and vice versa.

Law 3.16 Disjunction distributes through conjunction. For any propositions p, q, and r,

  $p \vee (q \wedge r) \Leftrightarrow (p \vee q) \wedge (p \vee r)$
  $(p \wedge q) \vee r \Leftrightarrow (p \vee r) \wedge (q \vee r)$

□

Law 3.17 Conjunction distributes through disjunction. For any propositions p, q, and r,

  $p \wedge (q \vee r) \Leftrightarrow (p \wedge q) \vee (p \wedge r)$
  $(p \vee q) \wedge r \Leftrightarrow (p \wedge r) \vee (q \wedge r)$

□

If the format of these laws appears difficult to comprehend at first, consider the distribution of multiplication through addition from arithmetic. Here, the term

  $3 \times (2 + 5)$

can be rewritten as

  $(3 \times 2) + (3 \times 5)$

which, of course, gives the same result as the original term: 21. The distribution of conjunction through disjunction (and, for that matter, vice versa) takes exactly the same form as this, more familiar, law of distribution.

Example 3.15 The proposition

  $true \vee (true \wedge false)$

is logically equivalent to

  $(true \vee true) \wedge (true \vee false)$

□

Exercise 3.16

1. Using the laws of disjunction seen so far, show that $p \wedge ((\neg p) \vee q)$ is equivalent to $p \wedge q$.

2. Using the laws of disjunction seen so far, show that $(p \wedge q) \vee (\neg p)$ is equivalent to $q \vee (\neg p)$.

□

3.6 Implication

Implication is the third of our logical operators, and, for most students of discrete mathematics, the least intuitive and most difficult to understand. The implication operator is written $\Rightarrow$, with the proposition $p \Rightarrow q$ being read "p implies q" or "if p, then q".

As an example, the statement "If it rains this afternoon, then Duncan will stay in" may be represented logically as $r \Rightarrow s$ (where r denotes "it rains this afternoon" and s denotes "Duncan will stay in"). If we were to think of this statement as if it were a contract, then there is only one circumstance under which we would be entitled to feel cheated: if it did rain this afternoon, and Duncan failed to stay in. If it did rain and Duncan stayed in, then we would be content that the contract hadn't been broken. Furthermore, if it failed to rain, then it shouldn't worry us too much what Duncan does with his afternoon, as this is quite clearly outside the boundaries of our contract. As such, $r \Rightarrow s$ is logically equivalent to *false* (i.e., the contract is broken) only when r is equivalent to *true* and s is equivalent to *false*; otherwise it is logically equivalent to *true*.

Exercise 3.17 Given that "Emily is happy" is denoted by e and "Rachel is happy" is denoted by r, translate each of the following natural language statements into propositional logic statements.

1. "If Emily is happy, then Rachel is happy."

2. "If Rachel is happy, it follows that Emily is happy."

3. "Emily is happy only if Rachel is happy."

4. "It is not the case that if Rachel is happy then Emily is not happy."

□

Exercise 3.18 Determine the truth or falsity of the following statements.

1. "If Marilyn Monroe was a man, then elephants can fly."

2. "If $1 + 1 = 2$, then Madrid is the capital of Spain."

3. "If Marilyn Monroe was a man, then $1 + 1 = 2$."

4. "If Madrid is the capital of Spain, then elephants can fly."

□

The truth table for implication is given below.

p	q	$\mathbf{p \Rightarrow q}$
true	*true*	**true**
true	*false*	**false**
false	*true*	**true**
false	*false*	**true**

Note that if p is equivalent to *false*, then the value of $p \Rightarrow q$ is always equivalent to *true*, no matter what the value of q is. On the other hand, if p is equivalent to *true*, then the value of $p \Rightarrow q$ is always equivalent to that of q.

Exercise 3.19 Write out the truth table for $(p \wedge q) \Rightarrow (p \vee q)$. $\square$

Exercise 3.20 Write out the truth table for $(p \vee q) \Rightarrow (p \wedge q)$. $\square$

One interpretation often used to motivate the implication operator is to consider a 'strength ordering' between *true* and *false*, in which *false* is stronger than *true*. What is more, *false* is at least as strong as any proposition (in fact, it is the strongest proposition) and *true* is at least as weak as any proposition (in fact, it is the weakest proposition). Using this technique, a proposition $p \Rightarrow q$ is equivalent to *false* if, and only if, p is weaker than q.

As any proposition p is always as strong as itself, *false* $\Rightarrow$ *false* and *true* $\Rightarrow$ *true* are both equivalent to *true*. Furthermore, as *false* is stronger than *true*, *false* $\Rightarrow$ *true* is equivalent to *true*. As an implication $p \Rightarrow q$ is only equivalent to *false* when the *antecedent* (the proposition appearing before the $\Rightarrow$) is weaker than the *consequent* (the proposition appearing after the $\Rightarrow$), *true* $\Rightarrow$ *false* is logically equivalent to *false*.

Example 3.16 We may use the above technique to determine the truth or falsity of

$$true \Rightarrow (p \Rightarrow (false \Rightarrow p))$$

First, *false* is at least as strong as any proposition, so

$$true \Rightarrow (p \Rightarrow (false \Rightarrow p))$$

is logically equivalent to

$$true \Rightarrow (p \Rightarrow true)$$

As *true* is at least as weak as any proposition, this is equivalent to

$$true \Rightarrow true$$

Finally, as *true* is no weaker than itself, this is equivalent to *true*. $\square$

None of the attractive properties which held of $\vee$ and $\wedge$ hold of implication; implication is not commutative, nor is it associative or idempotent. However, there is a very useful equivalence which we can use to transform an implication into a logically equivalent disjunction.

Law 3.18 For any propositions p and q,

$$(p \Rightarrow q) \Leftrightarrow ((\neg\, p) \vee q)$$

$\square$

Example 3.17 The implication "if it rains this afternoon, then Duncan will stay in" is logically equivalent to the disjunction "either it will not rain this afternoon or Duncan will stay in". $\square$

Exercise 3.21 Using the laws of propositional logic seen so far, calculate the truth values of the following propositions.

1. $p \Rightarrow p$
2. $(p \wedge q) \Rightarrow q$

$\square$

Exercise 3.22 Prove that the following properties do not hold of $\Rightarrow$.

1. Associativity.
2. Commutativity.
3. Idempotence.

$\square$

3.7 Equivalence

Our final propositional operator is *equivalence*. Throughout this chapter, we have referred to two propositions being *logically equivalent*, which we have denoted by $p \Leftrightarrow q$. For example, the statement "a score of above 70 is equivalent to a distinction" might be represented by the proposition $s \Leftrightarrow d$, where s denotes "a score of above 70" and d denotes "a distinction".

Essentially, two propositions p and q are logically equivalent if they have the same truth values, i.e., if p is true exactly when q is true. We write $p \Leftrightarrow q$ to denote the statement "p is logically equivalent to q" or "p if, and only if, q".

Example 3.18 The proposition $0 \le 1 \Leftrightarrow 1 \ge 1$ is true, whereas $0 \le 1 \Leftrightarrow 1 > 1$ is false. $\square$

Example 3.19 Given the propositions "Rick is a vegetarian", which is denoted by v, and "Rick eats chocolate", which is denoted by c, the proposition "Rick is a vegetarian if, and only if, Rick eats chocolate" is written $v \Leftrightarrow c$. $\square$

Exercise 3.23 Determine the truth or falsity of the following statements.

1. "Marilyn Monroe was a man if, and only if, elephants can fly."
2. "$1 + 1 = 2$ if, and only if, Madrid is the capital of Spain."

3. "Marilyn Monroe was a man if, and only if, $1 + 1 = 2$."

4. "Madrid is the capital of Spain if, and only if, elephants can fly."

☐

The truth table for equivalence is given below.

p	q	$p \Leftrightarrow q$
true	*true*	**true**
true	*false*	**false**
false	*true*	**false**
false	*false*	**true**

Exercise 3.24 Write out the truth table for $(p \Rightarrow q) \Leftrightarrow (q \Rightarrow p)$. ☐

The equivalence operator is both associative and commutative.

Law 3.19 Equivalence is associative. For any propositions p, q, and r,

$$((p \Leftrightarrow q) \Leftrightarrow r) \Leftrightarrow (p \Leftrightarrow (q \Leftrightarrow r))$$

☐

Law 3.20 Equivalence is commutative. For any propositions p and q,

$$(p \Leftrightarrow q) \Leftrightarrow (q \Leftrightarrow p)$$

☐

In addition, every proposition is equivalent to itself.

Law 3.21 For any proposition p,

$$(p \Leftrightarrow p) \Leftrightarrow true$$

☐

Conversely, no proposition is equivalent to its negation.

Law 3.22 For any proposition p,

$$(p \Leftrightarrow (\neg p)) \Leftrightarrow false$$

☐

Finally, to say that one proposition, p, is equivalent to another proposition, q, is the same as saying p implies q *and* q implies p.

Law 3.23 For any propositions p and q,

$$(p \Leftrightarrow q) \Leftrightarrow ((p \Rightarrow q) \wedge (q \Rightarrow p))$$

□

Example 3.20 As an example, the proposition "Clare gets a distinction if, and only if, Clare has scored above 70" is logically equivalent to "if Clare gets a distinction then Clare has scored above 70 and if Clare scores above 70 then Clare gets a distinction." □

Exercise 3.25 Using the laws of propositional logic seen so far, calculate the truth values of the following propositions.

1. $(p \Leftrightarrow true) \Leftrightarrow p$
2. $(p \Leftrightarrow false) \Leftrightarrow \neg p$

□

Exercise 3.26 Prove that $\Leftrightarrow$ is not idempotent. □

3.8 Precedence

As we have seen throughout this chapter, parentheses can be used to remove ambiguity from the meaning of a given proposition. For example, the proposition $\neg (p \wedge q)$ has a very different meaning to $(\neg p) \wedge q$. However, if we were simply to write $\neg p \wedge q$ then it might be unclear as to what we meant. The use of parentheses can help us to provide the necessary clarity.

However, sometimes, especially if we are dealing with a complex proposition, using parentheses in this way can often be rather cumbersome. For example, the proposition $\neg p \vee \neg q \vee \neg r$ has a fairly clear meaning, yet, if we were to be explicit about parenthesising this expression, we would have to write $(\neg p) \vee ((\neg q) \vee (\neg r))$ or $((\neg p) \vee (\neg q)) \vee (\neg r)$. (Of course, both of these propositions are logically equivalent; in this case, the parentheses are purely cosmetic.) While these latter, parenthesised, propositions have exactly the same logical meaning as the original, unparenthesised, proposition, they do appear somewhat clumsy by comparison.

To provide a happy medium between clarity and elegance, we may assign an order of precedence to our operators: $\neg$ has the highest priority (and, therefore, 'binds' tightest), followed by $\wedge$, then $\vee$, then $\Rightarrow$, and then $\Leftrightarrow$, which has the lowest priority (and, therefore, binds loosest). An operator that has a higher priority than a second operator effectively 'grabs' those propositions nearest to it before the second operator has the opportunity to 'grab' those propositions closest to it.

Example 3.21 If we were to write $\neg p \wedge q$, this would have the same meaning as $(\neg p) \wedge q$ rather than $\neg (p \wedge q)$, as the $\neg$ binds tighter than the $\wedge$. □

Example 3.22 The proposition $p \vee q \Rightarrow r$ would mean $(p \vee q) \Rightarrow r$ rather than $p \vee (q \Rightarrow r)$, as the $\vee$ binds tighter than the $\Rightarrow$. □

Exercise 3.27 Write the fully parenthesised versions of the following propositions.

1. $\neg\, p \lor q$

2. $p \lor q \land r$

3. $p \land q \Rightarrow r$

4. $\neg\, p \Leftrightarrow \neg\, q \land r \Rightarrow s$

□

3.9 Tautologies, contradictions, and contingencies

A proposition which is always logically equivalent to *true*—no matter what the values of its constituent atomic propositions are—is called a *tautology*. Examples of tautologies include $p \lor \neg\, p$, $p \Rightarrow p$, and $(p \land q) \lor (p \land \neg\, q) \lor (\neg\, p \land q) \lor (\neg\, p \land \neg\, q)$. No matter what the values of the individual atomic propositions are, in each case the truth value of the overall proposition is *true*.

Assuming that a given proposition consists of no more than a handful of atomic propositions, one easy method of determining whether or not it is a tautology is to substitute truth values for the atomic propositions. If the truth value of the proposition is *true* for all combinations of values attributed to the atomic propositions, then we may conclude that it is a tautology. Alternatively, if we can exhibit one combination of values for the atomic propositions for which the truth value of the proposition is *false*, then we may conclude that it is not a tautology.

Example 3.23 Given the proposition $p \lor \neg\, p$, if we substitute *true* for p, then the overall proposition is equivalent to *true*. Furthermore, if we substitute *false* for p, then the overall proposition is equivalent to *true*. Therefore, we can conclude that $p \lor \neg\, p$ is a tautology. □

Example 3.24 Consider the proposition $(p \lor \neg\, p) \Rightarrow q$. If the atomic proposition q assumes the value *false*, then—because $p \lor \neg\, p$ is equivalent to *true*—no matter what the value of p is, the value of the implication is *false*. As such, this proposition is not a tautology. □

Exercise 3.28 Which of the following propositions are tautologies?

1. $p \lor (\neg\, p \land q)$

2. $p \Rightarrow (q \Rightarrow p)$

3. $(p \Rightarrow q) \Rightarrow (q \Rightarrow p)$

4. $(p \Rightarrow q) \Leftrightarrow \neg\, (p \land \neg\, q)$

□

A proposition which is always *false* is called a *contradiction*. Again, we can substitute truth values for atomic propositions to determine whether or not a given proposition is a contradiction.

Example 3.25 Consider the proposition $\neg\, (p \Rightarrow p)$. If p is *true*, then the overall proposition is equivalent to *false*. Furthermore, if p is *false*, then the overall proposition is equivalent to *false*. As such, this proposition is a contradiction. □

Exercise 3.29 Which of the following propositions are contradictions?

1. $(p \land q) \land (\neg (p \lor q))$

2. $(p \lor \neg p) \Rightarrow (q \land \neg q)$

3. $(q \land \neg q) \Rightarrow (p \lor \neg p)$

□

If a proposition is neither a tautology nor a contradiction, then it is termed a *contingency*.

Example 3.26 Consider the proposition $p \Rightarrow q$. If p is *true* and q is *false*, then the implication is *false*. On the other hand, if both p and q are *true*, then the implication is *true*. As such, $p \Rightarrow q$ is a contingency. □

Exercise 3.30 Which of the following propositions are contingencies?

1. $p \Rightarrow (\neg p \Rightarrow p)$

2. $(p \lor q) \Rightarrow (p \land q)$

□

We have seen one method for determining whether a proposition is a tautology (i.e., logically equivalent to *true*) or a contradiction (i.e., logically equivalent to *false*) or a contingency (i.e., logically equivalent to neither *true* nor *false*). Unfortunately, this method only works for relatively simple propositions involving a small number of atomic propositions: while substituting possible truth values for a proposition consisting of many atomic propositions is feasible, it can be very time consuming. More general ways of establishing whether a proposition is a tautology (or a contradiction, or a contingency) are via truth tables, via natural deduction, and via equational reasoning. We shall study each of these techniques in turn.

3.10   Truth tables

Referring to Exercise 3.28, the proposition $p \Rightarrow (q \Rightarrow p)$ was seen to be a tautology, a fact which we established by simply substituting truth values for p and q. If we wished, we could also prove this fact via a truth table.

Example 3.27 The truth table for $p \Rightarrow (q \Rightarrow p)$ is given below.

p	q	$q \Rightarrow p$	$\mathbf{p \Rightarrow (q \Rightarrow p)}$
true	*true*	*true*	**true**
true	*false*	*true*	**true**
false	*true*	*false*	**true**
false	*false*	*true*	**true**

□

If we study the final column, we find that every entry is *true*; whenever this is the case for a proposition, we are free to conclude that it is a tautology. Conversely, if every entry in the final column of a given truth table is *false*, then that proposition is a contradiction. Finally, if the final column contains at least one *true* entry and at least one *false* entry, then it is a contingency.

Exercise 3.31 Determine, using truth tables, whether or not the following propositions are tautologies, contradictions, or contingencies.

1. $(p \wedge q) \vee (\neg (p \wedge q))$
2. $(p \Rightarrow q) \Leftrightarrow (\neg q \Rightarrow \neg p)$
3. $(\neg p \Rightarrow q) \Rightarrow (p \Rightarrow \neg q)$
4. $(p \Rightarrow q) \Rightarrow (\neg p \Rightarrow \neg q)$
5. $\neg (p \Rightarrow p) \wedge (\neg p \Rightarrow \neg p)$

□

Just as substituting values had its drawbacks, so does the use of truth tables. As we saw earlier, the size of a truth table for a proposition is directly related to the number of atomic propositions contained in that proposition. If there is only one atomic proposition contained in the proposition, then the truth table has two rows; if it has two atomic propositions, then the truth table has four rows; if it has three atomic propositions, then the truth table has eight rows; and so on. Realistically, if we were calculating truth tables by hand, three atomic propositions is an upper limit; any more than three atomic propositions would mean that the time taken to construct such a truth table would outweigh the benefits associated with the simplicity of the method. As with substitution, the task is feasible, but potentially very time consuming. In short, as was the case with substitution, the task may be appropriate for calculation by computers, but not for calculation by hand. As such, we need some more general techniques for determining if given propositions are tautologies; the first such technique that we consider is equational reasoning.

3.11   Equational reasoning

Throughout this chapter, we have determined the truth or falsity of propositions by substituting one logically equivalent proposition for another. When conducting this process, we have, at each stage, applied one of our laws of propositional logic. Because we have been strict about applying these laws, this apparently informal process has, in fact, been a rather formal one; we call this method of analysis *equational reasoning*.

Take, as an example, the solution to Exercise 3.21.1. Here, the truth value of $p \Rightarrow p$ was calculated as follows.

By Law 3.18, $p \Rightarrow p$ is equivalent to $(\neg p) \vee p$. By Law 3.15, this is equivalent to *true*.

A more formal (and far clearer) way of establishing that $p \Rightarrow p$ is equivalent to *true* is given below.

$$p \Rightarrow p$$
$$\Leftrightarrow (\neg p) \vee p \quad \text{[Law 3.18]}$$
$$\Leftrightarrow true \qquad \text{[Law 3.15]}$$

At each stage we substitute a logically equivalent proposition for the former proposition; this substitution is justified by the application of one of our laws of propositional logic. We can now have more confidence in this process, because it is clear what is being substituted for—and the justification for it is also clear—at each stage.

Example 3.28 We may prove that $p \Rightarrow (p \Rightarrow q)$ is logically equivalent to $p \Rightarrow q$ in the following way.

$$
\begin{aligned}
&(p \Rightarrow (p \Rightarrow q)) \Leftrightarrow (p \Rightarrow q) \\
\Leftrightarrow\ &(\neg p \vee (p \Rightarrow q)) \Leftrightarrow (p \Rightarrow q) && \text{[Law 3.18]} \\
\Leftrightarrow\ &(\neg p \vee (\neg p \vee q)) \Leftrightarrow (p \Rightarrow q) && \text{[Law 3.18]} \\
\Leftrightarrow\ &((\neg p \vee \neg p) \vee q) \Leftrightarrow (p \Rightarrow q) && \text{[Law 3.13]} \\
\Leftrightarrow\ &(\neg p \vee q) \Leftrightarrow (p \Rightarrow q) && \text{[Law 3.10]} \\
\Leftrightarrow\ &(p \Rightarrow q) \Leftrightarrow (p \Rightarrow q) && \text{[Law 3.18]} \\
\Leftrightarrow\ &true && \text{[Law 3.21]}
\end{aligned}
$$

□

Exercise 3.32 Exercise 3.25 asked you to determine the truth values of the following propositions.

1. $(p \Leftrightarrow true) \Leftrightarrow p$

2. $(p \Leftrightarrow false) \Leftrightarrow \neg p$

Rewrite the proofs using the more formal approach described above. □

When asked to prove—via equational reasoning—that two propositions are equivalent, there are two ways of establishing such a fact. The first approach, and that used in the above solutions, is to consider the whole proposition and establish that it is logically equivalent to *true*. A second approach is to start with the left-hand side of the equivalence, and—via equational reasoning—establish that this is logically equivalent to the right-hand side. Taking this approach, an alternative solution to the first of the above questions is as follows.

$$
\begin{aligned}
&p \Leftrightarrow true \\
\Leftrightarrow\ &(p \Rightarrow true) \wedge (true \Rightarrow p) && \text{[Law 3.23]} \\
\Leftrightarrow\ &((\neg p) \vee true) \wedge (true \Rightarrow p) && \text{[Law 3.18]} \\
\Leftrightarrow\ &((\neg p) \vee true) \wedge ((\neg true) \vee p) && \text{[Law 3.18]} \\
\Leftrightarrow\ &true \wedge ((\neg true) \vee p) && \text{[Law 3.12]} \\
\Leftrightarrow\ &true \wedge (false \vee p) && \text{[Law 3.1]} \\
\Leftrightarrow\ &true \wedge (p \vee false) && \text{[Law 3.14]} \\
\Leftrightarrow\ &true \wedge p && \text{[Law 3.11]} \\
\Leftrightarrow\ &p \wedge true && \text{[Law 3.7]} \\
\Leftrightarrow\ &p && \text{[Law 3.4]}
\end{aligned}
$$

Exercise 3.33 Prove, using equational reasoning, the following equivalences.

1. $((p \wedge \neg q) \Rightarrow q) \Leftrightarrow (\neg p \vee q)$

   2. $(p \Rightarrow q) \wedge p \Leftrightarrow q \wedge p$

$\square$

Exercise 3.34 Show, using equational reasoning, that the following propositions are equivalent to *true*.

   1. $(p \Rightarrow q) \Leftrightarrow (\neg q \Rightarrow \neg p)$

   2. $(p \Leftrightarrow true) \Rightarrow p$

$\square$

Exercise 3.35 Simply each of the following using equational reasoning.

   1. $(p \Leftrightarrow q) \Rightarrow (q \Leftrightarrow p)$

   2. $(p \wedge q) \Leftrightarrow (\neg p \vee \neg q)$

$\square$

3.12   Natural deduction

Earlier in this chapter, we referred to statements such as $p \vee \neg p$ and $p \Rightarrow p$ as *tautologies*: they are statements that are always true, no matter what the truth values of their constituent atomic propositions are. In another sense, we may refer to such statements as *theorems*: once we have established that the proposition is logically equivalent to *true*, we can use it to establish other tautologies—or theorems—of propositional logic.

    Natural deduction or *proof trees* can be used to prove the validity of theorems of propositional logic. As was the case with equational reasoning, such theorems can be proved by using certain rules to manipulate propositions. As was also the case with equational reasoning, the process is fairly mechanical and can—if desired—be automated.

    The derivation of theorems is guided by a set of well-formed rules. Each of these rules is of the form

$$\frac{premises}{conclusions} \; [\text{name}]$$

where the rule, called name, allows one or more *conclusions* to be reached on the basis of one or more *premises*.

    Although proof via proof trees (or natural deduction) is a generic method of establishing theorems of propositional logic, the rules that may be applied are often specific to a particular system. The system which we present is that associated with the formal description technique, Z, which is described in [WD96]. It should be noted that although we concentrate on a specific system, most of the rules associated with that system are true for most other systems of deduction. In addition, the interested reader is referred to [Hun84], in which tactics and strategy for applying natural deduction (although not the specific system introduced here) are described.

The first of our rules of natural deduction is that of *modus ponens*, or $\Rightarrow$ elimination. This is stated as follows.

$$\frac{p \Rightarrow q \quad p}{q} \ [\Rightarrow \text{elim}]$$

Essentially, this states that if we have established that q follows from p and we have also established that p is true, then we can conclude that q is true.

Example 3.29 The following is an example of the use of $\Rightarrow$ elimination.

$$\frac{\text{"If it is raining, then I will get wet"} \quad \text{"It is raining"}}{\text{"I will get wet"}} \ [\Rightarrow \text{elim}]$$

$\square$

Example 3.30 The following is another example of the use of $\Rightarrow$ elimination.

$$\frac{\text{"If I play with fire, then I will get burnt"} \quad \text{"I play with fire"}}{\text{"I will get burnt"}} \ [\Rightarrow \text{elim}]$$

$\square$

As is the case with all of our other binary operators, $\Rightarrow$ has a rule which 'introduces' it, as well as a rule which 'eliminates' it. The general form of this rule is given below.

$$\begin{array}{c} \lceil p \rceil_1 \\ \vdots \\ \frac{q}{p \Rightarrow q} \ [\Rightarrow \text{intro}_1] \end{array}$$

This rule states that if we have derived the proposition q from p through the use of one or more of our laws, then we can conclude that q follows from p, and, as such, $p \Rightarrow q$ is true. Note that we are free to use p as many times as necessary in the derivation of q.

The $\lceil$ and $\rceil$ surrounding p indicate that it was an *assumption* which has been *discharged*; this assumption is indexed—or identified—by the label 1. The root of a proof tree is the theorem that has been established—we start developing a proof tree by starting from this theorem and working backwards towards the leaves. The leaves of a proof tree are either other theorems that have been proved elsewhere, or assumptions. A theorem should have no undischarged assumptions.

Example 3.31 We can use our first two rules to establish that $(p \Rightarrow q) \Rightarrow (p \Rightarrow q)$ is a theorem of propositional logic. Here, $p \Rightarrow q$ and p are both assumptions that are discharged when our $\Rightarrow$ introduction rule is applied.

$$\frac{\dfrac{\dfrac{\lceil p \Rightarrow q \rceil_1 \quad \lceil p \rceil_2}{q} \ [\Rightarrow \text{elim}]}{p \Rightarrow q} \ [\Rightarrow \text{intro}_2]}{(p \Rightarrow q) \Rightarrow (p \Rightarrow q)} \ [\Rightarrow \text{intro}_1]$$

This tree is constructed as follows.

We start with the theorem that needs to be proved: $(p \Rightarrow q) \Rightarrow (p \Rightarrow q)$. As the dominant operator of this proposition is implication, we can guess that this could be arrived at by applying the $\Rightarrow$ introduction rule.

$$\frac{p \Rightarrow q}{(p \Rightarrow q) \Rightarrow (p \Rightarrow q)} \ [\Rightarrow \text{intro}_1]$$

Given the mechanics of the $\Rightarrow$ introduction rule, $p \Rightarrow q$ is now assumption 1 in our proof, and we are free to use it as many times as we like to derive $p \Rightarrow q$.

Moving up our tree, we now consider $p \Rightarrow q$. Again, the dominant operator here is implication, so we can guess that this could be arrived at by applying the $\Rightarrow$ introduction rule.

$$\frac{\dfrac{q}{p \Rightarrow q} \ [\Rightarrow \text{intro}_2]}{(p \Rightarrow q) \Rightarrow (p \Rightarrow q)} \ [\Rightarrow \text{intro}_1]$$

Again, given the mechanics of the $\Rightarrow$ introduction rule, p is now assumption 2 in our proof, and we are free to use it as many times as we like to derive q.

We are now in a position in which we have to prove q from p and $p \Rightarrow q$. There is, of course, a rule that allows us to do this: $\Rightarrow$ elimination. As such, we are in a position to complete our tree.

$$\frac{\dfrac{\dfrac{\lceil p \Rightarrow q \rceil_1 \quad \lceil p \rceil_2}{q} \ [\Rightarrow \text{elim}]}{p \Rightarrow q} \ [\Rightarrow \text{intro}_2]}{(p \Rightarrow q) \Rightarrow (p \Rightarrow q)} \ [\Rightarrow \text{intro}_1]$$

We have a tree with no undischarged assumptions, and, as such, we may conclude that

$$(p \Rightarrow q) \Rightarrow (p \Rightarrow q)$$

is a theorem of propositional logic. □

In the above, the subscript 1 indicates that the application of $\Rightarrow$ introduction to conclude $p \Rightarrow q$ is associated with the assumption of p and, in addition, the subscript 2 indicates that the application of $\Rightarrow$ introduction to conclude $(p \Rightarrow q) \Rightarrow (p \Rightarrow q)$ is associated with the assumption of $p \Rightarrow q$.

Recall that if a proposition q is logically equivalent to *true*, then, for any proposition p the implication $p \Rightarrow q$ is also logically equivalent to *true*. Similarly—in the context of natural deduction— if we have established via a proof tree that some proposition p is true, then we can conclude $q \Rightarrow p$ is also true for any proposition q.

Example 3.32 Recall that the statement "Pigs can fly $\Rightarrow 1 + 1 = 2$" is logically equivalent to *true* because $(p \Rightarrow true) \Leftrightarrow true$ is a theorem of propositional logic. We can prove that this is a theorem of propositional logic as follows.

$$\frac{\dfrac{}{1 + 1 = 2} \ [\text{mathematics}]}{\text{Pigs can fly} \Rightarrow 1 + 1 = 2} \ [\Rightarrow \text{intro}]$$

□

In the above, we can claim that $1 + 1 = 2$ is a theorem of mathematics. As such, we need no premises or assumptions to establish its validity. Even though the statement "Pigs can fly" was not used in the proof of $1+1 = 2$, we are free to invoke the $\Rightarrow$ introduction rule due to the fact that $p \Rightarrow true$ is logically equivalent to *true* for all propositions p. We shall refer to such assumptions as *vacuous* and, in such cases, we shall not subscript the corresponding rule. It is as though we have assumed "Pigs can fly" for our proof of $1+1 = 2$, but we have not had to use it at any stage.

Our next rule is called true introduction: it simply states that *true* is a theorem of propositional logic.

$$\frac{}{true} \text{ [true intro]}$$

Exercise 3.36 Using the proof rules seen so far, prove that *false* $\Rightarrow$ *true* is a theorem of propositional logic. $\square$

Our next rules concern the conjunction operator. If we have managed to prove that $p \wedge q$ is true, then it follows that p is true.

$$\frac{p \wedge q}{p} \text{ [}\wedge \text{ elim1]}$$

Similarly, if we have managed to prove that $p \wedge q$ is true, then it follows that q is true.

$$\frac{p \wedge q}{q} \text{ [}\wedge \text{ elim2]}$$

Note that the 1 and 2 in the names $\wedge$ elim1 and $\wedge$ elim2 are not associated with assumptions, but are part of the names of the rules and indicate which conjunct forms the conclusion.

Conversely, if we have managed to prove that p and q are both true, then it follows that $p \wedge q$ is *true*.

$$\frac{p \quad q}{p \wedge q} \text{ [}\wedge \text{ intro]}$$

Example 3.33 The above rules allow us to establish that

$$((p \wedge q) \wedge r) \Rightarrow (p \wedge (q \wedge r))$$

is a theorem of propositional logic. This is demonstrated below.

$$\frac{\dfrac{\dfrac{\lceil (p \wedge q) \wedge r \rceil_1}{p \wedge q} \text{ [}\wedge \text{ elim1]}}{p} \text{ [}\wedge \text{ elim1]} \qquad \dfrac{\dfrac{\dfrac{\lceil (p \wedge q) \wedge r \rceil_1}{p \wedge q} \text{ [}\wedge \text{ elim1]}}{q} \text{ [}\wedge \text{ elim2]} \qquad \dfrac{\lceil (p \wedge q) \wedge r \rceil_1}{r} \text{ [}\wedge \text{ elim2]}}{q \wedge r} \text{ [}\wedge \text{ intro]}}{\dfrac{p \wedge (q \wedge r)}{((p \wedge q) \wedge r) \Rightarrow (p \wedge (q \wedge r))} \text{ [}\Rightarrow \text{ intro}_1\text{]}}$$

$\square$

Exercise 3.37 Prove, using the introduction and elimination rules for conjunction, that

$$(p \wedge q) \Rightarrow (q \wedge p)$$

is a theorem of propositional logic. □

The rules for $\vee$ introduction are also intuitive: if we have established p, then we can conclude that $p \vee q$ is *true*.

$$\frac{p}{p \vee q} \;[\vee \text{ intro1}]$$

Similarly, if we have established q, then we can also conclude that $p \vee q$ is *true*.

$$\frac{q}{p \vee q} \;[\vee \text{ intro2}]$$

Exercise 3.38 Establish, using the natural deduction rules seen so far, that the following are theorems of propositional logic.

1. $p \wedge q \Rightarrow p \vee q$

2. *false* $\Rightarrow$ (*true* $\vee$ *false*)

□

The rule of $\vee$ elimination, though intuitive, can seem at first appearance to be far more complicated than it actually is.

First, assume that we have established—via natural deduction—that r follows from p, and also that r follows from q. Second, assume that we have established that $p \vee q$ holds. If we know that either p is true or q is true and also that r can be established from both p and q, then we may conclude that r is true.

The rule is given formally as follows.

$$\frac{\begin{array}{ccc} \lceil p \rceil_1 & \lceil q \rceil_1 & \\ \vdots & \vdots & \\ r & r & p \vee q \end{array}}{r} \;[\vee \text{ elim}_1]$$

Example 3.34 Assume that we have established that either Italy or Spain will win a particular football tournament. In addition, we also know—via Geography—that if Italy wins the tournament then there will be a European winner, and we know—again via Geography—that if Spain wins the tournament there will be a European winner. As such, we can conclude that a European team will win the tournament. This derivation may be represented formally as follows.

$$\frac{\dfrac{\lceil \text{Italy win} \rceil_1}{\text{Euro win}} \,[\text{geog}] \quad \dfrac{\lceil \text{Spain win} \rceil_1}{\text{Euro win}} \,[\text{geog}] \quad \dfrac{}{\text{Italy win} \vee \text{Spain win}} \,[\text{theorem}]}{\text{Euro win}} \;[\vee \text{ elim}_1]$$

□

Exercise 3.39 Using the rules of natural deduction seen thus far, prove that

$$((p \Rightarrow r) \land (q \Rightarrow r)) \Rightarrow ((p \lor q) \Rightarrow r)$$

is a theorem of propositional logic. □

Our deduction rules for $\Leftrightarrow$ are relatively intuitive. First, if we have established that $p \Leftrightarrow q$ is true, then it follows that $p \Rightarrow q$ and $q \Rightarrow p$ are also both true.

$$\frac{p \Leftrightarrow q}{p \Rightarrow q} \ [\Leftrightarrow \text{elim1}]$$

$$\frac{p \Leftrightarrow q}{q \Rightarrow p} \ [\Leftrightarrow \text{elim2}]$$

Furthermore, if we have established that both $p \Rightarrow q$ and $q \Rightarrow p$ are true, then it follows that $p \Leftrightarrow q$ is true.

$$\frac{p \Rightarrow q \quad q \Rightarrow p}{p \Leftrightarrow q} \ [\Leftrightarrow \text{intro}]$$

Exercise 3.40 Using the rules of natural deduction seen thus far, prove that

$$(p \Leftrightarrow q) \Rightarrow ((p \Rightarrow q) \land (q \Rightarrow p))$$

is a theorem of propositional logic. □

Exercise 3.41 Extend the solution to Exercise 3.37 to prove that

$$(p \land q) \Leftrightarrow (q \land p)$$

is a theorem of propositional logic. □

As proof trees are concerned with proving theorems, i.e., statements which are always *true*, if we ever establish *false* in a proof tree, then we can conclude that we have made an invalid assumption, which is akin to 'taking a wrong turning' in our derivation. This sentiment is captured by the following rule, which states that if we have managed to establish *false* from p, then p must be equivalent to *false*, i.e., $\neg p$ must be equivalent to *true*.

$$\begin{array}{c} p \\ \vdots \\ \dfrac{false}{\neg p} \ [\neg \text{intro}] \end{array}$$

Similarly, if we have managed to establish *false* from $\neg p$, then $\neg p$ must be *false*, i.e., p must be *true.*

$$\neg\, p$$
$$\vdots$$
$$\frac{false}{p} \ [\neg\, \text{elim}]$$

Finally, it can never be the case that both p and $\neg\, p$ are true. If we ever reach a stage where we have established both p and $\neg\, p$, then we are forced to conclude *false*.

$$\frac{p \quad \neg\, p}{false} \ [\text{false intro}]$$

Exercise 3.42 Using the rules of natural deduction, prove that the following are theorems of propositional logic.

1. $(p \land \neg\, p) \Rightarrow \neg\, (p \land \neg\, p)$
2. $\neg\, (p \lor \neg\, p) \Rightarrow (p \lor \neg\, p)$

$\square$

3.13   Further exercises

Exercise 3.43 Assume that p represents the statement "Jon is happy" and q represents the statement "Steve is in pain". What natural language statements do the following propositions represent?

1. $\neg\, p$
2. $p \land q$
3. $p \lor \neg\, q$
4. $p \Rightarrow q$
5. $p \Leftrightarrow q$

$\square$

Exercise 3.44 Assume that p represents the statement "Jon is happy" and q represents the statement "Steve is in pain". Write propositions to represent the following natural language statements.

1. "Steve is not in pain."
2. "Steve is not in pain and Jon is not happy."
3. "Steve is in pain or Jon is not happy."

4. "If Steve is in pain, then Jon is happy."

5. "Jon is not happy if, and only if, Steve is in pain."

□

Exercise 3.45 Assume that p represents the statement "Jon is happy", q represents the statement "Steve is in pain", and r represents the statement "Ali is angry". What natural language statements do the following propositions represent?

1. $(p \land q) \Rightarrow r$

2. $p \lor (q \land \neg r)$

3. $p \Leftrightarrow (q \land r)$

□

Exercise 3.46 Assume that p represents the statement "Jon is happy", q represents the statement "Steve is in pain", and r represents the statement "Ali is angry". Write propositions to represent the following natural language statements.

1. "Jon is happy and it is not the case that Steve is in pain and Ali is angry."

2. "If Ali is angry, then either Jon is happy or Steve is in pain."

3. "Steve is in pain if, and only if, Jon is happy and Ali is not angry."

□

Exercise 3.47 Write out truth tables for each of the following propositions.

1. $p \lor \neg q$

2. $\neg p \land q$

3. $p \Rightarrow (p \land q)$

4. $\neg p \Leftrightarrow q$

□

Exercise 3.48 Prove, using equational reasoning, that $p \Rightarrow (p \land q)$ is equivalent to $p \Rightarrow q$. □

Exercise 3.49 Prove that

$$(\neg p \Rightarrow p) \Leftrightarrow p$$

is a tautology using equational reasoning. □

Exercise 3.50 Prove, using a proof tree, that

$$p \Rightarrow (q \Rightarrow p)$$

is a theorem of propositional logic. □

Exercise 3.51 Prove, using a proof tree, that

$$(p \wedge \neg p) \Rightarrow (true \Rightarrow false)$$

is a theorem of propositional logic. □

Exercise 3.52 A *conjunct* is a collection of propositions combined via conjunction. For example, $p \wedge q \wedge r$ is a conjunct. A *disjunct* is a collection of propositions combined via disjunction. For example, $p \vee q \vee r$ is a disjunct. A proposition is said to be in *conjunctive normal form* when each of the following conditions are met.

- The operators $\Rightarrow$ and $\Leftrightarrow$ do not appear in the proposition.

- All occurrences of $\neg$ are associated with atomic propositions.

- If the proposition involves occurrences of $\wedge$ and $\vee$, then the proposition is a conjunct of disjuncts.

So, for example, $p \vee q$, $p \wedge \neg q$, and $(p \vee q) \wedge (p \vee r)$ are all in conjunctive normal form. The propositions $\neg (p \wedge q)$ and $(p \wedge q) \vee (p \wedge r)$, on the other hand, are not in conjunctive normal form.

Using equational reasoning, convert each of the following propositions to conjunctive normal form.

1. $p \Rightarrow (q \vee r)$

2. $(p \wedge q) \Rightarrow (q \vee r)$

□

Exercise 3.53 A proposition is said to be in *disjunctive normal form* when each of the following conditions are met.

- The operators $\Rightarrow$ and $\Leftrightarrow$ do not appear in the proposition.

- All occurrences of $\neg$ are associated with atomic propositions.

- If the proposition involves occurrences of $\wedge$ and $\vee$, then the proposition is a disjunct of conjuncts.

So, for example, $p \vee q$, $p \wedge \neg q$, and $(p \wedge q) \vee (p \wedge r)$ are all in disjunctive normal form. The propositions $\neg (p \wedge q)$ and $(p \vee q) \wedge (p \vee r)$, on the other hand, are not in disjunctive normal form.

Using equational reasoning, convert each of the following propositions to disjunctive normal form.

1. $(p \wedge (q \vee r)) \vee p$

2. $(p \vee (q \wedge r)) \wedge p$

□

3.14 Solutions

Solution 3.1

1. *true*

2. *true*

3. *false*

□

Solution 3.2 We can attach truth values to all but the first statement: there is insufficient information (i.e., *which* Jon) to be able to determine whether statement 1 is true or false. Statement 2 is false. Finally, statements 3 and 4 are both true. □

Solution 3.3

1. "It is not snowing."

2. "Jon does not like Ali."

3. "x is less than or equal to y."

□

Solution 3.4

1. *false*

2. *false*

3. *true*

□

Solution 3.5

1. *false*

2. *false*

3. *true*

□

Solution 3.6 The truth table for $(\neg\,p) \wedge (\neg\,q)$ is given below.

p	q	$\neg\,p$	$\neg\,q$	$(\neg\,\mathbf{p}) \wedge (\neg\,\mathbf{q})$
true	*true*	*false*	*false*	**false**
true	*false*	*false*	*true*	**false**
false	*true*	*true*	*false*	**false**
false	*false*	*true*	*true*	**true**

□

Solution 3.7 The truth table for $\neg\,(p \wedge q)$ is given below.

p	q	$p \wedge q$	$\neg\,(\mathbf{p} \wedge \mathbf{q})$
true	*true*	*true*	**false**
true	*false*	*false*	**true**
false	*true*	*false*	**true**
false	*false*	*false*	**true**

□

Solution 3.8

1. This truth table will have eight rows, as there eight possible combinations, as illustrated below.

p	q	r
true	*true*	*true*
true	*true*	*false*
true	*false*	*true*
true	*false*	*false*
false	*true*	*true*
false	*true*	*false*
false	*false*	*true*
false	*false*	*false*

2. The truth table for $p \wedge q$ has four rows; the truth table for $p \wedge q \wedge r$ has eight rows; the truth table for $p \wedge q \wedge r \wedge s$ has sixteen rows. In each case, each atomic proposition may take one of two possible values: *true* or *false*. As such, the truth table for a proposition with n atomic propositions will have 2^n rows.

3. Such a truth table will have one row, as the proposition *true* can only ever be *true* and the proposition *false* can only ever be *false*, i.e., there is only one possible combination of values.

□

Solution 3.9

1. By Law 3.2, $\neg\,\neg\,(true \wedge false)$ is equivalent to *true* $\wedge$ *false*. By Law 3.5, *true* $\wedge$ *false* is equivalent to *false*. As such, the proposition is logically equivalent to *false*.

2. By Law 3.5, $p \wedge false$ is equivalent to *false*. Again, by Law 3.5, *true* $\wedge$ *false* is equivalent to *false*. As such, the proposition is logically equivalent to *false*.

3. By Law 3.5, $p \wedge false$ is equivalent to *false*. By Law 3.1, $\neg\,false$ is equivalent to *true*. By either Law 3.3 or Law 3.4, *true* $\wedge$ *true* is equivalent to *true*. As such, the proposition is logically equivalent to *true*.

□

Solution 3.10

1. By Law 3.8, (*false* ∧ *p*) ∧ *q* is equivalent to *false* ∧ (*p* ∧ *q*). By Law 3.7, this is equivalent to (*p* ∧ *q*) ∧ *false*. By Law 3.5, this is equivalent to *false*. As such, the proposition is logically equivalent to *false*.

2. By Law 3.7, *p* ∧ (*false* ∧ *q*) is equivalent to *p* ∧ (*q* ∧ *false*). By Law 3.5, this is equivalent to *p* ∧ *false*. Again, by Law 3.5, this is equivalent to *false*. As such, the proposition is logically equivalent to *false*.

□

Solution 3.11

1. *false*

2. *true*

3. *true*

□

Solution 3.12 The truth table for ¬ (*p* ∨ *q*) is given below.

p	*q*	*p* ∨ *q*	¬ (**p** ∨ **q**)
true	*true*	*true*	**false**
true	*false*	*true*	**false**
false	*true*	*true*	**false**
false	*false*	*false*	**true**

□

Solution 3.13 The truth table for (¬ *p*) ∨ (¬ *q*) is given below.

p	*q*	¬ *p*	¬ *q*	(¬ **p**) ∨ (¬ **q**)
true	*true*	*false*	*false*	**false**
true	*false*	*false*	*true*	**true**
false	*true*	*true*	*false*	**true**
false	*false*	*true*	*true*	**true**

□

Solution 3.14

1. By Law 3.11, *p* ∨ *false* is equivalent to *p*. By Law 3.12, *p* ∨ *true* is equivalent to *true*. As such, the proposition is logically equivalent to *true*.

2. By Law 3.5, *p* ∧ *false* is equivalent to *false*. By Law 3.12, *false* ∨ *true* is equivalent to *true*. By Law 3.1, ¬ *true* is equivalent to *false*. As such, the proposition is logically equivalent to *false*.

□

Solution 3.15

1. Starting with the left conjunct, by Law 3.14, *true* ∨ *p* is equivalent to *p* ∨ *true*. By Law 3.12, *p* ∨ *true* is equivalent to *true*, and by Law 3.1, ¬ *true* is equivalent to *false*. Moving to the right conjunct, by Law 3.11, *p* ∨ *false* is equivalent to *p*. Now considering the overall proposition, by Law 3.7, *false* ∧ *p* is equivalent to *p* ∧ *false*. Finally, by Law 3.5, *p* ∧ *false* is equivalent to *false*. As such, the proposition is logically equivalent to *false*.

2. By Law 3.14, *true* ∨ *q* is equivalent to *q* ∨ *true*. By Law 3.13, *p* ∨ (*q* ∨ *true*) is equivalent to (*p* ∨ *q*) ∨ *true*. Finally, by Law 3.12, this is equivalent to *true*. As such, the proposition is logically equivalent to *true*.

☐

Solution 3.16

1. By Law 3.17, *p* ∧ ((¬ *p*) ∨ *q*) is equivalent to (*p* ∧ (¬ *p*)) ∨ (*p* ∧ *q*). By Law 3.6, this statement is equivalent to *false* ∨ (*p* ∧ *q*). By Law 3.14, this is equivalent to (*p* ∧ *q*) ∨ *false*. Finally, by Law 3.11, this is equivalent to *p* ∧ *q*.

2. By Law 3.16, (*p* ∧ *q*) ∨ ¬ *p* is equivalent to (*p* ∨ (¬ *p*)) ∧ (*q* ∨ (¬ *p*)). By Law 3.15, this is equivalent to *true* ∧ (*q* ∨ (¬ *p*)). By Law 3.7, this is equivalent to (*q* ∨ (¬ *p*)) ∧ *true*. Finally, by Law 3.4, this is equivalent to *q* ∨ (¬ *p*).

☐

Solution 3.17

1. $e \Rightarrow r$

2. $r \Rightarrow e$

3. $e \Rightarrow r$

4. $\neg (r \Rightarrow (\neg e))$

☐

Solution 3.18

1. *false* ⇒ *false*, which is equivalent to *true*.

2. *true* ⇒ *true*, which is equivalent to *true*.

3. *false* ⇒ *true*, which is equivalent to *true*.

4. *true* ⇒ *false*, which is equivalent to *false*.

☐

Solution 3.19 The truth table for $(p \wedge q) \Rightarrow (p \vee q)$ is given below.

p	q	$p \wedge q$	$p \vee q$	$(\mathbf{p} \wedge \mathbf{q}) \Rightarrow (\mathbf{p} \vee \mathbf{q})$
true	*true*	*true*	*true*	**true**
true	*false*	*false*	*true*	**true**
false	*true*	*false*	*true*	**true**
false	*false*	*false*	*false*	**true**

□

Solution 3.20 The truth table for $(p \vee q) \Rightarrow (p \wedge q)$ is given below.

p	q	$p \vee q$	$p \wedge q$	$(\mathbf{p} \vee \mathbf{q}) \Rightarrow (\mathbf{p} \wedge \mathbf{q})$
true	*true*	*true*	*true*	**true**
true	*false*	*true*	*false*	**false**
false	*true*	*true*	*false*	**false**
false	*false*	*false*	*false*	**true**

□

Solution 3.21

1. By Law 3.18, $p \Rightarrow p$ is equivalent to $(\neg p) \vee p$. By Law 3.15, this is equivalent to *true*.

2. By Law 3.18, $(p \wedge q) \Rightarrow q$ is equivalent to $(\neg (p \wedge q)) \vee q$. By Law 3.9, this is equivalent to $((\neg p) \vee (\neg q)) \vee q$. By Law 3.13, this is equivalent to $(\neg p) \vee ((\neg q) \vee q)$. By Law 3.15, this is equivalent to $(\neg p) \vee$ *true*. By Law 3.12, this is equivalent to *true*.

□

Solution 3.22

1. For $\Rightarrow$ to be associative, it must be the case that $p \Rightarrow (q \Rightarrow r)$ is logically equivalent to $(p \Rightarrow q) \Rightarrow r$ for all propositions p, q, and r. Assume that p is *false*, q is *true*, and r is *false*. Here, the former proposition is logically equivalent to *false* $\Rightarrow$ *false*, which is *true*, and the latter proposition is logically equivalent to *true* $\Rightarrow$ *false*, which is *false*. As such, $\Rightarrow$ is not associative.

2. For $\Rightarrow$ to be commutative, it must be the case that $p \Rightarrow q$ is logically equivalent to $q \Rightarrow p$ for all propositions p and q. Assume that p is *true* and q is *false*; it is certainly not the case that *true* $\Rightarrow$ *false* is logically equivalent to *false* $\Rightarrow$ *true*. As such, $\Rightarrow$ is not commutative.

3. For $\Rightarrow$ to be idempotent, it must be the case that $p \Rightarrow p$ is logically equivalent to p for all propositions p. Assume that p is *false*. Here, $p \Rightarrow p$ is equivalent to *true*. As such, $\Rightarrow$ is not idempotent.

□

Solution 3.23

1. *false* $\Leftrightarrow$ *false*, which is equivalent to *true*.

2. *true* $\Leftrightarrow$ *true*, which is equivalent to *true*.

3. *false* $\Leftrightarrow$ *true*, which is equivalent to *false*.

4. *true* $\Leftrightarrow$ *false*, which is equivalent to *false*.

$\square$

Solution 3.24 The truth table for $(p \Rightarrow q) \Leftrightarrow (q \Rightarrow p)$ is given below.

p	q	$p \Rightarrow q$	$q \Rightarrow p$	$(\mathbf{p \Rightarrow q}) \Leftrightarrow (\mathbf{q \Rightarrow p})$
true	*true*	*true*	*true*	**true**
true	*false*	*false*	*true*	**false**
false	*true*	*true*	*false*	**false**
false	*false*	*true*	*true*	**true**

$\square$

Solution 3.25

1. By Law 3.23, $p \Leftrightarrow$ *true* is equivalent to $(p \Rightarrow true) \wedge (true \Rightarrow p)$. By Law 3.18, this statement is equivalent to $((\neg p) \vee true) \wedge ((\neg true) \vee p)$. By Law 3.12, this is equivalent to $true \wedge ((\neg true) \vee p)$. By Law 3.1, we get $true \wedge (false \vee p)$. By Law 3.14, we get $true \wedge (p \vee false)$. By Law 3.11, this is equivalent to $true \wedge p$. By Law 3.7, this is equivalent to $p \wedge true$. By Law 3.4, this is equivalent to p. As the left-hand side of the equivalence is logically equivalent to p and the right-hand side of the equivalence is p, by Law 3.21, this proposition is equivalent to *true*.

2. By Law 3.23, $p \Leftrightarrow$ *false* is equivalent to $(p \Rightarrow false) \wedge (false \Rightarrow p)$. By Law 3.18, this statement is equivalent to $((\neg p) \vee false) \wedge ((\neg false) \vee p)$. By Law 3.11, this is equivalent to $(\neg p) \wedge ((\neg false) \vee p)$. By Law 3.1, we get $(\neg p) \wedge (true \vee p)$. By Law 3.14, we get $(\neg p) \wedge (p \vee true)$. By Law 3.12, this is equivalent to $(\neg p) \wedge true$. By Law 3.4, this is equivalent to $\neg p$. As the left-hand side of the equivalence is logically equivalent to $\neg p$ and the right-hand side of the equivalence is $\neg p$, by Law 3.21, this proposition is equivalent to *true*.

$\square$

Solution 3.26 For $\Leftrightarrow$ to be idempotent, we require that $p \Leftrightarrow p$ is logically equivalent to p for any proposition p. If p is *false* then $p \Leftrightarrow p$ is equivalent to *true*. As such, $\Leftrightarrow$ is not idempotent. $\square$

Solution 3.27

1. $(\neg p) \vee q$

2. $p \vee (q \wedge r)$

3. $(p \wedge q) \Rightarrow r$

4. $(\neg p) \Leftrightarrow (((\neg q) \wedge r) \Rightarrow s)$

$\square$

Solution 3.28

1. If p is *false* and q is *false*, then the overall proposition is *false*. Therefore, it is not a tautology.

2. No matter what combination of truth values we assign to p and q, the proposition is *true*. Therefore, it is a tautology.

3. If q is *true* and p is *false*, then the overall proposition is *false*. Therefore, it is not a tautology.

4. No matter what combination of truth values we assign to p and q, the proposition is *true*. Therefore, it is a tautology.

□

Solution 3.29

1. No matter what combination of truth values we assign to p and q, the proposition is *false*. Therefore, it is a contradiction.

2. No matter what combination of truth values we assign to p and q, the proposition is *false*. Therefore, it is a contradiction.

3. No matter what values we assign to p and q, the first argument of the implication will be *false*. As such, the proposition will always be *true*. Therefore, it is not a contradiction.

□

Solution 3.30

1. No matter what combination of truth values we assign to p and q, the proposition is *true*. Therefore, it is not a contingency.

2. If p is *true* and q is *false* (or vice versa), then the proposition is *false*. On the other hand, if p and q are both *true* (or both *false*), then the proposition is *true*. Therefore, it is a contingency.

□

Solution 3.31

1.

p	q	$p \wedge q$	$\neg (p \wedge q)$	$(\mathbf{p} \wedge \mathbf{q}) \vee \neg (\mathbf{p} \wedge \mathbf{q})$
true	*true*	*true*	*false*	**true**
true	*false*	*false*	*true*	**true**
false	*true*	*false*	*true*	**true**
false	*false*	*false*	*true*	**true**

Every entry in the final column is *true*, so the proposition is a tautology.

2.

p	q	$p \Rightarrow q$	$\neg q \Rightarrow \neg p$	$(\mathbf{p} \Rightarrow \mathbf{q}) \Leftrightarrow (\neg \mathbf{q} \Rightarrow \neg \mathbf{p})$
true	*true*	*true*	*true*	**true**
true	*false*	*false*	*false*	**true**
false	*true*	*true*	*true*	**true**
false	*false*	*true*	*true*	**true**

Every entry in the final column is *true*, so the proposition is a tautology.

3.

p	q	$\neg p \Rightarrow q$	$p \Rightarrow \neg q$	$(\neg \mathbf{p} \Rightarrow \mathbf{q}) \Rightarrow (\mathbf{p} \Rightarrow \neg \mathbf{q})$
true	*true*	*true*	*false*	**false**
true	*false*	*true*	*true*	**true**
false	*true*	*true*	*true*	**true**
false	*false*	*false*	*true*	**true**

There is one *false* entry and three *true* entries in the final column, so the proposition is a contingency.

4.

p	q	$p \Rightarrow q$	$\neg p \Rightarrow \neg q$	$(\mathbf{p} \Rightarrow \mathbf{q}) \Rightarrow (\neg \mathbf{p} \Rightarrow \neg \mathbf{q})$
true	*true*	*true*	*true*	**true**
true	*false*	*false*	*true*	**true**
false	*true*	*true*	*false*	**false**
false	*false*	*true*	*true*	**true**

There is one *false* entry and three *true* entries in the final column, so the proposition is a contingency.

5.

p	$p \Rightarrow p$	$\neg (p \Rightarrow p)$	$\neg p \Rightarrow \neg p$	$\neg (\mathbf{p} \Rightarrow \mathbf{p}) \wedge (\neg \mathbf{p} \Rightarrow \neg \mathbf{p})$
true	*true*	*false*	*true*	**false**
false	*true*	*false*	*true*	**false**

Every entry in the final column is *false*, so the proposition is a contradiction.

□

Solution 3.32

1. $(p \Leftrightarrow true) \Leftrightarrow p$
   $\Leftrightarrow ((p \Rightarrow true) \land (true \Rightarrow p)) \Leftrightarrow p$ [Law 3.23]
   $\Leftrightarrow (((\neg\, p) \lor true) \land ((\neg\, true) \lor p)) \Leftrightarrow p$ [Law 3.18]
   $\Leftrightarrow (true \land ((\neg\, true) \lor p)) \Leftrightarrow p$ [Law 3.12]
   $\Leftrightarrow (true \land (false \lor p)) \Leftrightarrow p$ [Law 3.1]
   $\Leftrightarrow (true \land (p \lor false)) \Leftrightarrow p$ [Law 3.14]
   $\Leftrightarrow (true \land p) \Leftrightarrow p$ [Law 3.11]
   $\Leftrightarrow (p \land true) \Leftrightarrow p$ [Law 3.7]
   $\Leftrightarrow p \Leftrightarrow p$ [Law 3.4]
   $\Leftrightarrow true$ [Law 3.21]

2. $(p \Leftrightarrow false) \Leftrightarrow (\neg\, p)$
   $\Leftrightarrow ((p \Rightarrow false) \land (false \Rightarrow p)) \Leftrightarrow (\neg\, p)$ [Law 3.23]
   $\Leftrightarrow (((\neg\, p) \lor false) \land ((\neg\, false) \lor p)) \Leftrightarrow (\neg\, p)$ [Law 3.18]
   $\Leftrightarrow ((\neg\, p) \land ((\neg\, false) \lor p)) \Leftrightarrow (\neg\, p)$ [Law 3.11]
   $\Leftrightarrow ((\neg\, p) \land (true \lor p)) \Leftrightarrow (\neg\, p)$ [Law 3.1]
   $\Leftrightarrow ((\neg\, p) \land (p \lor true)) \Leftrightarrow (\neg\, p)$ [Law 3.14]
   $\Leftrightarrow ((\neg\, p) \land true) \Leftrightarrow (\neg\, p)$ [Law 3.12]
   $\Leftrightarrow (\neg\, p) \Leftrightarrow (\neg\, p)$ [Law 3.4]
   $\Leftrightarrow true$ [Law 3.21]

□

Solution 3.33

1. $(p \land \neg\, q) \Rightarrow q$
   $\Leftrightarrow \neg\, (p \land \neg\, q) \lor q$ [Law 3.18]
   $\Leftrightarrow \neg\, p \lor q \lor q$ [Law 3.9]
   $\Leftrightarrow \neg\, p \lor q$ [Law 3.10]

2. $(p \Rightarrow q) \land p$
   $\Leftrightarrow (\neg\, p \lor q) \land p$ [Law 3.18]
   $\Leftrightarrow (\neg\, p \land p) \lor (q \land p)$ [Law 3.17]
   $\Leftrightarrow \neg\, (p \lor \neg\, p) \lor (q \land p)$ [Law 3.9]
   $\Leftrightarrow \neg\, (\neg\, p \lor p) \lor (q \land p)$ [Law 3.13]
   $\Leftrightarrow \neg\, true \lor (q \land p)$ [Law 3.15]
   $\Leftrightarrow false \lor (q \land p)$ [Law 3.1]
   $\Leftrightarrow (q \land p) \lor false$ [Law 3.14]
   $\Leftrightarrow q \land p$ [Law 3.11]

□

Solution 3.34

1. $(p \Rightarrow q) \Leftrightarrow (\neg\, q \Rightarrow \neg\, p)$
    $\Leftrightarrow (\neg\, p \vee q) \Leftrightarrow (\neg\, q \Rightarrow \neg\, p)$ [Law 3.18]
    $\Leftrightarrow (\neg\, p \vee q) \Leftrightarrow (\neg\,\neg\, q \vee \neg\, p)$ [Law 3.18]
    $\Leftrightarrow (\neg\, p \vee q) \Leftrightarrow (q \vee \neg\, p)$ [Law 3.2]
    $\Leftrightarrow (q \vee \neg\, p) \Leftrightarrow (q \vee \neg\, p)$ [Law 3.13]
    $\Leftrightarrow true$ [Law 3.21]

2. $(p \Leftrightarrow true) \Rightarrow p$
    $\Leftrightarrow p \Rightarrow p$ [Exercise 3.32]
    $\Leftrightarrow true$ [Exercise 3.21]

□

Solution 3.35

1. $(p \Leftrightarrow q) \Rightarrow (q \Leftrightarrow p)$
    $\Leftrightarrow (p \Leftrightarrow q) \Rightarrow (p \Leftrightarrow q)$ [Law 3.20]
    $\Leftrightarrow true$ [Exercise 3.32]

2. $(p \wedge q) \Leftrightarrow (\neg\, p \vee \neg\, q)$
    $\Leftrightarrow (p \wedge q) \Leftrightarrow \neg\, (p \wedge q)$ [Law 3.9]
    $\Leftrightarrow false$ [Law 3.22]

□

Solution 3.36

$$\dfrac{\dfrac{}{true}\;[\text{true intro}]}{false \Rightarrow true}\;[\Rightarrow \text{intro}]$$

□

Solution 3.37

$$\dfrac{\dfrac{\lceil p \wedge q \rceil_1}{q}\;[\wedge\text{ elim2}] \quad \dfrac{\lceil p \wedge q \rceil_1}{p}\;[\wedge\text{ elim1}]}{\dfrac{q \wedge p}{(p \wedge q) \Rightarrow (q \wedge p)}\;[\Rightarrow \text{intro}_1]}\;[\wedge\text{ intro}]$$

□

Solution 3.38

1.
$$\dfrac{\dfrac{\dfrac{\lceil p \wedge q \rceil_1}{p}\;[\wedge\text{ elim1}]}{p \vee q}\;[\vee\text{ intro1}]}{p \wedge q \Rightarrow p \vee q}\;[\Rightarrow \text{intro}_1]$$

2.

$$\frac{\dfrac{\overline{\rule{1.5cm}{0.4pt}}\ \text{[true intro]}}{true}}{\dfrac{true \lor false}{false \Rightarrow (true \lor false)}\ \text{[}\Rightarrow\text{ intro]}}\ \text{[}\lor\text{ intro1]}$$

□

Solution 3.39 Starting from the bottom of the tree, we have the following.

$$\frac{\dfrac{\dfrac{r \quad r \quad \lceil p \lor q \rceil_2}{r}\ \text{[}\lor\text{ elim}_3\text{]}}{(p \lor q) \Rightarrow r}\ \text{[}\Rightarrow\text{ intro}_2\text{]}}{((p \Rightarrow r) \land (q \Rightarrow r)) \Rightarrow ((p \lor q) \Rightarrow r)}\ \text{[}\Rightarrow\text{ intro}_1\text{]}$$

We are now left with two tasks: to establish r, given the assumptions p and $(p \Rightarrow r) \land (q \Rightarrow r)$, and to establish r, given the assumptions q and $(p \Rightarrow r) \land (q \Rightarrow r)$. We establish each of these in turn:

$$\frac{\dfrac{\lceil (p \Rightarrow r) \land (q \Rightarrow r) \rceil_1}{p \Rightarrow r}\ \text{[}\land\text{ elim1]} \qquad \lceil p \rceil_3}{r}\ \text{[}\Rightarrow\text{ elim]}$$

$$\frac{\dfrac{\lceil (p \Rightarrow r) \land (q \Rightarrow r) \rceil_1}{q \Rightarrow r}\ \text{[}\land\text{ elim2]} \qquad \lceil q \rceil_3}{r}\ \text{[}\Rightarrow\text{ elim]}$$

□

Solution 3.40

$$\frac{\dfrac{\dfrac{\lceil p \Leftrightarrow q \rceil_1}{p \Rightarrow q}\ \text{[}\Leftrightarrow\text{ elim1]} \quad \dfrac{\lceil p \Leftrightarrow q \rceil_1}{q \Rightarrow p}\ \text{[}\Leftrightarrow\text{ elim2]}}{(p \Rightarrow q) \land (q \Rightarrow p)}\ \text{[}\land\text{ intro]}}{(p \Leftrightarrow q) \Rightarrow ((p \Rightarrow q) \land (q \Rightarrow p))}\ \text{[}\Rightarrow\text{ intro}_1\text{]}$$

□

Solution 3.41

$$\frac{\dfrac{\dfrac{\dfrac{\lceil p \land q \rceil_1}{q}\ \text{[}\land\text{ elim2]} \quad \dfrac{\lceil p \land q \rceil_1}{p}\ \text{[}\land\text{ elim1]}}{q \land p}\ \text{[}\land\text{ intro]}}{(p \land q) \Rightarrow (q \land p)}\ \text{[}\Rightarrow\text{ intro}_1\text{]} \qquad \dfrac{\dfrac{\dfrac{\lceil q \land p \rceil_2}{p}\ \text{[}\land\text{ elim2]} \quad \dfrac{\lceil q \land p \rceil_2}{q}\ \text{[}\land\text{ elim1]}}{p \land q}\ \text{[}\land\text{ intro]}}{(q \land p) \Rightarrow (p \land q)}\ \text{[}\Rightarrow\text{ intro}_2\text{]}}{(p \land q) \Leftrightarrow (q \land p)}\ \text{[}\Leftrightarrow\text{ intro]}$$

□

Solution 3.42

1.

$$\cfrac{\cfrac{\lceil p \wedge \neg p \rceil_1}{p} \text{ [}\wedge \text{ elim1]} \qquad \cfrac{\lceil p \wedge \neg p \rceil_1}{\neg p} \text{ [}\wedge \text{ elim2]}}{\cfrac{\cfrac{false}{\neg (p \wedge \neg p)} \text{ [}\neg \text{ intro]}}{(p \wedge \neg p) \Rightarrow \neg (p \wedge \neg p)} \text{ [}\Rightarrow \text{ intro}_1\text{]}} \text{ [false intro]}$$

2.

$$\cfrac{\cfrac{\cfrac{\lceil \neg (p \vee \neg p) \rceil_1}{\neg p \wedge \neg \neg p} \text{ [de Morgan]}}{\neg p} \text{ [}\wedge \text{ elim1]} \qquad \cfrac{\cfrac{\lceil \neg (p \vee \neg p) \rceil_1}{\neg p \wedge \neg \neg p} \text{ [de Morgan]}}{\neg \neg p} \text{ [}\wedge \text{ elim2]}}{\cfrac{\cfrac{false}{p \vee \neg p} \text{ [}\neg \text{ elim]}}{\neg (p \vee \neg p) \Rightarrow (p \vee \neg p)} \text{ [}\Rightarrow \text{ intro}_1\text{]}} \text{ [false intro]}$$

□

Solution 3.43

1. "Jon is not happy."
2. "Jon is happy and Steve is in pain."
3. "Jon is happy or Steve is not in pain."
4. "If Jon is happy, then Steve is in pain."
5. "Jon is happy if, and only if, Steve is in pain."

□

Solution 3.44

1. $\neg q$
2. $\neg q \wedge \neg p$
3. $q \vee \neg p$
4. $q \Rightarrow p$
5. $\neg p \Leftrightarrow q$

□

Solution 3.45

1. "If Jon is happy and Steve is in pain, then Ali is angry."
2. "Either Jon is happy, or Steve is in pain and Ali is not angry."

3. "Jon is happy if, and only if, Steve is in pain and Ali is angry."

□

Solution 3.46

1. $p \wedge \neg (q \wedge r)$
2. $r \Rightarrow (p \vee q)$
3. $q \Leftrightarrow p \wedge \neg r$

□

Solution 3.47

1.

p	q	$\neg q$	$\mathbf{p} \vee \neg \mathbf{q}$
true	*true*	*false*	**true**
true	*false*	*true*	**true**
false	*true*	*false*	**false**
false	*false*	*true*	**true**

2.

p	q	$\neg p$	$\neg \mathbf{p} \wedge \mathbf{q}$
true	*true*	*false*	**false**
true	*false*	*false*	**false**
false	*true*	*true*	**true**
false	*false*	*true*	**false**

3.

p	q	$p \wedge q$	$\mathbf{p} \Rightarrow (\mathbf{p} \wedge \mathbf{q})$
true	*true*	*true*	**true**
true	*false*	*false*	**false**
false	*true*	*false*	**true**
false	*false*	*false*	**true**

4.

p	q	$\neg p$	$\neg \mathbf{p} \Leftrightarrow \mathbf{q}$
true	*true*	*false*	**false**
true	*false*	*false*	**true**
false	*true*	*true*	**true**
false	*false*	*true*	**false**

□

Solution 3.48

$$
\begin{aligned}
& p \Rightarrow (p \wedge q) \\
\Leftrightarrow\ & \neg\, p \vee (p \wedge q) && \text{[Law 3.18]} \\
\Leftrightarrow\ & (\neg\, p \vee p) \wedge (\neg\, p \vee q) && \text{[Law 3.16]} \\
\Leftrightarrow\ & true \wedge (\neg\, p \vee q) && \text{[Law 3.15]} \\
\Leftrightarrow\ & (\neg\, p \vee q) \wedge true && \text{[Law 3.8]} \\
\Leftrightarrow\ & \neg\, p \vee q && \text{[Law 3.4]} \\
\Leftrightarrow\ & p \Rightarrow q && \text{[Law 3.18]}
\end{aligned}
$$

□

Solution 3.49

$$
\begin{aligned}
& (\neg\, p \Rightarrow p) \Leftrightarrow p \\
\Leftrightarrow\ & ((\neg\, p \Rightarrow p) \Rightarrow p) \wedge (p \Rightarrow (\neg\, p \Rightarrow p)) && \text{[Law 3.23]} \\
\Leftrightarrow\ & (\neg\, (\neg\, p \Rightarrow p) \vee p) \wedge (\neg\, p \vee (\neg\, p \Rightarrow p)) && \text{[Law 3.18]} \\
\Leftrightarrow\ & (\neg\, (p \vee p) \vee p) \wedge (\neg\, p \vee (p \vee p)) && \text{[Law 3.18]} \\
\Leftrightarrow\ & (\neg\, p \vee p) \wedge (\neg\, p \vee p) && \text{[Law 3.10]} \\
\Leftrightarrow\ & true \wedge true && \text{[Law 3.15]} \\
\Leftrightarrow\ & true && \text{[Law 3.4]}
\end{aligned}
$$

□

Solution 3.50

$$
\dfrac{\dfrac{\lceil p \rceil_1}{q \Rightarrow p}\ [\Rightarrow \text{intro}]}{p \Rightarrow (q \Rightarrow p)}\ [\Rightarrow \text{intro}_1]
$$

□

Solution 3.51

$$
\dfrac{\dfrac{\dfrac{\lceil p \wedge \neg\, p \rceil_1}{p}\ [\wedge\ \text{elim1}] \quad \dfrac{\lceil p \wedge \neg\, p \rceil_1}{\neg\, p}\ [\wedge\ \text{elim2}]}{\dfrac{false}{true \Rightarrow false}\ [\Rightarrow \text{intro}]}\ [\text{false intro}]}{(p \wedge \neg\, p) \Rightarrow (true \Rightarrow false)}\ [\Rightarrow \text{intro}_1]
$$

□

Solution 3.52

1. $p \Rightarrow (q \vee r)$
     $\Leftrightarrow \neg\, p \vee (q \vee r)$ [Law 3.18]
     $\Leftrightarrow \neg\, p \vee q \vee r$ [Associativity of $\vee$]

2. $(p \land q) \Rightarrow (q \lor r)$
    $\Leftrightarrow \lnot\, (p \land q) \lor (q \lor r)$ [Law 3.18]
    $\Leftrightarrow (\lnot\, p \lor \lnot\, q) \lor (q \lor r)$ [Law 3.9]
    $\Leftrightarrow \lnot\, p \lor \lnot\, q \lor q \lor r$ [Associativity of $\lor$]

□

Solution 3.53

1. $(p \land (q \lor r)) \lor p$
    $\Leftrightarrow ((p \land q) \lor (p \land r)) \lor p$ [Law 3.16]
    $\Leftrightarrow (p \land q) \lor (p \land r) \lor p$ [Associativity of $\lor$]

2. $(p \lor (q \land r)) \land p$
    $\Leftrightarrow (p \land p) \lor (p \land (q \land r))$ [Law 3.17]
    $\Leftrightarrow p \lor (p \land (q \land r))$ [Law 3.3]
    $\Leftrightarrow p \lor (p \land q \land r)$ [Associativity of $\land$]

□

Chapter 4

Set theory

The two fundamental building blocks of discrete mathematics are propositional logic and set theory. We discussed the former in Chapter 3, and we present an introduction to the latter in this chapter.

Readers who are keen to learn more about this topic are referred to [End77], [Hen86], or [CLP01].

4.1 Sets

A set is simply an unordered collection of objects. For example, the natural numbers, which we met in Chapter 2, form a set, which we denote by $\mathbb{N}$, as do the real numbers, which we denote by $\mathbb{R}$. In addition, we can think of all of the countries in the world as forming a set, as well as a set containing all the people in the world; we might refer to the former as *Countries*, and the latter as *People*.

In each of the above examples, we have referred to a set by name and in each case we have some idea of what is contained in the set: in each case, we have some notion of the set's *elements* (or *members*).

Sometimes we need to be more concrete; sometimes, it is not sufficient to refer to an abstract entity, such as *People* representing all of the people in the world. In such cases, we may be interested in defining a set in terms of its elements; maybe, for example, we are interested in defining a set containing the name of the countries that have won football's World Cup.

Our means of doing this is to list all of the set's elements, and surround them by 'curly brackets' (or 'curly braces'): { and }. Every element is listed between these brackets, with each element being separated by a comma. Thus, we may define our set of World Cup winners in the following fashion.

$$Winners = \{\, argentina, brazil, england, france, germany, italy, uruguay \,\}$$

This is referred to as defining the set *Winners* by *extension* or by *listing*.

Example 4.1 Graham owns three compact discs—one by Abba, one by Westlife, and one by Madonna. The set containing these CDs is called *graham_cds* and is given by

$$graham\_cds = \{\, abba, westlife, madonna \,\}$$

□

Example 4.2 The balls associated with a game of pool (with the exception of the cue ball) are labelled 1 to 15. As such, we might define the set *pool_balls* in the following way.

$$pool\_balls = \{1, 2, 3, 4, 5, 6, 7, 8, 9, 10, 11, 12, 13, 14, 15\}$$

□

Exercise 4.1 List the elements of the following sets.

1. The set of vowels contained in the English alphabet.
2. The set of even prime numbers greater than two.
3. The set of letters that appear in the phrase "the set of letters".

□

Exercise 4.2 Define the following sets by extension.

1. The set of natural numbers less than 6.
2. The set of natural numbers less than 6 that are odd.
3. The set of vowels that appear in the phrase "the set of vowels".

□

Any element can occur at most once in a given set. As such, the sets $\{1, 2, 3, 3\}$ and $\{1, 2, 2, 3\}$ are equal. Indeed, they are both equal to the set $\{1, 2, 3\}$.

Example 4.3 The information contained in the set *fruitbowl* tells us only which fruit is available, not how many pieces of each fruit there are.

$$fruitbowl = \{banana, pear, apple, orange, peach\}$$

This set is equal to, for example,

$$\{banana, banana, pear, pear, apple, orange, peach\}$$

□

Furthermore, the order in which elements occur in a set is of no importance whatsoever. As such, the sets $\{3, 1, 2\}$, $\{1, 2, 3\}$, and $\{2, 1, 3\}$ are all equal.

Example 4.4 The information contained in the set *eaten* tells us only which types of fruit have been eaten, not the order in which they were eaten.

$$eaten = \{banana, orange, pear\}$$

This set is equal to, for example,

$$\{pear, orange, banana\}$$

□

Example 4.5 Each of the following sets are equal to $\{jack, jill\}$.

$$\{jill, jack\}, \{jack, jack, jill\}, \{jack, jill, jack\},$$
$$\{jack, jill, jill\}, \{jill, jack, jack\}, \{jill, jack, jill\}, \{jill, jill, jack\}$$

☐

Two sets are said to be *equal* when they contain exactly the same elements; we write $S = T$ when it is the case that two sets are equal. For example, we have seen that sets have no concept of order.

As such, $\{a, b\}$ and $\{b, a\}$ are equal, and so we may write

$$\{a, b\} = \{b, a\}$$

In addition, we have seen that sets have no notion of repetition. As such,

$$\{a, b, b, a\} = \{a, b\}$$

Exercise 4.3 Which of the following sets are equal?

1. The set of letters contained in the word "spear".

2. The set of letters contained in the word "pears".

3. The set of letters contained in the word "spares".

4. The set of letters contained in the word "spears".

☐

4.2 Singleton sets

If a set contains just one member, then we say that it is a *singleton* set. As an example, the set $\{a\}$ is a singleton set, as is the set $\{b\}$. The set $\{a, b\}$, on the other hand, is not a singleton set, as it contains two elements.

Example 4.6 The set of prime numbers between 24 and 30 is a singleton set: it is the set $\{29\}$.
☐

Example 4.7 The set of even numbers between 24 and 30 is not a singleton set: it is the set $\{26, 28\}$. ☐

Exercise 4.4 Which of the following sets are singleton sets?

1. The divisors of 9, other than 1 and itself.

2. The divisors of 11, other than 1 and itself.

3. The divisors of 16, other than 1 and itself.

☐

4.3   The empty set

As we saw in Exercise 4.4, it is sometimes the case that sets do not have any elements; such sets are equal to the empty set, i.e., the set that contains no elements. We may denote this set either by $\emptyset$ or by $\{\}$.

Example 4.8 The set of prime numbers between 32 and 36 is empty, as is the set of even prime numbers greater than 2. $\square$

Exercise 4.5 Which of these sets are equivalent to the empty set?

1. The set of letters of the English alphabet that follow Z.

2. The set of countries in the world that are not members of the United Nations.

3. The set of natural numbers that do not have a successor.

$\square$

4.4   Set membership

Given a set S, it is sometimes desirable to reason about whether or not a particular element is contained in S. Given an element s, if the element s appears in the set S, then we may say that *s is a member of* S. This is written $s \in S$.

Example 4.9 The number 1 is an element of the natural numbers. As such, we may write $1 \in \mathbb{N}$. $\square$

Example 4.10 Referring to Example 4.3, the set *fruitbowl* contains the element *banana*. As such, we may write *banana* $\in$ *fruitbowl*. $\square$

If an element t does not appear in a set S, then we may say that *t is not a member of* S; this is written $t \notin S$.

Example 4.11 Referring again to the natural numbers, the number -1 is not an element of the natural numbers. As such, we may write $-1 \notin \mathbb{N}$. $\square$

Example 4.12 Referring to Example 4.3, the set *fruitbowl* does not contain the element *lemon*. As such, we may write *lemon* $\notin$ *fruitbowl*. $\square$

It follows that $\notin$ is the complement of $\in$.

Law 4.1 For any set S and any element s,

$$\neg\,(s \in S) \Leftrightarrow s \notin S$$

$\square$

In addition, no element appears in the empty set. As such, the statement $x \in \emptyset$, for any element x, is logically equivalent to *false*.

Law 4.2 For any element x,

$$x \in \emptyset \Leftrightarrow false$$

□

Exercise 4.6 Assuming that *Primes* represents the set of all prime numbers, and *Evens* represents the set of all even numbers, which of the following statements are true and which are false?

1. $6 \in Primes$
2. $6 \notin Evens$
3. $7 \in Primes$
4. $7 \notin Evens$

□

For the moment, we allow ourselves to write $x \in X$ only if x is an *element* and X is a *set* (we shall formalise this restriction later). So, for example, if S is a set containing some natural numbers, such that $S = \{2, 3, 4\}$, we may write $2 \in S$ and $5 \notin S$, but not $\{2\} \in S$ or $\{5\} \notin S$. The statement $2 \in S$ is true and the statement $5 \in S$ is false. However, the values of the statements $\{2\} \in S$ and $\{5\} \notin S$ are both undefined, as $\{2\}$ and $\{5\}$ are not elements—they are sets.

Exercise 4.7 Assume that S is a set of letters drawn from the English alphabet, such that $S = \{a, b, c\}$. Which of the following statements are true, which are false, and which are undefined?

1. $a \in S$
2. $d \in S$
3. $\emptyset \in S$
4. $\{d\} \in S$

□

4.5   Subsets

Sometimes we may wish to reason about one set being contained within another set. For example, in Chapter 2 we saw that every element of the natural numbers is contained in the set of integers, and, in turn, every element of the integers is contained in the set of real numbers.

Given two sets, S and T, if every element of S is contained in T, then we may say that S is a *subset* of T and write $S \subseteq T$. For example, if $S = \{1, 3, 5\}$ and $T = \{1, 3, 5, 7\}$ then $S \subseteq T$ clearly holds. This state of affairs is illustrated by the following *Venn diagram*.[1]

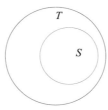

Here, the whole of the set S is contained within T.

Example 4.13 As every element of the natural numbers is contained in the set of integers and every integer is contained in the set of real numbers, we may write $\mathbb{N} \subseteq \mathbb{Z}$ and $\mathbb{Z} \subseteq \mathbb{R}$. □

Example 4.14 The set of all vegetarians, which is denoted *Vegetarian*, is a subset of the set of all people, which is denoted *People*. As such, *Vegetarian* $\subseteq$ *People*. □

Recall that if two sets S and T contain exactly the same elements, then we can say that they are equal, and write $S = T$. In such circumstances, it is always the case that $S \subseteq T$ and $T \subseteq S$.

Law 4.3 For any sets S and T,

$$(S \subseteq T \land T \subseteq S) \Leftrightarrow S = T$$

□

Example 4.15 Sets A and B are defined as follows.

$$A = \{barry, john\}$$
$$B = \{john, barry\}$$

Here, $A \subseteq B$ and $B \subseteq A$. As such, it follows that $A = B$. □

The empty set is a subset of every other set.

Law 4.4 For any set S,

$$\emptyset \subseteq S$$

□

---

[1]Venn diagrams are abstract, graphical representations with which we can represent relationships between different sets: the subset relation being a prime example. See [Hun96] for a thorough introduction to the topic of Venn diagrams.

Every set is a subset of itself.

Law 4.5 For any set S,

$$S \subseteq S$$

□

Exercise 4.8 Which of the following statements are true and which are false?

1. $\emptyset \subseteq \{cat, dog\}$
2. $\{cat\} \subseteq \{cat, dog\}$
3. $\{cat, dog\} \subseteq \{cat, dog\}$
4. $\{cat, dog\} \subseteq \emptyset$
5. $\{cat, dog\} \subseteq \{cat\}$

□

If it is the case that some of the elements of one set are not contained in another set, then we can say that the former is not a subset of the latter.

Example 4.16 In the following Venn diagram, the set R is not a subset of S, although it is a subset of T.

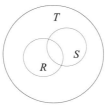

In this case, we may write $R \not\subseteq S$ and $R \subseteq T$. Furthermore, for this example, we may also write $S \subseteq T$, $S \not\subseteq R$, $T \not\subseteq R$, and $T \not\subseteq S$. □

The following law captures the relationship between $\subseteq$ and $\not\subseteq$.

Law 4.6 For any sets S and T,

$$\neg (S \subseteq T) \Leftrightarrow S \not\subseteq T$$

□

Example 4.17 The set of integers is not a subset of the natural numbers: $\mathbb{Z} \not\subseteq \mathbb{N}$. □

Exercise 4.9 Which of the following statements are true and which are false?

1. $\emptyset \not\subseteq \emptyset$
2. $\emptyset \not\subseteq \{a\}$

   3. $\{a\} \not\subseteq \emptyset$

   4. $\{a\} \not\subseteq \{a\}$

□

If a set S is completely contained within another set T, and T contains some elements that do not appear in S, then we may say that S is a *proper subset* of T, and write $S \subset T$.

Example 4.18 Recall that the sets *eaten* and *fruitbowl* were defined as follows.

$$fruitbowl = \{banana, pear, apple, orange, peach\}$$
$$eaten = \{banana, orange, pear\}$$

As such, *eaten* $\subset$ *fruitbowl*. □

The following law captures the relationship between $\subseteq$, $\subset$, and $=$.

Law 4.7 For any sets S and T,

$$S \subseteq T \Leftrightarrow (S \subset T \vee S = T)$$

□

Exercise 4.10 Which of the following statements are true and which are false?

   1. $\emptyset \subset \{cat, dog\}$

   2. $\{cat\} \subset \{cat, dog\}$

   3. $\{cat, dog\} \subset \{cat, dog\}$

   4. $\{cat, dog\} \subset \emptyset$

   5. $\{cat, dog\} \subset \{cat\}$

□

If S is not a strict subset of T, then we may write $S \not\subset T$. The following law captures the relationship between $\subset$ and $\not\subset$.

Law 4.8 For any sets S and T,

$$S \not\subset T \Leftrightarrow \neg (S \subset T)$$

□

Given any set S, it is always the case that S is not a strict subset of itself.

Law 4.9 For any set S,

$$S \not\subset S$$

□

In addition, if one set S is a proper subset of another set T, then T cannot also be a proper subset of S.

Law 4.10 Given any two sets S and T,

$$S \subset T \Rightarrow T \not\subset S$$

□

Example 4.19 Each of the following statements is true.

$$\{1,2\} \not\subset \{1\}$$
$$\{1,2\} \not\subset \{1,2\}$$

□

Exercise 4.11 Which of the following statements are true and which are false?

1. $\emptyset \not\subset \{cat, dog\}$

2. $\{cat\} \not\subset \{cat, dog\}$

3. $\{cat, dog\} \not\subset \{cat, dog\}$

4. $\{cat, dog\} \not\subset \emptyset$

5. $\{cat, dog\} \not\subset \{cat\}$

□

Exercise 4.12 Let $S = \{1, 2, 3\}$. Which of the following statements are true and which are false?

1. $\{1\} \subseteq S$

2. $\{1, 2, 3\} \subseteq S$

3. $\{1, 2, 3\} \subset S$

4. $\{1\} \subset S$

□

Exercise 4.13 Given the set S of the previous exercise, list all of the subsets of S. □

Exercise 4.14 Assuming the set S of Exercise 4.12, list all of the proper subsets of S. □

Exercise 4.15 Prove the following assertion.

$$S \subseteq \emptyset \Rightarrow S = \emptyset$$

□

As was the case with $\in$, we must be strict with regards to the circumstances under which we may use $\subseteq$ (or any of its derivatives). We may only write $S \subseteq T$ if S and T are both sets: we may not write $S \subseteq T$ if S or T are elements.[2] So, for example, we may write $\{1\} \subseteq \{1,2,3\}$—which is true—and $\{1,2,3\} \subseteq \{1\}$—which is false—but we may not write $1 \subseteq \{1,2,3\}$ or $\{1,2,3\} \subseteq 1$. The values of these last two statements are undefined.

Exercise 4.16 Assume $S = \{a,b,c\}$. Which of the following are true, which are false, and which are undefined?

1. $a \subseteq S$
2. $d \subseteq S$
3. $\emptyset \subseteq S$
4. $\{d\} \not\subseteq S$

$\square$

4.6 Supersets

The superset relationship is the complement of the subset relationship: a set S is a superset of another set T exactly when T is a subset of S. For example, the set $\{1\}$ is a subset of the set $\{1,2\}$. As such, $\{1,2\}$ is a superset of $\{1\}$. In this case, we write $\{1,2\} \supseteq \{1\}$.

Law 4.11 For any sets S and T,

$$S \supseteq T \Leftrightarrow T \subseteq S$$

$\square$

Here, the relationship between $\subseteq$ and $\supseteq$ is similar to that which holds between $\leq$ and $\geq$. Recall that $m \leq n$ holds for two natural numbers m and n if, and only if, $n \geq m$. Similarly, $S \subseteq T$ holds for two sets S and T if, and only if, $T \supseteq S$.

As was the case with $\subseteq$, there are a number of derived relationships associated with $\supseteq$: $\not\supseteq$, $\supset$, and $\not\supset$.

Example 4.20 All of the following statements are true.

$\emptyset \not\supseteq \{a,b\}$
$\{a\} \not\supseteq \{a,b\}$
$\{a,b\} \supset \emptyset$
$\{a,b\} \supset \{a\}$
$\{a\} \not\supset \{a\}$
$\emptyset \not\supset \{a\}$

$\square$

---

[2]We formalise this rule in Chapter 6.

Exercise 4.17 Which of the following statements are true and which are false?

1. $\{jack\} \supseteq \{jack, jill\}$

2. $\{jack\} \supset \{jack, jill\}$

3. $\{jack\} \not\supseteq \{jack\}$

4. $\{jack\} \not\supset \{jack\}$

5. $\{jack\} \supseteq \emptyset$

6. $\{jack\} \supset \emptyset$

□

4.7 Set union

Thus far, we have introduced a number of means for reasoning about the relationships that hold between sets. We can, for example, determine whether a given element appears in a set, or whether one set is a subset of another set. We are now in a position to introduce some *operators* on sets—these provide us with means of combining two or more sets to produce a new one.

The first operator we concern ourselves with is *set union*. This involves taking two sets and combining them to form a bigger set which contains all of the elements of both sets—remembering, of course, that any element can occur at most once in any given set. Given two sets S and T, the set union of S and T is denoted $S \cup T$.

Referring to the Venn diagram given below, the elements contained in the union of sets S and T are those elements contained in the highlighted area of the diagram.

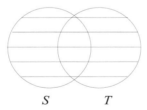

Example 4.21 If $S = \{1, 2, 3\}$ and $T = \{2, 3, 4\}$, then $S \cup T = \{1, 2, 3, 4\}$. □

Formally, an element a appears in the union of two sets S and T if, and only if, it appears in at least one of S and T.

Law 4.12 For any element a, and any sets S and T,

$$a \in S \cup T \Leftrightarrow (a \in S \vee a \in T)$$

□

Example 4.22 Graham and Christine have decided to merge their CD collections. Graham's CD collection is given by

$$graham\_cds = \{abba, westlife, madonna\}$$

Christine also has three CDs. This CD collection is given by

$$christine\_cds = \{\,britney\_spears,\, spice\_girls,\, the\_beatles\,\}$$

The union of their collections is given by $graham\_cds \cup christine\_cds$, is equivalent to the set

$$\{\,abba,\, westlife,\, madonna,\, britney\_spears,\, spice\_girls,\, the\_beatles\,\}$$

□

There are a number of laws associated with set union. First, the empty set contributes nothing to a set union.

Law 4.13 For any set S,

$$S \cup \emptyset = S$$

□

Example 4.23

$$\{a, b\} \cup \emptyset = \{a, b\}$$

□

Second, the union of any set S with itself produces S.

Law 4.14 For any set S,

$$S \cup S = S$$

□

Example 4.24

$$\{a, b\} \cup \{a, b\} = \{a, b\}$$

□

Set union also possesses two properties that we first met in Chapter 3—commutativity and associativity.

Law 4.15 For any sets S and T,

$$S \cup T = T \cup S$$

□

Law 4.16 For any sets R, S and T,

$$R \cup (S \cup T) = (R \cup S) \cup T$$

□

Example 4.25

$$\{a, b\} \cup (\{b, c\} \cup \{c, d\}) = (\{a, b\} \cup \{b, c\}) \cup \{c, d\}$$

□

Finally, the union of two sets is always at least as big as the individual sets.

Law 4.17 For any sets S and T,

$$S \subseteq S \cup T$$

□

Example 4.26

$$\{a, b\} \subseteq \{a, b\} \cup \{b, c\}$$

□

Exercise 4.18 Calculate each of the following.

1. $\{1, 2, 3\} \cup \emptyset$
2. $\{1, 2, 3\} \cup \{1, 2, 3\}$
3. $\{1, 2, 3\} \cup \{4\}$
4. $\{1, 2, 3\} \cup \{1, 2, 3, 4\}$
5. $\{1, 2\} \cup \{3, 4\}$

□

4.8   Set intersection

Sometimes we are interested in elements which occur in two different sets. For example, we might be interested in gathering information pertaining to people who both read *The Guardian* newspaper and subscribe to cable television; if a person satisfies only one of these criteria, then we are not interested in his or her details. Given two sets S and T, those elements which appear both in S and in T are referred to as the *intersection* of the two sets; we denote the intersection of S and T by $S \cap T$.

Referring to the Venn diagram given below, the elements contained in the intersection of sets S and T are those elements contained in the highlighted area of the diagram.

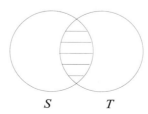

S T

Example 4.27 If $S = \{1, 2, 3\}$ and $T = \{2, 3, 4\}$, then $S \cap T = \{2, 3\}$. □

Formally, an element a appears in the intersection of two sets S and T if, and only if, it appears in both S and T.

Law 4.18 For any element a, and any sets S and T,

$$a \in S \cap T \Leftrightarrow (a \in S \wedge a \in T)$$

□

Example 4.28 Graham and Christine have no CDs in common. Recall that their CD collections were given by

$$graham\_cds = \{abba, westlife, madonna\}$$

and

$$christine\_cds = \{britney\_spears, spice\_girls, the\_beatles\}$$

respectively. As such, the intersection of these two sets is given by

$$graham\_cds \cap christine\_cds = \emptyset$$

□

As was the case with set union, there are a number of laws associated with set intersection. First, the intersection of any set with the empty set is equal to the empty set.

Law 4.19 For any set S,

$$S \cap \emptyset = \emptyset$$

□

Example 4.29

$$\{a, b\} \cap \emptyset = \emptyset$$

□

Second, the intersection of any set S and itself produces S.

Law 4.20 For any set S,

$$S \cap S = S$$

□

Example 4.30

$$\{a, b\} \cap \{a, b\} = \{a, b\}$$

□

As was the case with set union, set intersection is both commutative and associative.

Law 4.21 For any sets S and T,

$$S \cap T = T \cap S$$

□

Law 4.22 For any sets, R, S and T,

$$R \cap (S \cap T) = (R \cap S) \cap T$$

□

Example 4.31

$$\{a, b\} \cap (\{b, c\} \cap \{c, d\}) = (\{a, b\} \cap \{b, c\}) \cap \{c, d\}$$

□

Furthermore, the intersection of two sets is always at least as small as the individual sets.

Law 4.23 For any sets S and T,

$$S \cap T \subseteq S$$

□

Example 4.32

$$\{a, b\} \cap \{b, c\} \subseteq \{a, b\}$$

□

Finally, $\cup$ distributes through $\cap$, and vice versa.

Law 4.24 For any sets R, S and T,

$$R \cup (S \cap T) = (R \cup S) \cap (R \cup T)$$
$$R \cap (S \cup T) = (R \cap S) \cup (R \cap T)$$

□

We use the following example to illustrate the first of these laws.

Example 4.33 Assume $R = \{a, b\}$, $S = \{b, c\}$, and $T = \{c, d\}$. Here, the left-hand side of

$$R \cup (S \cap T) = (R \cup S) \cap (R \cup S)$$

can be calculated as follows.

$$\begin{aligned}
R \cup (S \cap T) &= \{a, b\} \cup (\{b, c\} \cap \{c, d\}) \\
&= \{a, b\} \cup \{c\} \\
&= \{a, b, c\}
\end{aligned}$$

Next, the right-hand side of

$$R \cup (S \cap T) = (R \cup S) \cap (R \cup S)$$

can be calculated as follows.

$$\begin{aligned}
(R \cup S) \cap (R \cup S) &= (\{a, b\} \cup \{b, c\}) \cap (\{a, b\} \cup \{c, d\}) \\
&= \{a, b, c\} \cap \{a, b, c, d\} \\
&= \{a, b, c\}
\end{aligned}$$

As can be seen, the two sides are indeed equal. □

Exercise 4.19 Calculate each of the following.

1. $\{1, 2, 3\} \cap \emptyset$
2. $\{1, 2, 3\} \cap \{1, 2, 3\}$
3. $\{1, 2, 3\} \cap \{4\}$
4. $\{1, 2, 3\} \cap \{1, 2, 3, 4\}$
5. $\{1, 2\} \cap \{3, 4\}$

□

4.9 Set difference

The set difference operator allows us to consider elements which appear in a set S and do not appear in another set T; this is denoted $S \setminus T$. For example, we might be interested in the details of people who read *The Guardian* newspaper and who do not subscribe to cable television. Given relevant sets G and C, we may denote this set of people by $G \setminus C$.

Referring to the Venn diagram given below, the elements contained in the set difference $S \setminus T$ are those elements contained in the highlighted area of the diagram.

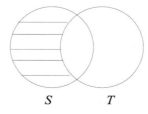

$S \qquad T$

Example 4.34 If $S = \{1, 2, 3\}$ and $T = \{2, 3, 4\}$, then $S \setminus T = \{1\}$. □

Example 4.35 Consider the sets $S = \{a, b, c\}$ and $T = \{b, c, d\}$. Here,

$$S \setminus T = \{a\}$$
$$T \setminus S = \{d\}$$

□

Formally, an element a appears in the set $S \setminus T$ if, and only if, it appears in S and it does not appear in T.

Law 4.25 For any element a, and any sets S and T,

$$a \in S \setminus T \Leftrightarrow (a \in S \wedge a \notin T)$$

□

Example 4.36 Given the sets $S = \{jon, jim, dave\}$ and $T = \{jim\}$, the set $S \setminus T$ is given by $\{jon, dave\}$: the elements jon and $dave$ both appear in S, but do not appear in T. □

Again, there are a number of laws associated with this operator. First, given any set S, removing the elements of the empty set from S will have no effect.

Law 4.26 For any set S,

$$S \setminus \emptyset = S$$

□

Example 4.37

$$\{a, b\} \setminus \emptyset = \{a, b\}$$

□

Second, given any set S, removing all the elements of S will leave the empty set.

Law 4.27 For any set S,

$$S \setminus S = \emptyset$$

□

Example 4.38

$$\{a, b\} \setminus \{a, b\} = \emptyset$$

□

In addition, no elements can be removed from the empty set.

Law 4.28 For any set S,

$$\emptyset \setminus S = \emptyset$$

$\square$

Example 4.39

$$\emptyset \setminus \{a, b\} = \emptyset$$

$\square$

Set difference also has the following interesting properties.

Law 4.29 For any sets R, S and T,

$$R \setminus (S \cup T) = (R \setminus S) \cap (R \setminus T)$$
$$R \setminus (S \cap T) = (R \setminus S) \cup (R \setminus T)$$

$\square$

We use the following example to illustrate that

$$R \setminus (S \cup T) = (R \setminus S) \cap (R \setminus T)$$

holds.

Example 4.40 Assume $R = \{a, b\}$, $S = \{b, c\}$, and $T = \{c, d\}$. Here, the left-hand side of

$$R \setminus (S \cup T) = (R \setminus S) \cap (R \setminus T)$$

is calculated as follows.

$$\begin{aligned}
R \setminus (S \cup T) &= \{a, b\} \setminus (\{b, c\} \cup \{c, d\}) \\
&= \{a, b\} \setminus \{b, c, d\} \\
&= \{a\}
\end{aligned}$$

Next, the right-hand side of

$$R \setminus (S \cup T) = (R \setminus S) \cap (R \setminus T)$$

is calculated as follows.

$$\begin{aligned}
(R \setminus S) \cap (R \setminus T) &= (\{a, b\} \setminus \{b, c\}) \cap (\{a, b\} \setminus \{c, d\}) \\
&= \{a\} \cap \{a, b\} \\
&= \{a\}
\end{aligned}$$

As can be seen, the two sides are indeed equal. $\square$

We use the following example to illustrate that

$$R \setminus (S \cap T) = (R \setminus S) \cup (R \setminus T)$$

holds.

Example 4.41 Assume $R = \{a, b\}$, $S = \{b, c\}$, and $T = \{c, d\}$. Here, the left-hand side of

$$R \setminus (S \cap T) = (R \setminus S) \cup (R \setminus T)$$

is calculated as follows.

$$
\begin{aligned}
R \setminus (S \cap T) &= \{a, b\} \setminus (\{b, c\} \cap \{c, d\}) \\
&= \{a, b\} \setminus \{c\} \\
&= \{a, b\}
\end{aligned}
$$

Next, the right-hand side of

$$R \setminus (S \cap T) = (R \setminus S) \cup (R \setminus T)$$

is calculated as follows.

$$
\begin{aligned}
(R \setminus S) \cup (R \setminus T) &= (\{a, b\} \setminus \{b, c\}) \cup (\{a, b\} \setminus \{c, d\}) \\
&= \{a\} \cup \{a, b\} \\
&= \{a, b\}
\end{aligned}
$$

As can be seen, the two sides are indeed equal. □

Finally, removing elements from a set S will result in a set that is at least as small as S.

Law 4.30 For any sets S and T,

$$S \setminus T \subseteq S$$

□

Example 4.42

$$\{1, 2, 3\} \setminus \{1, 2\} \subseteq \{1, 2, 3\}$$

□

Exercise 4.20 Assuming the sets $A = \{1, 2, 3, 4\}$, $B = \{2, 4, 6, 8\}$, and $C = \{3, 4, 5, 6\}$, calculate the following.

1. $A \setminus B$

2. $C \setminus A$

3. $B \setminus C$

4. $B \setminus B$

5. $(A \setminus B) \setminus C$

6. $A \setminus (B \setminus C)$

$\square$

Exercise 4.21 Prove that $\setminus$ is not commutative. $\square$

Exercise 4.22 Prove that $\setminus$ is not associative. $\square$

Exercise 4.23 Prove that $\setminus$ is not idempotent. $\square$

4.10   Reasoning about sets

We have already exhibited a number of laws about set theory, some of which describe exactly what it means for an element to be in a set of the form $S \cap T$ or a set of the form $S \setminus T$. In addition, we have stated that two sets are equal exactly when they have the same elements; this can be characterised by the following law.

Law 4.31 Given two sets, S and T, $S = T$ if, and only if, for every element x,

$$x \in S \Leftrightarrow x \in T$$

$\square$

It follows that the laws provided in this chapter, together with the concept of equational reasoning which was discussed in the previous chapter, provide us with a means of determining if two sets are equal.

Suppose we wanted to prove, using propositional logic, that the commutativity of $\cap$ really is a valid property, i.e., that $S \cap T = T \cap S$ for all sets S and T. In this case, our proof might take the following form.

$$
\begin{aligned}
& x \in S \cap T \\
\Leftrightarrow{} & x \in S \wedge x \in T \quad [\text{Law } 4.18] \\
\Leftrightarrow{} & x \in T \wedge x \in S \quad [\text{Law } 3.7] \\
\Leftrightarrow{} & x \in T \cap S \qquad [\text{Law } 4.18]
\end{aligned}
$$

Here, it is assumed that x is an arbitrary element; as it is an arbitrary element, we can claim that there is nothing special about x, and, as such,

$$x \in S \cap T \Leftrightarrow x \in T \cap S$$

holds for every element.[3]

---

[3]We shall put a more subtle restriction on x in later chapters, but, for now, this interpretation will suffice.

Exercise 4.24 Prove the following laws of set theory.

1. $S \cap S = S$

2. $S \cup S = S$

3. $S \setminus S = \emptyset$

4. $S \cap \emptyset = \emptyset$

5. $S \cup \emptyset = S$

6. $S \setminus \emptyset = S$

□

4.11   Cardinality

The size, or *cardinality*, of a set S refers to the number of elements contained in that set. We denote the cardinality of a set S by $\# S$.

As an example, assume that the sets *Friends* and *Enemies* are defined as follows.

$$Friends = \{fred, felicity\}$$
$$Enemies = \{ewan, elizabeth\}$$

Here, it is clear that the number of elements contained in *Friends* is two, and also that the number of elements contained in *Enemies* is two. As such, we may write $\# \, Friends = 2$ and $\# \, Enemies = 2$. Furthermore, $\# \, (Friends \cup Enemies) = 4$, and $\# \, (Friends \cap Enemies) = 0$.

From the above, we can see that the cardinality of the empty set is 0.

Law 4.32

$$\# \, \emptyset = 0$$

□

Furthermore, the operators $\cap$ and $\setminus$ restrict cardinality, whereas $\cup$ increases it.

Law 4.33 For any sets S and T,

$$\# \, (S \cap T) = \# \, S - \# (S \setminus T)$$

□

Law 4.34 For any sets S and T,

$$\# \, (S \cup T) = \# \, S + \# \, T - \# \, (S \cap T)$$

□

Law 4.35 For any sets S and T,

$$\#(S \setminus T) = \#S - \#(S \cap T)$$

□

Example 4.43 Assume the following sets.

$$S = \{2,3,4,5\}$$
$$T = \{3,4,5,6\}$$

Here, $\#S = 4$ and $\#T = 4$. Furthermore,

$$S \cup T = \{2,3,4,5,6\}$$
$$S \cap T = \{3,4,5\}$$
$$S \setminus T = \{2\}$$

As such, we have the following.

$$\#(S \cup T) = 5$$
$$\#(S \cap T) = 3$$
$$\#(S \setminus T) = 1$$

Clearly, Law 4.33 holds for this example, as $3 = 4 - 1$. In addition, Law 4.35 holds for this example, as $1 = 4 - 3$. Finally, Law 4.34 holds for this example, as $5 = 4 + 4 - 3$. □

Exercise 4.25 Give the cardinality of each of the following sets.

1. The set of letters in the word "hello".

2. The set of prime numbers between 16 and 25.

3. The set of divisors of 16.

4. $\{1,2,3\} \cup \{2,3,4\}$

5. $\{1,2,3\} \cap \{2,3,4\}$

6. $\{1,2,3\} \setminus \{2,3,4\}$

□

4.12   Finite and infinite sets

In Exercise 4.25, we considered a number of sets for which the elements could be enumerated and then counted. For example, the elements contained in the set of letters in the word "hello" are h, e, l, and o. When it is possible for us to enumerate the elements of a given set, count the elements, and *guarantee that this process will eventually stop*, we may say that that set is *finite*. For example, the set of divisors of 16 is finite. On the other hand, if this process will never stop for a given set—i.e., if counting the elements will continue forever—then we may say that that set is *infinite*. For example, the set of natural numbers, $\mathbb{N}$, is infinite.

Exercise 4.26 Which of the following sets are finite and which are infinite?

1. The empty set.

2. The set of integers.

3. The set of odd natural numbers.

4. The letters contained in the English alphabet.

5. The number of words with less than seven letters, constructed from the letters of the English alphabet.

6. The set of all strings of letters drawn from the English alphabet.

7. The people who have ever lived on earth.

□

4.13 Power sets

Earlier in this chapter, we described how we can consider the subsets of any set S. For example, if $T = \{a, b\}$, then the subsets of T are given by $\emptyset$, $\{a\}$, $\{b\}$, and $\{a, b\}$.

It is possible to enumerate the process of listing all subsets of a given set in the following way. First, consider the subsets of cardinality 0. In all cases, there is only one: the empty set. Next, consider all subsets of cardinality 1. In the above case, this gives the subsets $\{a\}$ and $\{b\}$. Next, consider all subsets of cardinality 2. This process continues until subsets of cardinality equal to that of the original set are to be considered. Again, there will only be one subset of this cardinality: the original set itself.

If we apply this process to the set $\{1, 2, 3, 4\}$ we obtain the following.

subsets of cardinality 0 : $\emptyset$
subsets of cardinality 1 : $\{1\}, \{2\}, \{3\}, \{4\}$
subsets of cardinality 2 : $\{1, 2\}, \{1, 3\}, \{1, 4\}, \{2, 3\}, \{2, 4\}, \{3, 4\}$
subsets of cardinality 3 : $\{1, 2, 3\}, \{1, 2, 4\}, \{1, 3, 4\}, \{2, 3, 4\}$
subsets of cardinality 4 : $\{1, 2, 3, 4\}$

It is natural to think of these subsets as themselves forming a set. Thus, given a set S, we denote the set of all subsets of S—or the *power set* of S—by $\mathbb{P}\,S$.[4] Thus, for our above examples, we have

$$\mathbb{P}\,T = \{\emptyset, \{a\}, \{b\}, \{a, b\}\}$$

and

$$\mathbb{P}\,\{1, 2, 3, 4\} = \{\emptyset, \{1\}, \{2\}, \{3\}, \{4\}, \{1, 2\}, \{1, 3\}, \{1, 4\}, \{2, 3\}, \{2, 4\}, \{3, 4\},$$
$$\{1, 2, 3\}, \{1, 2, 4\}, \{1, 3, 4\}, \{2, 3, 4\}, \{1, 2, 3, 4\}\}$$

---

[4]Note that in some texts, the power set of a set S is denoted $P(S)$ or 2^S rather than $\mathbb{P}\,S$.

Example 4.44 The set of sets of numbers between 4 and 6 is given by $\mathbb{P}\{4, 5, 6\}$, which is equivalent to

$$\{\emptyset, \{4\}, \{5\}, \{6\}, \{4, 5\}, \{4, 6\}, \{5, 6\}, \{4, 5, 6\}\}$$

□

Example 4.45 As we saw earlier in this chapter, the empty set is a subset of *every* set (including itself). As such, $\mathbb{P}\,\emptyset$ is simply the set containing the empty set, i.e.,

$$\mathbb{P}\,\emptyset = \{\emptyset\}$$

□

The relationship between $\in$ and $\subseteq$ is captured by the following law.

Law 4.36 For any sets S and T,

$$S \in \mathbb{P}\,T \Leftrightarrow S \subseteq T$$

□

Example 4.46 Given the set $\{a, b\}$, it is the case that $\{a\} \in \mathbb{P}\{a, b\}$, and—equivalently—$\{a\} \subseteq \{a, b\}$. □

Furthermore, as the empty set is a subset of every set, it follows that it is an element of the power set of every set.

Law 4.37 For any set S,

$$\emptyset \in \mathbb{P}\,S$$

□

Finally, as every set is a subset of itself, it follows that every set is an element of its power set.

Law 4.38 For any set S,

$$S \in \mathbb{P}\,S$$

□

Exercise 4.27 Calculate each of the following.

1. $\mathbb{P}\{x\}$
2. $\mathbb{P}\{x, y\}$
3. $\mathbb{P}\{x, y, z\}$

□

Exercise 4.28 The power set of $\{4, 5, 6\}$ has eight elements. How many elements does the power set of $\{4, 5\}$ have? In general, if a set has n elements, how many elements will its power set have? □

Law 4.39 For any set S,

$$\# \left(\mathbb{P} \, S \right) = 2^{\#(S)}$$

□

The reason that the above law holds is that—for any subset T of S, and for any element $s \in S$—there are two possibilities for s: it is either in T or it is not. As an example, assume that S contains two elements: a and b. As such, there are $2 \times 2 = 4$ possibilities (or possible subsets): a is in and b is in (the set $\{a, b\}$); a is in and b is not in (the set $\{a\}$); a is not in and b is in (the set $\{b\}$); and a is not in and b is not in (the set $\emptyset$). This argument generalises to sets of any (finite) size.

Exercise 4.29 Calculate the following.

1. $\mathbb{P} \{1, 2\}$

2. $\mathbb{P} \{\emptyset\}$

3. $\mathbb{P} \{\emptyset, \{1, 2\}\}$

4. $\mathbb{P} \, \mathbb{P} \{1, 2\}$

□

In considering the power sets of the previous exercise, one must bear in mind that sets that contain sets as elements are being generated. Although at first glance these sets may appear complicated, through experience the generation of such power sets becomes a lot less daunting.

Consider for a moment the set $\{a\}$. This is a set containing a single element: a. As such, this set has two subsets: $\emptyset$ and $\{a\}$. Therefore, the power set of this set is given by

$$\{\emptyset, \{a\}\}$$

Now consider the set $\{\emptyset\}$. Again, this is a set containing a single element: $\emptyset$. As such, this set has two subsets: $\emptyset$ and $\{\emptyset\}$. Therefore, the power set of this set is given by

$$\{\emptyset, \{\emptyset\}\}$$

The method of determining the power set is identical in both cases, with the only difference in the power sets being the element contained in the singleton set: in the first power set it is the element, a; in the second power set it is the empty set, $\emptyset$.

4.14 Generalised operations

We have seen how, when given two sets S and T, $S \cup T$ denotes the union of the two sets. Furthermore, we have seen that $R \cup S \cup T$ denotes the union of R, S, and T. We may, of course, go further, and talk about such sets as $Q \cup R \cup S \cup T$ or $P \cup Q \cup R \cup S \cup T$. Of course, this approach to representing the union of a number of sets can become rather cumbersome if that number is relatively large.

The *generalised union* operator provides us with a more elegant means of representing the union of a number of sets. Given a collection of sets, the generalised union of these sets is the set containing each element which appears in *at least one of the sets*.

Example 4.47 Given the sets R, S and T, such that $R = \{1, 2, 3\}$, $S = \{1, 4\}$ and $T = \{1, 5\}$, the generalised union of these sets—equivalent, of course, to $R \cup S \cup T$—is given by $\{1, 2, 3, 4, 5\}$. □

We write $\bigcup \{R, S, T\}$ to represent the generalised union of the sets R, S, and T. As we have seen, the resulting set is equivalent $R \cup S \cup T$, but writing this set in terms of the generalised union operator can have the advantage of being slightly more elegant.

Law 4.40 For any set of sets X and any element x, $x \in \bigcup X$ if, and only if, there is some set $S \in X$ such that $x \in S$. □

Exercise 4.30 Assume the following sets.

$$X = \{jack, jill\}$$
$$Y = \{jack, richard\}$$
$$Z = \{margaret, emma\}$$

Given the sets X, Y, and Z, calculate the following.

1. $\bigcup \{X, Y\}$

2. $\bigcup \{X\}$

3. $\bigcup \{X, Y, Z\}$

4. $\bigcup \{X, X\}$

5. $\bigcup \{X, \emptyset\}$

□

The *generalised intersection* operator works in an analogous fashion to the generalised union operator. Given a collection of sets, the generalised intersection of these sets is the set containing each element which appears *in every one of the sets*.

Example 4.48 Given the sets R, S and T, such that $R = \{1, 2, 3\}$, $S = \{1, 4\}$ and $T = \{1, 5\}$, the generalised intersection of these sets—equivalent, of course, to $R \cap S \cap T$—is given by $\{1\}$. □

We write $\bigcap \{R, S, T\}$ to represent the generalised intersection of R, S and T .

Law 4.41 For any set of sets X and any element x, $x \in \bigcap X$ if, and only if, for every set $S \in X$ it is the case that $x \in S$. □

Exercise 4.31 Assume the following sets.

$X = \{jack, jill\}$
$Y = \{jack, richard\}$
$Z = \{margaret, emma\}$

Given the sets X, Y, and Z, calculate the following.

1. $\bigcap \{X, Y\}$
2. $\bigcap \{X\}$
3. $\bigcap \{X, Y, Z\}$
4. $\bigcap \{X, X\}$
5. $\bigcap \{X, \emptyset\}$

□

4.15 Further exercises

Exercise 4.32 Define the following sets by extension.

1. All natural numbers between 7 and 13 (inclusive).

2. All odd natural numbers between 7 and 13 (inclusive).

3. All natural numbers x satisfying the equation

$$7 + (2 \times x) = 13$$

4. All natural numbers x satisfying the equation

$$7 + (2 \times x) < 13$$

□

Exercise 4.33 Which of the following sets are equal?

$A = \{a, b\}$
$B = \{b, a\}$
$C = \{a, b, c\}$
$D = \{c, a, b\}$
$E = \{a, b, a, b\}$
$F = \{a, b, c, a\}$

□

Exercise 4.34 Which (if any) of the following sets are equal: $\emptyset$, $\{0\}$, and $\{\emptyset\}$? □

Exercise 4.35 Which of the following are singleton sets?

1. $\emptyset$
2. $\{0\}$
3. $\{\emptyset\}$

$\square$

Exercise 4.36 Give the power sets of the following sets.

1. $\emptyset$
2. $\{jon, richard\}$
3. $\{\{jon\}, \{richard\}\}$
4. $\{\{jon, richard\}\}$

$\square$

Exercise 4.37 Assume $S = \{r, s, t\}$. Which of the following statements are defined?

1. $r \in S$
2. $r \subseteq S$
3. $\{r\} \in S$
4. $\{r\} \subseteq S$

$\square$

Exercise 4.38 Consider the following sets.

$$R = \{jack\}$$
$$S = \{jill\}$$
$$T = \{jack, jill\}$$

Which of the following statements are true and which are false?

1. $\emptyset \subset R$
2. $S \supseteq \emptyset$
3. $R \subseteq S$
4. $T \supset T$
5. $T \not\subseteq R$
6. $\emptyset \not\supseteq S$

7. $R \not\subseteq \emptyset$

8. $S \not\supseteq T$

□

Exercise 4.39 Prove each of the following statements.

1. $\{1, 2, 3\} \not\subseteq \{2, 3, 4\}$

2. $\{1, 2, 3\} \subset \{1, 2, 3, 4\}$

□

Exercise 4.40 Illustrate the following sets using Venn diagrams.

1. $(R \cap S) \cap T$

2. $(R \cap S) \setminus T$

3. $(R \setminus S) \cup T$

4. $(R \cap S) \cup T$

□

Exercise 4.41   Assume the following sets.

$$A = \{1, 2, 3, 4, 5\}$$
$$B = \{3, 4, 5, 6, 7\}$$
$$C = \{2, 4, 6\}$$
$$D = \{1, 3, 5, 7\}$$

Calculate each of the following.

1. $(A \cap C) \setminus (B \cap D)$

2. $(A \cup D) \setminus (B \cup C)$

3. $\bigcap \{A, B, D\}$

4. $(\bigcup \{A, D\}) \cap (\bigcup \{B, D\})$

□

Exercise 4.42 Given the sets of Exercise 4.41, calculate the following.

1. $A \setminus B$

2. $A \setminus (B \setminus C)$

3. $(A \setminus B) \setminus C$

4. $(A \setminus B) \setminus (C \setminus D)$

□

Exercise 4.43 Given that $A = \{1, 2, 3, 4\}$, calculate the elements of the sets B and C, given the following facts.

$$A = B \cap C$$
$$C \setminus A = \{5, 6\}$$
$$(B \cup A) \setminus C = \{7\}$$

□

Exercise 4.44 Prove the following statement.

$$(B \cup C) \cap A = (B \cap A) \cup (C \cap A)$$

□

Exercise 4.45 Prove the following statement.

$$(B \cap C) \cup A = (B \cup A) \cap (C \cup A)$$

□

Exercise 4.46 How many elements do the following sets have?

1. $\{a, b\}$
2. $\mathbb{P}\{a, b\}$
3. $\mathbb{P}\,\mathbb{P}\{a, b\}$
4. $\emptyset$
5. $\mathbb{P}\,\emptyset$
6. $\mathbb{P}\,\mathbb{P}\,\emptyset$

□

Exercise 4.47 List the elements of the following sets.

1. $\mathbb{P}\{a, b\}$
2. $\mathbb{P}\,\emptyset$
3. $\mathbb{P}\,\mathbb{P}\,\emptyset$
4. $\mathbb{P}\{\emptyset, \{a, b\}\}$

□

Exercise 4.48 Which of the following sets are finite, and which are infinite?

1. The set of months of the year.
2. The set of ways in which the numbers between 1 and 1,000,000 can be ordered.

3. The set of integers that divide 7.

4. The set of natural numbers containing the digit 0.

☐

Exercise 4.49 Assuming the sets $A = \{1, 2\}$ and $B = \{2, 3\}$, calculate the following.

1. $\mathbb{P}\, A$

2. $A \cap B$

3. $A \cup B$

4. $A \setminus B$

5. $\mathbb{P}\,((A \setminus B) \cap B)$

☐

Exercise 4.50 Prove that for any sets A and B,

$$(A \setminus B) \cap B = \emptyset$$

☐

Exercise 4.51 Prove that for any sets A and B,

$$(A \setminus B) \cup B = A \cup B$$

☐

Exercise 4.52 Calculate the following.

1. $\bigcap \{\emptyset\}$

2. $\bigcup \{\emptyset\}$

3. $\bigcap \{\{1\}, \{3\}, \{5\}\}$

4. $\bigcup \{\{1\}, \{3\}, \{5\}\}$

5. $\bigcap \{\{1, 3\}, \{3, 5\}\}$

6. $\bigcup \{\{1, 3\}, \{3, 5\}\}$

☐

Exercise 4.53 Calculate the generalised union of the following sets.

1. $\{john, graham\}$, $\{andy, john\}$, and $\{richard, john\}$

2. $\{1\}$ and $\{1, 2\}$

   3. $\{1\}$ and $\emptyset$

   4. $\{\{1\}\}$ and $\{\{2\}\}$

$\square$

Exercise 4.54 Calculate the generalised intersection of the sets of the previous question. $\square$

Exercise 4.55 When is it the case that $\bigcup S = \bigcap S$? $\square$

4.16   Solutions

Solution 4.1

   1. The elements are: a, e, i, o, and u.

   2. This set has no elements.

   3. The elements are: t, h, e, s, o, f, l, and r.

$\square$

Solution 4.2

   1. $\{0, 1, 2, 3, 4, 5\}$

   2. $\{1, 3, 5\}$

   3. $\{e, o\}$

$\square$

Solution 4.3 They are all equal. $\square$

Solution 4.4 Only 1 is a singleton set. 2 has no elements, while 3 has three elements: 2, 4, and 8. $\square$

Solution 4.5 1 and 3 are equal to the empty set, whereas 2 is not. $\square$

Solution 4.6 1 and 2 are false, and 3 and 4 are true. $\square$

Solution 4.7 1 is true, 2 is false, and 3 and 4 are undefined. $\square$

Solution 4.8 1, 2 and 3 are true, and 4 and 5 are false. $\square$

Solution 4.9 Only the third of these statements is true. $\square$

Solution 4.10 The first two statements are true, and the last three are false. $\square$

Solution 4.11 The first two statements are false, and the last three are true. $\square$

Solution 4.12 1, 2 and 4 are true, and 3 is false. $\square$

Solution 4.13 $\emptyset, \{1\}, \{2\}, \{3\}, \{1,2\}, \{1,3\}, \{2,3\}$, and $\{1,2,3\}$. $\square$

Solution 4.14 $\emptyset, \{1\}, \{2\}, \{3\}, \{1,2\}, \{1,3\}$, and $\{2,3\}$. $\square$

Solution 4.15 By Law 4.4, $\emptyset$ is a subset of every set, so $\emptyset$ is a subset of S. By Law 4.3, as S is also a subset of $\emptyset$, the two sets must be equal. $\square$

Solution 4.16 1 and 2 are undefined, and 3 and 4 are true. $\square$

Solution 4.17 The first three are false, and the last three are true. $\square$

Solution 4.18

1. $\{1,2,3\}$
2. $\{1,2,3\}$
3. $\{1,2,3,4\}$
4. $\{1,2,3,4\}$
5. $\{1,2,3,4\}$

$\square$

Solution 4.19

1. $\emptyset$
2. $\{1,2,3\}$
3. $\emptyset$
4. $\{1,2,3\}$
5. $\emptyset$

$\square$

Solution 4.20

1. $\{1,3\}$
2. $\{5,6\}$
3. $\{2,8\}$
4. $\emptyset$
5. $\{1\}$
6. $\{1,3,4\}$

$\square$

Solution 4.21 For \ to be commutative, we would require that

$$S \setminus T = T \setminus S$$

for any sets S and T. Consider the sets $S = \{0, 1\}$ and $T = \{0\}$. Here $S \setminus T = \{1\}$ and $T \setminus S = \emptyset$. As such, \ is not commutative. □

Solution 4.22 For \ to be associative, we would require that

$$R \setminus (S \setminus T) = (R \setminus S) \setminus T$$

for any sets R, S, and T. Consider the sets $R = \{1, 2, 3\}$, $S = \{1, 2\}$, and $T = \{3\}$. Here,

$$\begin{aligned}
R \setminus (S \setminus T) &= \{1, 2, 3\} \setminus (\{1, 2\} \setminus \{3\}) \\
&= \{1, 2, 3\} \setminus \{1, 2\} \\
&= \{3\}
\end{aligned}$$

and

$$\begin{aligned}
(R \setminus S) \setminus T &= (\{1, 2, 3\} \setminus \{1, 2\}) \setminus \{3\} \\
&= \{3\} \setminus \{3\} \\
&= \emptyset
\end{aligned}$$

As such, \ is not associative. □

Solution 4.23 For \ to be idempotent, we would require that

$$S \setminus S = S$$

for any set S. Consider the set $S = \{1, 2\}$. Here,

$$\begin{aligned}
S \setminus S &= \{1, 2\} \setminus \{1, 2\} \\
&= \emptyset
\end{aligned}$$

As such, \ is not idempotent. □

Solution 4.24

1. $x \in S \cap S$
   $\Leftrightarrow x \in S \land x \in S$ [Law 4.18]
   $\Leftrightarrow x \in S$ [Law 3.7]

2. $x \in S \cup S$
   $\Leftrightarrow x \in S \lor x \in S$ [Law 4.12]
   $\Leftrightarrow x \in S$ [Law 3.14]

3. $x \in S \setminus S$
   $\Leftrightarrow x \in S \land x \notin S$ [Law 4.25]
   $\Leftrightarrow x \in S \land \lnot (x \in S)$ [Law 4.1]
   $\Leftrightarrow \mathit{false}$ [Law 3.6]
   $\Leftrightarrow x \in \emptyset$ [Law 4.2]

4. $x \in S \cap \emptyset$
    $\Leftrightarrow x \in S \land x \in \emptyset$ [Law 4.18]
    $\Leftrightarrow x \in S \land \text{\textit{false}}$ [Law 4.2]
    $\Leftrightarrow \text{\textit{false}}$ [Law 3.5]
    $\Leftrightarrow x \in \emptyset$ [Law 4.2]

5. $x \in S \cup \emptyset$
    $\Leftrightarrow x \in S \lor x \in \emptyset$ [Law 4.12]
    $\Leftrightarrow x \in S \lor \text{\textit{false}}$ [Law 4.2]
    $\Leftrightarrow x \in S$ [Law 3.11]

6. $x \in S \setminus \emptyset$
    $\Leftrightarrow x \in S \land x \notin \emptyset$ [Law 4.25]
    $\Leftrightarrow x \in S \land \lnot (x \in \emptyset)$ [Law 4.1]
    $\Leftrightarrow x \in S \land \lnot \text{\textit{false}}$ [Law 4.2]
    $\Leftrightarrow x \in S \land \text{\textit{true}}$ [Law 3.1]
    $\Leftrightarrow x \in S$ [Law 3.4]

$\square$

Solution 4.25

1. $\# \{h, e, l, o\} = 4$
2. $\# \{17, 19, 23\} = 3$
3. $\# \{1, 2, 4, 8, 16\} = 5$
4. $\# \{1, 2, 3, 4\} = 4$
5. $\# \{2, 3\} = 2$
6. $\# \{1\} = 1$

$\square$

Solution 4.26 2, 3, and 6 are infinite, while the others are finite. $\square$

Solution 4.27

1. $\{\emptyset, \{x\}\}$
2. $\{\emptyset, \{x\}, \{y\}, \{x, y\}\}$
3. $\{\emptyset, \{x\}, \{y\}, \{z\}, \{x, y\}, \{x, z\}, \{y, z\}, \{x, y, z\}\}$

$\square$

Solution 4.28 The power set of $\{4, 5\}$ has four elements: $\emptyset$, $\{4\}$, $\{5\}$, and $\{4, 5\}$. In general, if a set S has n elements then $\mathbb{P}\, S$ has 2^n elements. $\square$

Solution 4.29

1. $\{\emptyset, \{1\}, \{2\}, \{1, 2\}\}$

2. $\{\emptyset, \{\emptyset\}\}$

3. $\{\emptyset, \{\emptyset\}, \{\{1, 2\}\}, \{\emptyset, \{1, 2\}\}\}$

4. $\{\emptyset, \{\emptyset\}, \{\{1\}\}, \{\{2\}\}, \{\{1, 2\}\}, \{\emptyset, \{1\}\}, \{\emptyset, \{2\}\}, \{\emptyset, \{1, 2\}\}, \{\{1\}, \{2\}\},$
   $\{\{1\}, \{1, 2\}\}, \{\{2\}, \{1, 2\}\}, \{\emptyset, \{1\}, \{2\}\}, \{\emptyset, \{1\}, \{1, 2\}\}, \{\emptyset, \{2\}, \{1, 2\}\},$
   $\{\{1\}, \{2\}, \{1, 2\}\}, \{\emptyset, \{1\}, \{2\}, \{1, 2\}\}\}$

□

Solution 4.30

1. $\{jack, jill, richard\}$

2. $\{jack, jill\}$

3. $\{jack, jill, richard, margaret, emma\}$

4. $\{jack, jill\}$

5. $\{jack, jill\}$

□

Solution 4.31

1. $\{jack\}$

2. $\{jack, jill\}$

3. $\emptyset$

4. $\{jack, jill\}$

5. $\emptyset$

□

Solution 4.32

1. $\{7, 8, 9, 10, 11, 12, 13\}$

2. $\{7, 9, 11, 13\}$

3. $\{3\}$

4. $\{0, 1, 2\}$

□

Solution 4.33 $A = B = E$ and $C = D = F$. □

Solution 4.34 None of them. They are all different. □

Solution 4.35 The second and third sets are singleton sets. □

Solution 4.36

1. $\{\emptyset\}$

2. $\{\emptyset, \{jon\}, \{richard\}, \{jon, richard\}\}$

3. $\{\emptyset, \{\{jon\}\}, \{\{richard\}\}, \{\{jon\}, \{richard\}\}\}$

4. $\{\emptyset, \{\{jon, richard\}\}\}$

□

Solution 4.37 1 and 4 are legal; 2 and 3 are illegal. □

Solution 4.38 All except 3 and 4 are true. □

Solution 4.39

1. For $\{1, 2, 3\} \not\subseteq \{2, 3, 4\}$ to be true, we need to exhibit an element which appears in the former set and does not appear in the latter set. 1 is such an element. As such, $\{1, 2, 3\} \not\subseteq \{2, 3, 4\}$ is true.

2. For $\{1, 2, 3\} \subset \{1, 2, 3, 4\}$ to be true, we need to demonstrate that every element which appears in the former set appears in the latter set, and that there is at least one element in the latter set which does not appear in the former. Every element of $\{1, 2, 3\}$ appears in $\{1, 2, 3, 4\}$. Furthermore, 4 appears in the latter, but not in the former. As such, $\{1, 2, 3\} \subset \{1, 2, 3, 4\}$ is true.

□

Solution 4.40

1.

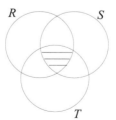

2.

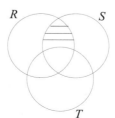

3.

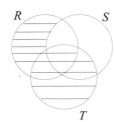

4.

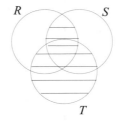

□

Solution 4.41

1. $\{2, 4\}$
2. $\{1\}$
3. $\{3, 5\}$
4. $\{1, 3, 4, 5, 7\}$

□

Solution 4.42

1. $\{1, 2\}$
2. $\{1, 2, 4\}$
3. $\{1\}$
4. $\{1\}$

□

Solution 4.43

$$B = \{1, 2, 3, 4, 7\}$$

and

$$C = \{1, 2, 3, 4, 5, 6\}$$

□

Solution 4.44

$$x \in (B \cup C) \cap A$$
$$\Leftrightarrow (x \in B \cup C) \wedge x \in A \qquad \text{[Law 4.18]}$$
$$\Leftrightarrow (x \in B \vee x \in C) \wedge x \in A \qquad \text{[Law 4.12]}$$
$$\Leftrightarrow (x \in B \wedge x \in A) \vee (x \in C \wedge x \in A) \quad \text{[Law 3.9]}$$
$$\Leftrightarrow (x \in B \cap A) \vee (x \in C \cap A) \qquad \text{[Law 4.18]}$$
$$\Leftrightarrow x \in (B \cap A) \cup (C \cap A) \qquad \text{[Law 4.12]}$$

□

Solution 4.45

$$x \in (B \cap C) \cup A$$
$$\Leftrightarrow (x \in B \cap C) \vee x \in A \qquad \text{[Law 4.12]}$$
$$\Leftrightarrow (x \in B \wedge x \in C) \vee x \in A \qquad \text{[Law 4.18]}$$
$$\Leftrightarrow (x \in B \vee x \in A) \wedge (x \in C \vee x \in A) \quad \text{[Law 3.9]}$$
$$\Leftrightarrow (x \in B \cup A) \wedge (x \in C \cup A) \qquad \text{[Law 4.12]}$$
$$\Leftrightarrow x \in (B \cup A) \cap (C \cup A) \qquad \text{[Law 4.18]}$$

□

Solution 4.46

1. 2
2. $2^2 = 4$
3. $2^{2^2} = 2^4 = 16$
4. 0
5. $2^0 = 1$
6. $2^{2^0} = 2^1 = 2$

□

Solution 4.47

1. $\{\emptyset, \{a\}, \{b\}, \{a, b\}\}$
2. $\{\emptyset\}$
3. $\{\emptyset, \{\emptyset\}\}$
4. $\{\emptyset, \{\emptyset\}, \{\{a, b\}\}, \{\emptyset, \{a, b\}\}\}$

□

Solution 4.48 They are all finite, except for 4. □

Solution 4.49

1. $\{\emptyset, \{1\}, \{2\}, \{1,2\}\}$

2. $\{2\}$

3. $\{1,2,3\}$

4. $\{1\}$

5. $\{\emptyset\}$

□

Solution 4.50

$$
\begin{aligned}
&x \in (A \setminus B) \cap B \\
\Leftrightarrow\ & (x \in A \setminus B) \wedge x \in B && \text{[Law 4.18]} \\
\Leftrightarrow\ & (x \in A \wedge x \notin B) \wedge x \in B && \text{[Law 4.25]} \\
\Leftrightarrow\ & x \in A \wedge \neg\,(x \in B) \wedge x \in B && \text{[Law 4.1]} \\
\Leftrightarrow\ & x \in A \wedge \textit{false} && \text{[Law 3.6]} \\
\Leftrightarrow\ & \textit{false} && \text{[Law 3.5]} \\
\Leftrightarrow\ & x \in \emptyset && \text{[Law 4.2]}
\end{aligned}
$$

□

Solution 4.51

$$
\begin{aligned}
&x \in (A \setminus B) \cup B \\
\Leftrightarrow\ & x \in (A \setminus B) \vee x \in B && \text{[Law 4.12]} \\
\Leftrightarrow\ & (x \in A \wedge x \notin B) \vee x \in B && \text{[Law 4.25]} \\
\Leftrightarrow\ & (x \in A \vee x \in B) \wedge (x \notin B \vee x \in B) && \text{[Law 3.9]} \\
\Leftrightarrow\ & (x \in A \vee x \in B) \wedge (\neg\,(x \in B) \vee x \in B) && \text{[Law 4.1]} \\
\Leftrightarrow\ & (x \in A \vee x \in B) \wedge \textit{true} && \text{[Law 3.15]} \\
\Leftrightarrow\ & x \in A \vee x \in B && \text{[Law 3.4]} \\
\Leftrightarrow\ & x \in A \cup B && \text{[Law 4.12]}
\end{aligned}
$$

□

Solution 4.52

1. $\emptyset$

2. $\emptyset$

3. $\emptyset$

4. $\{1,3,5\}$

    5. $\{3\}$

    6. $\{1, 3, 5\}$

☐

Solution 4.53

    1. $\{john, graham, andy, richard\}$

    2. $\{1, 2\}$

    3. $\{1\}$

    4. $\{\{1\}, \{2\}\}$

☐

Solution 4.54

    1. $\{john\}$

    2. $\{1\}$

    3. $\emptyset$

    4. $\emptyset$

☐

Solution 4.55 When S is empty, or when S contains exactly one non-empty set. ☐

Chapter 5

Boolean algebra

5.1 Introduction

Both set theory and propositions have similar properties. These properties can be used to define a mathematical structure called *Boolean algebra*.[1] For example, the idempotence laws for set theory and propositional logic look very similar, as can be seen from the following table.

Set theory	Propositional logic
$S \cup S = S$	$p \vee p = p$
$S \cap S = S$	$p \wedge p = p$

This is also true of the commutativity laws for set theory and propositional logic, as can be seen from the following table.

Set theory	Propositional logic
$S \cup T = T \cup S$	$p \vee q = q \vee p$
$S \cap T = T \cap S$	$p \wedge q = q \wedge p$

Exercise 5.1 Give the distributivity laws for set theory and propositional logic. □

The fact that we can identify rules which are common to both set theory and propositional logic means that we can apply results from one area to the other. In addition, the notion of a Boolean algebra underpins the design of digital circuits—an integral part of computer science that we shall meet in Chapter 14. This relationship is explored fully in [Gre98].

In this chapter we concentrate on the rules which constitute a Boolean algebra. Before studying Boolean algebra in detail, we revisit the topics of propositional logic and set theory in the context of this chapter.

For a comprehensive introduction to the topic of Boolean algebra, the interested reader is referred to [Mon89].

---

[1]Named after George Boole (1815 - 1864). The interested reader is referred to [GGB97] for a collection of works by George Boole.

5.2   Propositional logic revisited

Recall that every proposition can be associated with a truth value—*true* or *false*. In addition, we have seen that, via the equivalence relation, $\Leftrightarrow$, there are, in effect, three classes of propositions: those which are logically equivalent to *true* (the set of tautologies), those which are logically equivalent to *false* (the set of contradictions), and those which are logically equivalent to neither *true* nor *false* (the set of contingencies). We may denote the set of tautologies by T and the set of contradictions by F.

In addition, propositional logic involved three basic operators: $\neg$, $\wedge$, and $\vee$ (recall that we can define $\Rightarrow$ and $\Leftrightarrow$ in terms of these basic operators), together with a number of laws, such as the commutativity and distributivity of $\wedge$ and $\vee$, that propositional logic satisfied.

In summary, we may conclude that propositional logic consists of the classes T and F, the operators, $\wedge$, $\vee$ and $\neg$, and a number of laws of propositional logic.

5.3   Set theory revisited

Consider a set S, together with its power set, $\mathbb{P}\,S$. The set theory operators, $\cap$ and $\cup$, may be viewed as being operations on $\mathbb{P}\,S$, as, given any two elements of $\mathbb{P}\,S$, the resulting union or intersection will also be an element of $\mathbb{P}\,S$.

Now consider some set $X \in \mathbb{P}\,S$. It is, of course, always the case that $X \cap S = X$. We capture this formally below.

Law 5.1   For any set $X \in \mathbb{P}\,S$,

$$X \cap S = X$$

$\square$

Example 5.1   Consider the set S, such that $S = \{1, 2, 3\}$. Here,

$$\{1, 2\} \cap \{1, 2, 3\} = \{1, 2\}$$

$\square$

We may define a further operation on $\mathbb{P}\,S$—the complement operator—as follows. Given any set, $X \in \mathbb{P}\,S$, the complement of X, denoted X^-, is given by $S \setminus X$.

Law 5.2   For any set $X \in \mathbb{P}\,S$,

$$X^- = S \setminus X$$

$\square$

Example 5.2   Consider again the set $S = \{1, 2, 3\}$. Here,

$$\{1, 2\}^- = \{3\}$$

$\square$

Exercise 5.2 Given the set $S = \{a, b, c, d, e\}$, calculate the following.

1. $\{a, d\}^-$
2. $\{a, b, c, d\}^-$
3. $\{a\}^- \cap \{b\}^-$
4. $\emptyset^-$
5. $\{a, b, c, d, e\}^-$

□

Given the nature of the complement operator, we may introduce the following two laws.

Law 5.3 For any set $X \in \mathbb{P} \ S$,

$$X \cup X^- = S$$

□

Law 5.4 For any set $X \in \mathbb{P} \ S$,

$$X \cap X^- = \emptyset$$

□

We may conclude that set theory consists of a set $\mathbb{P} \ S$, the operators, $\cap$, $\cup$ and $^-$ that operate on $\mathbb{P} \ S$, and a number of laws of set theory.

Thus, we can see that the abstract structures of set theory and propositional logic are remarkably similar. By studying Boolean algebra, we can focus on the laws governing this general structure, i.e., laws governing both propositional logic and set theory, rather than propositional logic or set theory specifically. Thus, any subsequent laws discovered are applicable to both propositional logic and set theory and, of course, other structures that satisfy the laws of Boolean algebra.

Exercise 5.3 Prove Law 5.1 using equational reasoning. □

Exercise 5.4 Prove Law 5.3 using equational reasoning. □

Exercise 5.5 Prove Law 5.4 using equational reasoning. □

5.4 Fundamentals

The concept of a Boolean algebra describes a mathematical structure which encompasses the laws of both propositional logic and set theory. Such a structure is defined as follows.

Definition 5.1 Assume a set, B, with two binary operators, denoted $+$ and $*$, and a unary operator, denoted $'$, defined on it. In addition, let 0 and 1 denote two distinct elements of B. The sextuplet

$$\langle B, +, *, ', 0, 1 \rangle$$

is called a *Boolean algebra* if the following four axioms hold for any elements a, b, and c of the set B.

Commutativity: $a + b = b + a$
$$a * b = b * a$$

Distributivity: $a + (b * c) = (a + b) * (a + c)$
$$a * (b + c) = (a * b) + (a * c)$$

Identity: $a + 0 = a$
$$a * 1 = a$$

Complement: $a + a' = 1$
$$a * a' = 0$$

□

That is, any structure which satisfies the above laws can be regarded as a Boolean algebra.

Example 5.3 A Boolean algebra, B, consists of two elements—0 and 1—with operations, $'$, $+$ and $*$, which are defined as follows.

$0' = 1$
$1' = 0$

$0 + 0 = 0$
$0 + 1 = 1$
$1 + 0 = 1$
$1 + 1 = 1$

$0 * 0 = 0$
$0 * 1 = 0$
$1 * 0 = 0$
$1 * 1 = 1$

For this collection of elements and operations to be a Boolean algebra, it must satisfy the laws of commutativity, distributivity, identity, and complement. □

We now consider the general structure of Boolean algebras, and, doing so, we shall adopt the following terminology.

First, the element 0 is called the *zero* element, so-called because it behaves in a similar fashion to 0 in arithmetic. For example, $n * 0 = 0$ holds in both Boolean algebra and arithmetic.

Second, the element 1 is called the *unit* element, again, so-called because it behaves in a similar fashion to 1 in arithmetic. For example $n * 1 = n$ holds in both Boolean algebra and arithmetic.

Next, a' is called the *complement* of a, as $0' = 1$ and $1' = 0$.

Finally, the results produced by the $*$ operation and the $+$ operation are called *products* and *sums* respectively.

There are a number of other laws which follow as a consequence from those stated above; some of these may look familiar.

First, the product and sum operators are both associative.

Law 5.5 For any elements a, b, and c,

$$a + (b + c) = (a + b) + c$$
$$a * (b * c) = (a * b) * c$$

□

Next, the complement of the complement of a is equal to a.

Law 5.6 For any element a,

$$a'' = a$$

□

This is analogous to the equivalence $(\neg \ \neg \ p) \Leftrightarrow p$ of Chapter 3.

Our next law states that the sum and product operators are both idempotent.

Law 5.7 For any element a,

$$a + a = a$$
$$a * a = a$$

□

Law 5.8 tells us that de Morgan's laws are true of all Boolean algebras, not just the Boolean algebras of propositional logic and set theory.

Law 5.8 For any elements a and b,

$$(a + b)' = a' * b'$$
$$(a * b)' = a' + b'$$

□

Just as the disjunction of any proposition with *true* returns the result *true* and the conjunction of any proposition with *false* returns the result *false*, so the sum of any Boolean algebra term and 1 and the product of any Boolean algebra term and 0 return the results 1 and 0 respectively.

Law 5.9 For any element a,

$$a + 1 = 1$$
$$a * 0 = 0$$

□

Our next law may be motivated by considering its propositional logic and set theory incarnations.

First consider two atomic propositions, p and q. The proposition $p \lor (p \land q)$ is true if, and only if, at least one disjunct is true. However, both disjuncts require p to be true. As such,

$$p \lor (p \land q) \Leftrightarrow p$$

is a theorem of propositional logic.

Next consider two sets, S and T. The union of S and T, $S \cup T$, contains all of the elements of S and T. The intersection of S with $S \cup T$ contains all of the elements of S. As such,

$$S \cap (S \cup T) = S$$

is true of all sets, S and T.

Both of these facts are true of all Boolean algebras and are captured by the following law.

Law 5.10 For any elements a and b,

$$a + (a * b) = a$$
$$a * (a + b) = a$$

□

Finally, the following law's incarnation in propositional logic is given by the logical equivalences $(\neg \text{ false}) \Leftrightarrow \text{true}$ and $(\neg \text{ true}) \Leftrightarrow \text{false}$.

Law 5.11

$$0' = 1$$
$$1' = 0$$

□

Exercise 5.6 Prove, using the axioms of Boolean algebra, together with Laws 5.6 and 5.8, that $0' = 1$ holds. □

Exercise 5.7 Prove, using the axioms of Boolean algebra, that for any element a,

$$a + 0 = a * 1$$

holds. □

Exercise 5.8 Prove, using the axioms of Boolean algebra, that for any element a,

$$a + a = a$$

holds. □

Exercise 5.9 Use the result of Exercise 5.8 and Law 5.5, together with the axioms of Boolean algebra, to prove that for any element a,

$$a + 1 = 1$$

holds. □

5.5 Conventions

We may, if we wish, drop the product symbol, and use juxtaposition as a shorthand notation. That is, we may write ab instead of $a * b$ or abc instead of $a * b * c$.

Example 5.4 The distributivity laws of Boolean algebra may be written as follows.

$$a(b + c) = (ab) + (ac)$$
$$a + (bc) = (a + b)(a + c)$$

□

Exercise 5.10 Rewrite the following using this shorthand notation.

1. $a * b$
2. $a * (b + c)$
3. $a + (b * c)$
4. $a * b'$
5. $(a * b)'$

□

5.6 Precedence

Just as we enforced a precedence ordering on our propositional logic operators, so a precedence order exists with respect to the operators of Boolean algebra. Here, $'$ has precedence over $*$, and $*$ has precedence over $+$. That is, $'$ binds strongest and $+$ binds weakest.

Example 5.5 $a + b * c$ means $a + (b * c)$ and not $(a + b) * c$. □

Example 5.6 $a * b'$ means $a * (b')$ and not $(a * b)'$. □

Exercise 5.11 Write the following Boolean expressions in their fully parenthesised form.

1. ab'
2. $ab + c$
3. $a + b'$
4. $ab + cd + ef'$

□

5.7 The Boolean algebra of sets

We are now in a position to consider the Boolean algebra of sets.

Assume a set S, together with its power set, $\mathbb{P}\, S$. The set $\mathbb{P}\, S$, and the union, intersection and complement operators on $\mathbb{P}\, S$, form a Boolean algebra. This Boolean algebra has the empty set, $\emptyset$, as the zero element and the set S as the unit element.

Exercise 5.12 Prove that $\mathbb{P}\, S$ is a Boolean algebra. □

5.8   The Boolean algebra of propositions

Let Π be the set of propositions. Then Π, together with the propositional operators $\vee$, $\wedge$ and $\neg$, is a Boolean algebra. This Boolean algebra has *true* as the unit element and *false* as the zero element.

Exercise 5.13 Prove that Π is a Boolean algebra. $\square$

5.9   Isomorphic Boolean algebras

Two Boolean algebras B and $\overline{B}$ are said to be *isomorphic* if there exists an operation, f, which transforms elements of B to elements of $\overline{B}$.[2]

Essentially, such an operation converts the zero and unit elements of B to the zero and unit elements of $\overline{B}$, and 'preserves' the three operations, i.e., it ensures that, for any elements $a, b \in B$, the following hold.

$$f\,(a + b) = f\,(a)\,\overline{+}\,f\,(b)$$
$$f\,(a * b) = f\,(a)\,\overline{*}\,f\,(b)$$
$$f\,(a') = f\,(a)'$$

Here, $\overline{+}$ and $\overline{*}$ represent the sum and product operators respectively of $\overline{B}$.

Example 5.7   We may establish that the Boolean algebras of set theory and propositional logic are isomorphic by defining an operation, g, such that $g\,(true) = S$ and $g\,(false) = \emptyset$, and by establishing that

$$g\,(a \vee b) = g\,(a) \cup g\,(b)$$
$$g\,(a \wedge b) = g\,(a) \cap g\,(b)$$
$$g\,(\neg\,a) = g\,(a)^-$$

for all propositions a and b. $\square$

Exercise 5.14 Apply the operation of Example 5.7 to each of the following propositions.

1. *true* $\vee$ *false*

2. $\neg\,(true \wedge (true \vee false))$

3. $(true \wedge (true \vee false)) \vee false$

$\square$

Exercise 5.15 Assume an operation, h, which is defined as follows.

$$h\,(\emptyset) = false$$
$$h\,(S) = true$$
$$h\,(a \cup b) = h\,(a) \vee h\,(b)$$
$$h\,(a \cap b) = h\,(a) \wedge h\,(b)$$
$$h\,(a^-) = \neg\,h\,(a)$$

---

[2]We shall meet the subject of functions formally in Chapter 9, where we shall also be able to tighten our definition of what it means for two Boolean algebras to be isomorphic.

Apply h to each of the following terms.

1. $\emptyset^-$
2. $\emptyset \cap S^-$
3. $\emptyset \cap (S \cup \emptyset)$

□

Exercise 5.16 Recall that the Boolean algebra of Example 5.3 was defined as having two elements—0 and 1—with operations, $'$, $+$ and $*$, defined as follows.

$0' = 1$
$1' = 0$

$0 + 0 = 0$
$0 + 1 = 1$
$1 + 0 = 1$
$1 + 1 = 1$

$0 * 0 = 0$
$0 * 1 = 0$
$1 * 0 = 0$
$1 * 1 = 1$

How might we establish that the Boolean algebra of sets and the Boolean algebra of Example 5.3 are isomorphic? □

Exercise 5.17 How might we establish that the Boolean algebra of propositions and the Boolean algebra of Example 5.3 are isomorphic? □

5.10   Duality

In a Boolean algebra, the dual of any statement s is the statement obtained by interchanging the operations $+$ and $*$, and also interchanging the corresponding identity elements 0 and 1. That is, for example, the dual of $0 + 1$ is $1 * 0$.

Example 5.8 The dual of $(a * 1) * (0 + a') = 0$ is given by

$(a + 0) + (1 * a') = 1$

□

Example 5.9 The dual of $a + a'b = a + b$ is given by

$a * (a' + b) = a * b$

□

Example 5.10 In the Boolean algebra of propositional logic the dual of $p \wedge true$ is given by $p \vee false$. □

Example 5.11 In the Boolean algebra of set theory the dual of $R \cap T$ is given by $R \cup T$. □

An important consequence of duality is that the dual of any theorem in a Boolean algebra is also a theorem. For example, we have seen that $p \vee true \Leftrightarrow true$ is a theorem of propositional logic—it is always true, no matter what the value of p is. Its dual, $p \wedge false \Leftrightarrow false$ is also a theorem of propositional logic. Again, it is always true, no matter what the value of p is.

Exercise 5.18 Give the dual of each of the following Boolean algebra equations.

1. $(a1)(0 + a') = 0$
2. $a + (a'b) = a + b$
3. $a(a' + b) = ab$
4. $(a + 1)(a + 0) = a$
5. $(a + b)(b + c) = ac + b$

□

Exercise 5.19 Give the dual of each of the following theorems of propositional logic.

1. $\neg p \wedge p \Leftrightarrow false$
2. $p \wedge q \Leftrightarrow q \wedge p$
3. $p \vee (q \vee r) \Leftrightarrow (p \vee q) \vee r$
4. $p \wedge (p \vee q) \Leftrightarrow p$

□

5.11   Further exercises

Exercise 5.20 Assuming that the propositions p, q and r represent the Boolean algebra terms a, b and c, give the propositional logic incarnations of the following Boolean algebra terms.

1. $a + b$
2. $(a + b)'$
3. $(a' * b') + c'$

□

Exercise 5.21 Assuming that the sets R, S and T represent the Boolean algebra terms a, b and c, represent the Boolean algebra terms of Exercise 5.20 as set theory expressions. □

Exercise 5.22 Simplify the following Boolean expressions.

1. $a * (a * b')'$

2. $b * (a + b)'$

3. $((a * b) + (a' * c)) + (b * c')$

$\square$

Exercise 5.23 Consider again the operation f of Solution 5.16, which is defined in the following way.

$$f(S) = 1$$
$$f(\emptyset) = 0$$
$$f(a \cup b) = f(a) + f(b)$$
$$f(a \cap b) = f(a) * f(b)$$
$$f(a^-) = f(a)'$$

Apply this operation to the term $S \cap (\emptyset \cup S^-)$. $\square$

Exercise 5.24 Consider again the operation f of Solution 5.16 and the operation g of Example 5.7, which were defined as follows.

$$f(S) = 1$$
$$f(\emptyset) = 0$$
$$f(a \lor b) = f(a) + f(b)$$
$$f(a \land b) = f(a) * f(b)$$
$$f(\neg a) = f(a)'$$

$$g(true) = S$$
$$g(false) = \emptyset$$
$$g(a \lor b) = g(a) \cup g(b)$$
$$g(a \land b) = g(a) \cap g(b)$$
$$g(\neg a) = g(a)^-$$

Apply f to the result of applying g to the proposition $true \land (false \lor \neg\, true)$. $\square$

Exercise 5.25 Give the duals of each of the following laws of set theory.

1. $X \cap (Y \cup Z) = (X \cap Y) \cup (X \cap Z)$

2. $X \cap \emptyset = \emptyset$

3. $X \cup X^- = S$

$\square$

Exercise 5.26 In Exercise 5.9, we proved that $a + 1 = 1$ holds. Give the proof of its dual rule. $\square$

Exercise 5.27 In Exercise 5.8, we proved that $a + a = a$ holds. Give the proof of its dual rule. $\square$

Exercise 5.28 Rewrite each of the following Boolean terms so that no parentheses or product symbols appear.

1. $(a * b) * (a' * c)$
2. $(a * (b * c)) * b$
3. $(a * b) * (c' * (a * (b * a)))$
4. $(a * b) * (c' * (a * (b' * c')))$

□

Exercise 5.29 In Boolean algebra, a literal is defined to be a variable or a complemented variable, such as a or a'. In addition, a fundamental product is defined to be a collection of literals in which no variable appears more than once. As an example, ab' is a fundamental product, but abb' is not. Reduce the following Boolean products either to 0 or to a fundamental product.

1. $abba'c$
2. $abcbc'$
3. $abc'abc'$
4. $a'b'c'abc$

□

Exercise 5.30 A fundamental product P is said to be included in another fundamental product Q if the literals of P are also literals of Q. For example, ac, a and bc are all included in abc, but ac is not included in ac'. If P and Q are fundamental products and P is included in Q, show that $P + Q = P$. □

5.12   Solutions

Solution 5.1 The distributivity laws for set theory are as follows.

$$R \cap (S \cup T) = (R \cap S) \cup (R \cap T)$$
$$R \cup (S \cap T) = (R \cup S) \cap (R \cup T)$$

The distributivity laws for propositional logic are as follows.

$$p \wedge (q \vee r) = (p \wedge q) \vee (p \wedge r)$$
$$p \vee (q \wedge r) = (p \vee q) \wedge (p \vee r)$$

□

Solution 5.2

1. $\{b, c, e\}$
2. $\{e\}$

3. $\{c, d, e\}$

4. $\{a, b, c, d, e\}$

5. $\emptyset$

□

Solution 5.3 For any element $x \in S$, we have the following.

$$x \in X \cap S \Leftrightarrow x \in X \land x \in S \quad \text{[Law 4.18]}$$
$$\Leftrightarrow x \in X \land \mathit{true} \quad \text{[assumption that } x \in S]$$
$$\Leftrightarrow x \in X \quad \text{[Law 3.4]}$$

□

Solution 5.4 For any element $x \in S$, we have the following.

$$x \in X \cup X' \Leftrightarrow x \in X \lor x \in X' \quad \text{[Law 4.12]}$$
$$\Leftrightarrow x \in X \lor x \in S \setminus X \quad \text{[Law 5.2]}$$
$$\Leftrightarrow x \in X \lor (x \in S \land x \notin X) \quad \text{[Law 4.25]}$$
$$\Leftrightarrow x \in X \lor (\mathit{true} \land x \notin X) \quad \text{[assumption that } x \in S]$$
$$\Leftrightarrow x \in X \lor (x \notin X \land \mathit{true}) \quad \text{[Law 3.7]}$$
$$\Leftrightarrow x \in X \lor x \notin X \quad \text{[Law 3.4]}$$
$$\Leftrightarrow \mathit{true} \quad \text{[Law 3.15]}$$
$$\Leftrightarrow x \in S \quad \text{[assumption that } x \in S]$$

□

Solution 5.5 For any element $x \in S$, we have the following.

$$x \in X \cap X' \Leftrightarrow x \in X \land x \in X' \quad \text{[Law 4.18]}$$
$$\Leftrightarrow x \in X \land x \in S \setminus X \quad \text{[Law 5.2]}$$
$$\Leftrightarrow x \in X \land x \in S \land x \notin X \quad \text{[Law 4.25]}$$
$$\Leftrightarrow x \in X \land x \notin X \land x \in S \quad \text{[Law 3.8]}$$
$$\Leftrightarrow \mathit{false} \land x \in S \quad \text{[Law 3.6]}$$
$$\Leftrightarrow x \in S \land \mathit{false} \quad \text{[Law 3.7]}$$
$$\Leftrightarrow \mathit{false} \quad \text{[Law 3.5]}$$
$$\Leftrightarrow x \in \emptyset \quad \text{[Law 4.2]}$$

□

Solution 5.6

$$0' = (a * a')' \quad \text{[Complement]}$$
$$= a' + a'' \quad \text{[Law 5.8]}$$
$$= a' + a \quad \text{[Law 5.6]}$$
$$= 1 \quad \text{[Complement]}$$

□

Solution 5.7

$$a + 0 = a \qquad \text{[Identity]}$$
$$= a * 1 \quad \text{[Identity]}$$

□

Solution 5.8

$$a + a = (a + a) * 1 \qquad \text{[Identity]}$$
$$= (a + a) * (a + a') \quad \text{[Complement]}$$
$$= a + (a * a') \qquad \text{[Distributivity]}$$
$$= a + 0 \qquad \text{[Complement]}$$
$$= a \qquad \text{[Identity]}$$

□

Solution 5.9

$$a + 1 = a + (a + a') \quad \text{[Complement]}$$
$$= (a + a) + a' \quad \text{[Law 5.5]}$$
$$= a + a' \qquad \text{[Solution 5.8]}$$
$$= 1 \qquad \text{[Complement]}$$

□

Solution 5.10

1. ab
2. $a(b + c)$
3. $a + (bc)$
4. ab'
5. $(ab)'$

□

Solution 5.11

1. $a * (b')$
2. $(a * b) + c$
3. $a + (b')$
4. $(a * b) + (c * d) + (e * (f'))$

□

Solution 5.12 To prove that $\mathbb{P}\,S$ forms a Boolean algebra, we need to establish that the axioms of commutativity, distributivity, identity and complement all hold. It follows that $\mathbb{P}\,S$ forms a Boolean algebra, as each of the following axioms hold for all sets $X, Y, Z \in \mathbb{P}\,S$.

$$
\begin{array}{ll}
X \cup Y = Y \cup X & \text{(Law 4.15)} \\
X \cap Y = Y \cap X & \text{(Law 4.21)} \\
X \cup (Y \cap Z) = (X \cup Y) \cap (X \cup Z) & \text{(Law 4.24)} \\
X \cap (Y \cup Z) = (X \cap Y) \cup (X \cap Z) & \text{(Law 4.24)} \\
X \cup \emptyset = X & \text{(Law 4.13)} \\
X \cap S = X & \text{(Law 5.1)} \\
X \cup X' = S & \text{(Law 5.4)} \\
X \cap X' = \emptyset & \text{(Law 5.3)}
\end{array}
$$

$\square$

Solution 5.13 To prove that Π forms a Boolean algebra, we need to establish that the axioms of commutativity, distributivity, identity and complement all hold. It follows that Π forms a Boolean algebra, as each of the following axioms hold for all propositions $p, q, r \in \Pi$.

$$
\begin{array}{ll}
p \vee q = q \vee p & \text{(Law 3.14)} \\
p \wedge q = q \wedge p & \text{(Law 3.7)} \\
p \vee (q \wedge r) = (p \vee q) \wedge (p \vee r) & \text{(Law 3.16)} \\
p \wedge (q \vee r) = (p \wedge q) \vee (p \wedge r) & \text{(Law 3.17)} \\
p \vee \textit{false} = p & \text{(Law 3.11)} \\
p \wedge \textit{true} = p & \text{(Law 3.4)} \\
p \vee \neg\, p = \textit{true} & \text{(Law 3.15)} \\
p \wedge \neg\, p = \textit{false} & \text{(Law 3.6)}
\end{array}
$$

$\square$

Solution 5.14

1. $\begin{aligned}[t]
g\,(\textit{true} \vee \textit{false}) &= g\,(\textit{true}) \cup g\,(\textit{false}) \\
&= S \cup g\,(\textit{false}) \\
&= S \cup \emptyset
\end{aligned}$

2. $\begin{aligned}[t]
g\,(\neg\,(\textit{true} \wedge (\textit{true} \vee \textit{false}))) &= (g\,(\textit{true} \wedge (\textit{true} \vee \textit{false})))^{-} \\
&= (g\,(\textit{true}) \cap g\,(\textit{true} \vee \textit{false}))^{-} \\
&= (g\,(\textit{true}) \cap (g\,(\textit{true}) \cup g\,(\textit{false})))^{-} \\
&= (g\,(\textit{true}) \cap (S \cup g\,(\textit{false})))^{-} \\
&= (g\,(\textit{true}) \cap (S \cup \emptyset))^{-} \\
&= (S \cap (S \cup \emptyset))'
\end{aligned}$

3. $g\left(\left(true \wedge \left(true \vee false\right)\right) \vee false\right)$

$\qquad = g\left(true \wedge \left(true \vee false\right)\right) \cup g\left(false\right)$

$\qquad = \left(g\left(true\right) \cap g\left(true \vee false\right)\right) \cup g\left(false\right)$

$\qquad = \left(S \cap g\left(true \vee false\right)\right) \cup g\left(false\right)$

$\qquad = \left(S \cap \left(g\left(true\right) \cup g\left(false\right)\right)\right) \cup g\left(false\right)$

$\qquad = \left(S \cap \left(S \cup g\left(false\right)\right)\right) \cup g\left(false\right)$

$\qquad = \left(S \cap \left(S \cup \emptyset\right)\right) \cup g\left(false\right)$

$\qquad = \left(S \cap \left(S \cup \emptyset\right)\right) \cup \emptyset$

$\square$

Solution 5.15

1. $h\left(\emptyset^-\right) = \neg\, h\left(\emptyset\right)$

$\qquad = \neg\, false$

2. $h\left(\emptyset \cap S^-\right) = h\left(\emptyset\right) \wedge h\left(S^-\right)$

$\qquad = false \wedge h\left(S^-\right)$

$\qquad = false \wedge \neg\, h\left(S\right)$

$\qquad = false \wedge \neg\, true$

3. $h\left(\emptyset \cap \left(S \cup \emptyset\right)\right) = h\left(\emptyset\right) \wedge h\left(S \cup \emptyset\right)$

$\qquad = false \wedge h\left(S \cup \emptyset\right)$

$\qquad = false \wedge \left(h\left(S\right) \vee h\left(\emptyset\right)\right)$

$\qquad = false \wedge \left(true \vee h\left(\emptyset\right)\right)$

$\qquad = false \wedge \left(true \vee false\right)$

$\square$

Solution 5.16 We may exhibit an operation f, which is defined as follows.

$$f\left(S\right) = 1$$
$$f\left(\emptyset\right) = 0$$
$$f\left(R \cup T\right) = f\left(R\right) + f\left(T\right)$$
$$f\left(R \cap T\right) = f\left(R\right) * f\left(T\right)$$
$$f\left(R^-\right) = f\left(R\right)'$$

$\square$

Solution 5.17 We may exhibit an operation f, which is defined as follows.

$$f\left(true\right) = 1$$
$$f\left(false\right) = 0$$
$$f\left(p \vee q\right) = f\left(p\right) + f\left(q\right)$$
$$f\left(p \wedge q\right) = f\left(p\right) * f\left(q\right)$$
$$f\left(\neg\, p\right) = f\left(p\right)'$$

$\square$

Solution 5.18

1. $(a + 0) + (1a') = 1$

2. $a(a' + b) = ab$

3. $a + (a'b) = a + b$

4. $(a0) + (a1) = a$

5. $(ab) + (bc) = (a + c)b$

□

Solution 5.19

1. $\neg\, p \vee p \Leftrightarrow true$

2. $p \vee q \Leftrightarrow q \vee p$

3. $p \wedge (q \wedge r) \Leftrightarrow (p \wedge q) \wedge r$

4. $p \vee (p \wedge q) \Leftrightarrow p$

□

Solution 5.20

1. $p \vee q$

2. $\neg\, (p \vee q)$

3. $(\neg\, p \wedge \neg\, q) \vee \neg\, r$

□

Solution 5.21

1. $R \cup S$

2. $(R \cup S)^-$

3. $(R^- \cap S^-) \cup T^-$

□

Solution 5.22

1. $\begin{aligned}
a * (a * b')' &= a * (a' + b'') \\
&= a * (a' + b) \\
&= (a * a') + (a * b) \\
&= 0 + (a * b) \\
&= a * b
\end{aligned}$

2. $b * (a + b)' = b * (a' * b')$
    $$= b * (b' * a')$$
    $$= (b * b') * a'$$
    $$= 0 * a'$$
    $$= 0$$

3. $((a * b) + (a' * c)) + (b * c') = ((a' * c) + (a * b)) + (b * c')$
    $$= (a' * c) + ((a * b) + (b * c'))$$
    $$= (a' * c) + ((b * a) + (b * c'))$$
    $$= (a' * c) + (b * (a + c'))$$
    $$= ((a' * c) + b) * ((a' * c) + (a + c'))$$
    $$= ((a' * c) + b) * ((a' * c) + (a'' + c'))$$
    $$= ((a' * c) + b) * ((a' * c) + (a' * c)')$$
    $$= ((a' * c) + b) * 1$$
    $$= (a' * c) + b$$

□

Solution 5.23

$$f(S \cap (\emptyset \cup S^-)) = f(S) * f(\emptyset \cup S^-)$$
$$= 1 * f(\emptyset \cup S^-)$$
$$= 1 * (f(\emptyset) + f(S^-))$$
$$= 1 * (f(\emptyset) + f(S)')$$
$$= 1 * (0 + f(S)')$$
$$= 1 * (0 + 1')$$

□

Solution 5.24 Applying g to *true* $\wedge$ (*false* $\vee \neg$ *true*) gives $S \cap (\emptyset \cup S^-)$. By the solution to the previous exercise, $f(S \cap (\emptyset \cup S^-))$ gives $1 * (0 + 1')$. □

Solution 5.25

1. $X \cup (Y \cap Z) = (X \cup Y) \cap (X \cup Z)$
2. $X \cup S = S$
3. $X \cap X^- = \emptyset$

□

Solution 5.26

$$a * 0 = a * (a * a') \quad \text{[Complement]}$$
$$= (a * a) * a' \quad \text{[Associativity]}$$
$$= a * a' \quad\quad\ \text{[Idempotence]}$$
$$= 0 \quad\quad\quad\ \ \text{[Complement]}$$

□

Solution 5.27

$$
\begin{aligned}
a * a &= (a * a) + 0 && \text{[Identity]} \\
&= (a * a) + (a * a') && \text{[Complement]} \\
&= a * (a + a') && \text{[Distributivity]} \\
&= a * 1 && \text{[Complement]} \\
&= a && \text{[Identity]}
\end{aligned}
$$

□

Solution 5.28

1. $aba'c$

2. $abcb$

3. $abc'aba$

4. $abc'ab'c'$

□

Solution 5.29

1. $abba'c = aa'bbc$
   $= 0bbc$
   $= 0$

2. $abcbc' = abbcc'$
   $= abb0$
   $= 0$

3. $abc'abc' = aabbc'c'$
   $= abc'$

4. $a'b'c'abc = aa'bb'cc'$
   $= 0bb'cc'$
   $= 0$

□

Solution 5.30 If P is included in Q, then $Q = P * R$ for some R. As such, we have the following.

$$
\begin{aligned}
P + Q &= P + (P * R) \\
&= (P * 1) + (P * R) && \text{[Identity]} \\
&= P * (R + 1) && \text{[Distributivity]} \\
&= P * 1 && \text{[Law 5.9]} \\
&= P && \text{[Identity]}
\end{aligned}
$$

□

Chapter 6

Typed set theory

Thus far we have been very careful about the types of statement we have used with regards to our set theory operators. In Example 4.18, we stated that *eaten* was a subset of *fruitbowl*: both of these sets were concerned with fruit. Furthermore, we have been careful not to combine, for example, sets of numbers with, say, sets of fruit when using the operators $\cap$, $\cup$, or $\setminus$. When using such operators, we must insist that both sets contain elements *of the same type*.

At first glance, statements such as $1 \in eaten$, $banana \in \mathbb{N}$, or $\mathbb{N} \subseteq People$ might appear to be untrue. If we negate these statements to arrive at $1 \notin eaten$, $banana \notin \mathbb{N}$ and $\mathbb{N} \nsubseteq People$, one might jump to the conclusion that these negated statements are true. In our approach to discrete mathematics, however, the second collection of statements is as nonsensical as the first set. This is because we concern ourselves with *typed set theory*.

Typed set theory asserts that every set (and, indeed, every element) must have an associated *type*. To reason about statements such as $s \in S$, S must be a set of the same type as that which is associated with s. Similarly, to reason about statements of the form $S \subseteq T$, or $S \cap T$, we must ensure that S and T are sets of the same type. For example, if S is a set of numbers then T must also be a set of numbers for the result of the intersection $S \cap T$ to be defined. This is consistent with our view of set theory as a Boolean algebra.

A major advantage of this approach is that, when sets are implemented in the 'real world', such as, for example, in the context of relational databases, all such entities must have an associated type. If we were to reason about information pertaining to the users of a sports hall and we wished to print out the names and addresses of all members, we would not expect to be presented with their ages and telephone numbers; the types of these objects are very different. Furthermore, while it would be sensible to merge two sets of membership numbers one would expect the result of merging a set of addresses with a set of membership numbers to produce an error. It is the concept of typed sets that provides this desired structure. As we shall see, there is a very clear relationship between the material of this and the following two chapters, and the apparently more applied subject of relational databases.

Thus, in this chapter, we outline the reasons for using typed set theory, and describe some further ways of defining sets in terms of this approach. An advanced text on this subject for the interested reader is [Sup60].

6.1 The need for types

In Chapter 4 we discussed the notion of a set. There, we described a set as being "an unordered collection of objects". Although we were careful to ensure that sets contained similar types of objects, and that distinct types of objects were kept apart, we were—at that stage—under no obligation to do so. However, if we were to continue with such an approach, then a number of *paradoxes* can arise; we discuss two of the most famous such paradoxes here.

Cantor's paradox

Consider the set of all sets, which we may denote by $\mathcal{S}$. Here,

$$X \in \mathcal{S} \Leftrightarrow X \text{ is a set}$$

Every subset of $\mathcal{S}$ is a set and, as such, every subset of $\mathcal{S}$ is an element of $\mathcal{S}$. It follows that $\mathbb{P}\, \mathcal{S} \subseteq \mathcal{S}$. However, as we saw in Chapter 4, it is a fact that the cardinality of the power set of any set X is always greater than the cardinality of X (recall that if X has n elements then $\mathbb{P}\, X$ has 2^n elements). As such, these two statements are contradictory. This is known as *Cantor's paradox*.[1]

Russell's paradox

It is possible for some sets to be members of themselves; for example, $\mathcal{S}$, the set of all sets is a member of itself. On the other hand, some sets are not members of themselves: the set of all dogs, for example, is not a dog. Assume that the set of all sets which are not members of themselves is denoted $\mathcal{R}$. Here,

$$X \in \mathcal{R} \Leftrightarrow X \notin X$$

This definition of $\mathcal{R}$ begs the question "is $\mathcal{R}$ a member of itself?" If $\mathcal{R} \in \mathcal{R}$, then $\mathcal{R}$ is a member of the set of all sets which are not members of themselves, and, as such, $\mathcal{R} \notin \mathcal{R}$. On the other hand, if $\mathcal{R} \notin \mathcal{R}$, then $\mathcal{R}$ is not a member of the set of all sets which are not members of themselves, and, as such, $\mathcal{R} \in \mathcal{R}$. These two statements are contradictory. This is known as *Russell's paradox*.[2]

Allowing ourselves a great deal of flexibility in the definition of sets—as we did in Chapter 4—can lead to the sort of situations described by the above paradoxes. *Axiomatic set theory*[3] introduces a number of *axioms*—some of which we have already seen—to prevent these, and other, paradoxes. Thus, a set is properly defined only if it satisfies such axioms.[4] In particular, the axioms guarantee that statements such as "the set of all sets" and "the set of all sets which are not members of themselves" are invalidated—we do not allow ourselves to reason about such sets.

As well as enforcing a number of laws, axiomatic set theory insists that a *type* be associated with each set. It follows that our set theory operators, such as $\cap$ and $\cup$, can only be applied to two sets if they are of the same type.

---

[1] Named after Georg Cantor (1845 - 1918). See [Dau90] for a description of Cantor's life and work.

[2] Named after Bertrand Russell (1872 - 1970). See [Gar92] for a description of the influence of Russell and Russell's paradox, and [Mon97] for an excellent biography of the first fifty years of Russell's life.

[3] Developed originally by Ernst Zermelo (1871 - 1953) and Abraham Fraenkel (1891 - 1965).

[4] We choose not to list the axioms here. Suffice to say that many of these axioms are captured in terms of some of the laws of Chapter 4.

Example 6.1 We may concern ourselves with the *type* of people, denoted *People*, and the *type* of cars, denoted *Cars*. While we are free to reason about the intersections, unions, and set differences of any sets $X, Y \subseteq People$, we cannot reason about the intersection or union of X and some set Z, such that $Z \subseteq Cars$, as their types are incompatible. □

Exercise 6.1 Assume that *People* and *Cars* are types, such that *People* denotes the set of all people and *Cars* denotes the set of all cars. Assume further that $A, B \subseteq People$ and $X, Y \subseteq Cars$. Which of the following applications of $\cap$ are legal?

1. $A \cap B$

2. $A \cap X$

3. $A \cap Y$

4. $A \cap People$

5. $A \cap Cars$

6. $A \cap \emptyset$

□

Recall that a set of people is a very different entity to a set of set of people—the type of element that may appear in the first set is different to the type of element that may appear in the second. For example, a typical element of the first set is $\{joe\}$ while a typical element of the second set is $\{\{joe\}\}$.

Example 6.2 Consider again the sets $A, B \subseteq People$. The intersections $A \cap B$ and $A \cap People$ are defined, but the intersections $A \cap \{B\}$ and $\{A\} \cap People$ are not. □

Exercise 6.2 Which of the following applications of $\cap$ are legal?

1. $\{1, 2, 3\} \cap \{\{1\}\}$

2. $\{1, 2, 3\} \cap \emptyset$

3. $\{1, 2, 3\} \cap \{\emptyset\}$

4. $\{1, 2, 3\} \cap \{1\}$

□

As well as restricting the use of operators which deal with entities of the same *type*, such as $\cap$, $\cup$ and $\setminus$, typed set theory also restricts the use of operators which deal with entities of different *orders*, such as $\in$ and $\subseteq$.

As a means of motivating this, consider a bucket containing a banana. Now consider inserting that first bucket into a second bucket. If one was asked what was contained in this second bucket, the answer would not be "a banana", but "a bucket containing a banana". Typed set theory enforces this principle: the elements of the set $\{\{1\}, \{2\}\}$ are $\{1\}$ and $\{2\}$, not 1 and 2, and, as such, the statement $\{1\} \in \{\{1\}, \{2\}\}$ is legitimate, whereas the statement $1 \in \{\{1\}, \{2\}\}$ is not. Provided that x is—in this sense—'one level lower' than X, then the value of $x \in X$ is defined.

Example 6.3 Assume that the type *People* represents the set of all people, and that *andy, duncan, john* and *richard* are all elements of *People*.

The following applications of $\in$ and $\subseteq$ are all legal and, furthermore, all have the value *true*.

$\emptyset \subseteq People$
$\emptyset \in \mathbb{P}\ People$
$andy \in People$
$\{andy, john\} \subseteq People$
$\{andy, john\} \subseteq \{andy, duncan, john, richard\}$

The following applications of $\in$, $\notin$ and $\subseteq$ are all legal, and all have the value *false*.

$People \subseteq \emptyset$
$\emptyset \notin \mathbb{P}\ People$
$andy \notin People$
$\{andy, john\} \subseteq \{john\}$
$\{andy, john\} \subseteq \{andy, duncan, richard\}$

The following applications of $\in$ and $\subseteq$ are all illegal.

$\emptyset \subseteq andy$
$\emptyset \in People$
$andy \subseteq People$
$\{andy, john\} \in People$
$\{andy, john\} \in \{andy, duncan, john, richard\}$

$\square$

Exercise 6.3 Assume the set $S = \{1, 2\}$. Which of the following statements are legal?

1. $S \in S$

2. $S \subseteq S$

3. $\{1\} \in S$

4. $1 \in S$

5. $\{1\} \subseteq S$

6. $1 \subseteq S$

7. $S \in \mathbb{P}\ S$

8. $\{S\} \in \mathbb{P}\ S$

9. $\{1\} \in \mathbb{P}\ S$

10. $1 \in \mathbb{P}\ S$

$\square$

6.2 The empty set revisited

The consideration of typed set theory raises some important questions regarding the empty set. Consider the term $\emptyset \cap \{1\}$. This is a legal application of $\cap$, and the resulting set is equal to $\emptyset$. Now consider the term $\{jack\} \setminus \{jack\}$. Again, this is a legal application of a set theory operator, with the resulting set being equal to $\emptyset$. Now, given these two terms, does it follow that

$$(\{jack\} \setminus \{jack\}) \cap \{1\}$$

is a legal term? The answer, of course, is no. Even though we have stated that both $\emptyset \cap \{1\}$ and $\{jack\} \setminus \{jack\}$ are legal applications of our set theory operators, the overall statement $(\{jack\} \setminus \{jack\}) \cap \{1\}$ is not. This is because the empty set of the statement $\emptyset \cap \{1\}$ is of a different type to the empty set which is the result of $\{jack\} \setminus \{jack\}$.

The reasoning behind this is that typed set theory requires that there is not one, unique, empty set, but many empty sets—infinitely many, in fact. Specifically, typed set theory requires that every type has an associated empty set. As such, the empty set of people is a different entity to the empty set of natural numbers. The former can be combined with sets of people, while the latter can be combined with sets of natural numbers. However, the empty set of people cannot be combined with any set of natural numbers—including the empty set of natural numbers.

In our above example, the empty set of $\emptyset \cap \{1\}$ is the empty set of natural numbers, whereas the empty set which is the result of $\{jack\} \setminus \{jack\}$ is the empty set of people. Although we cannot distinguish between the two empty sets *syntactically*, we can determine which empty set we are dealing with in each case by the context in which they are being used. Thus, the reason that $(\{jack\} \setminus \{jack\}) \cap \{1\}$ is an illegal statement is that we are trying to establish the intersection of the empty set of *people* and the set containing 1: this is clearly a type mismatch.

Exercise 6.4 Assuming the type *Car*, such that $\{ford, ferrari\} \subseteq Car$, which of the following terms are defined?

1. $\{ford, ferrari\} \cap \{1, 2\}$

2. $\{ford, ferrari\} \cap \emptyset$

3. $\{1, 2\} \cap \emptyset$

4. $\{ford, ferrari\} \cup (\{1, 2\} \cap \{3, 4\})$

□

6.3 Set comprehension

In Chapter 4, we saw how we might define sets as abstract entities—such as the set *People* representing all of the people in the world—or explicitly, by enumerating all of their members, such as

$$Union = \{england, northern\ ireland, scotland, wales\}$$

In this section, we demonstrate a means of defining new sets in terms of existing ones: this is referred to as *set comprehension*.

The form of such set comprehensions is that given by

$$\{declaration \mid predicate\}$$

where *declaration* involves the enumeration of entities of the appropriate type (or types), and *predicate* acts as a filter—keeping all those entities that satisfy the predicate, and removing all those that don't.

Example 6.4 The set of all people who own mobile phones may be defined as follows.

$$Mobile\_owners = \{p : People \mid p \text{ owns a mobile phone}\}$$

Here, the declaration $p : People$ indicates that we are concerned with all people, and the predicate "p owns a mobile phone" states that only those elements of *People* who own a mobile phone should be included in the set *Mobile_owners*. □

Example 6.5 The set of even natural numbers may be defined via set comprehension in the following way.

$$even\_naturals = \{n : \mathbb{N} \mid n \bmod 2 = 0\}$$

This set comprehension generates the set containing 0, 2, 4, 6, etc. □

Set comprehension can be thought of as a machine: the declaration provides the inputs, while the predicate checks whether each input satisfies the relevant property. For example, in the above, 0 is generated first. This is checked against the criterion: it passes the test and, as a result, is retained. 1 is the next number to be generated. This is checked against the criterion: it fails, and, as such, is rejected. 2 is the next number to be generated. Again, it is compared against the criterion: it passes, and, as a result, is retained. This process then continues (in this case, infinitely).

Exercise 6.5 List the elements of the following sets.

1. $A = \{n : \mathbb{N} \mid n < 7\}$
2. $B = \{n : \mathbb{N} \mid n < 10 \wedge n \bmod 3 = 0\}$
3. $C = \{n : \mathbb{N} \mid (4 \times n = 12) \vee (6 \times n = 12)\}$
4. $D = \{n : \mathbb{N} \mid n < 0\}$

□

Exercise 6.6 Assuming the types *People* and *Car* and appropriate predicates, define the following sets using set comprehension.

1. The set of all red cars.
2. The set of all vegetarians.
3. The set of all cars with sun roofs.

4. The set of all people who are rich.

□

Statements such as "*p* is rich" are—for the moment—necessarily informal: we have not specified formally how we represent such a statement mathematically. We shall see, in the following chapters, how to represent such statements formally in terms of *predicates*. For now, however, this semi-formal approach shall suffice.

In the previous exercise, we generated sets of cars and sets of people. Often, however, there are circumstances in which we require further information about those people who are rich; we might, for example, not be concerned with the people *per se*, but with their addresses (for whatever nefarious purposes). An extension of our set comprehension notation allows us to capture such information formally. In such a case, the form of the set comprehension is

$$\{ declaration \mid predicate \bullet term \}$$

Here, *term* generates a term for those entities enumerated by *declaration*, and filtered by *predicate*.

Example 6.6 If we wished to generate a set containing the addresses of all mobile phone owners, then we may define

$$Mobile\_addresses = \{ p : People \mid p \text{ owns a mobile phone} \bullet address(p) \}$$

□

In the above example, we assume that *address* is an operation which, when given a person as input, returns their address as output.

Example 6.7 The set of squares of all even naturals is given by

$$even\_natural\_squares = \{ n : \mathbb{N} \mid n \bmod 2 = 0 \bullet n^2 \}$$

This set comprehension generates the numbers 0, 4, 16, 36, etc. □

Again, we can think of set comprehension as being a machine. Step 1 generates entities; step 2 filters them; and step 3 performs some action on them, and then provides this as output. So, in the above example, the declaration generates the numbers 0, 1, 2, 3, 4, 5, and so on. The next stage—the predicate stage—filters these numbers: 0 is kept, 1 is rejected, 2 is kept, 3 is rejected, 4 is kept, 5 is rejected, and so on. Finally, the term part of the set comprehension readies these remaining inputs for output: 0 is squared to give 0, 2 is squared to give 4, 4 is squared to give 16, and so on.

Exercise 6.7 List the elements of the following sets.

1. $A = \{ n : \mathbb{N} \mid n < 7 \bullet n^2 \}$

2. $B = \{ n : \mathbb{N} \mid n < 10 \wedge n \bmod 3 = 0 \bullet n \}$

3. $C = \{ n : \mathbb{N} \mid (4 \times n = 12) \vee (6 \times n = 12) \bullet n \bmod 2 \}$

4. $D = \{n : \mathbb{N} \mid n < 0 \bullet n + 1\}$

□

Exercise 6.8 Assuming the types *People* and *Car*, define the following sets using set comprehension. In each case, you should assume the existence of relevant operations.

1. The set of registration numbers of all red cars.

2. The set of ages of all vegetarians.

3. The set of owners of all cars with sun roofs.

4. The set of addresses of all people who are rich.

□

Just as we might wish to define the set of registration numbers of all red cars, we might wish to be more general and define the set of registration numbers of all cars, whatever their colour. Under such circumstances, we need to use a filter which doesn't reject any element generated by the declaration; that is, a predicate which is always true. Of course, we have such a predicate in our mathematical language: *true*. So, the set comprehension

$$\{c : Car \mid true \bullet registration(c)\}$$

gives us the set that we require.

In such circumstances, we may freely drop the predicate part of the set comprehension as it contributes nothing to the overall result: in our machine analogy, we may assume that the second step has simply been switched off. So, in this example, we may also write

$$\{c : Car \bullet registration(c)\}$$

and achieve exactly the same result.

Example 6.8 The set of squares of all natural numbers may be defined as follows.

$$squares = \{n : \mathbb{N} \bullet n^2\}$$

This set comprehension generates the set containing 0, 1, 4, 9, 16, and so on. □

Exercise 6.9 List the first nine elements of the following sets.

1. $A = \{n : \mathbb{N} \bullet n^2\}$

2. $B = \{n : \mathbb{N} \bullet n\}$

3. $C = \{n : \mathbb{N} \bullet n \text{ div } 2\}$

4. $D = \{n : \mathbb{N} \bullet n + 1\}$

□

Exercise 6.10 Assuming the types *People* and *Car*, define the following sets using set comprehension. In each case, you may assume the existence of relevant operations.

1. The set of all car registration numbers.

2. The set of ages of people.

3. The set of all car owners.

4. The set of all addresses of people.

□

6.4 Characteristic tuples

Just as it is possible for us to omit the declaration part from a set comprehension, so it is possible for us to omit the term part from a set comprehension. Indeed, the first few examples of set comprehensions that we considered had no term part. In such cases, the type of elements produced by the set comprehension are determined not by the term part, but by the declaration. This information—known as the *characteristic tuple* of the set comprehension—is akin to enforcing a default term on such set comprehensions.

As an example, the set comprehensions

$$\{ c : Car \mid c \text{ is red} \}$$

and

$$\{ c : Car \mid c \text{ is red} \bullet c \}$$

are entirely equivalent. In this case, the characteristic tuple is c. As a second example, the set comprehensions

$$\{ n : \mathbb{N} \mid n < 7 \}$$

and

$$\{ n : \mathbb{N} \mid n < 7 \bullet n \}$$

are also equivalent. In this case, the characteristic tuple is n.

At this stage, all the characteristic tuples that we may meet are of this simple variety. However, the notion of a characteristic tuple will prove to be more relevant when we encounter Cartesian products[5] in Section 6.6.

Exercise 6.11 Give the characteristic tuples of the following sets.

1. $A = \{ n : \mathbb{N} \mid n < 7 \}$

2. $B = \{ m : \mathbb{N} \mid m < 10 \wedge m \bmod 3 = 0 \}$

□

---

[5] Named after René Descartes (1596 - 1650).

6.5   Abbreviations

Sometimes when dealing with a subset of the natural numbers, we may not wish to write out the elements of this subset in full; it may be more convenient to provide an abbreviated version. One way of accomplishing this is to use the .. operator, which is pronounced "up to". Given two natural numbers x and y such that $x \leq y$, we may write $x .. y$ to represent the set of natural numbers which occur between x and y (inclusive). Thus, for example, $1 .. 3$ denotes the set $\{1, 2, 3\}$.

Example 6.9 The set of natural numbers between 10 and 20 (inclusive) may be written $10 .. 20$. This is equivalent to the set

$$\{10, 11, 12, 13, 14, 15, 16, 17, 18, 19, 20\}$$

$\square$

Example 6.10 The set of *pool_balls* of Example 4.2 might have been defined by

$$pool\_balls = 1 .. 15$$

$\square$

Exercise 6.12 Write the following sets out in full.

1. $0 .. 5$

2. $1 .. 1$

3. $0 .. 5 \setminus 0 .. 3$

4. $\{n : \mathbb{N} \mid n \in 0 .. 3\}$

$\square$

When we wish to denote the fact that a series of numbers carries on indefinitely, we will not provide an upper bound on the abbreviation. However, in such cases, we will explicitly provide the first few numbers in the sequence to indicate how it progresses. So, for example, $3, 4, 5, \ldots$ indicates the set containing all natural numbers, from 3 upwards. This, of course, could equally well be written $\mathbb{N} \setminus \{0, 1, 2\}$. As a further example, $0, 2, 4, 6, 8, \ldots$ denotes the set of even natural numbers.

Exercise 6.13 Write the following sets out in full.

1. $\{n : 0 .. 10 \mid n \in 7, 8, 9, \ldots\}$

2. $\{n : 0 .. 10 \mid n \in 2, 4, 6, \ldots\}$

3. $\{n : 0 .. 10 \mid n \in 3, 6, 9, \ldots\}$

$\square$

6.6   Cartesian products

In Chapter 4 and in this chapter so far, we have tended to concentrate on relatively simple sets. Furthermore, all of the operators of set theory that we have seen have combined sets of the same simple types. In reality, however, we often have to reason about structures which incorporate different types. For example, in any substantial database there are entities of many different types that the underlying structure of the database must combine in some meaningful way. A record that consists of employees' names and salaries must combine information of different, simple, types and store the related information in structures of further, more complicated, types. If we are to reason about such entities formally, then we require an operator that allows us to structure information in an appropriate fashion.

The set theoretic operator which allows us to combine sets of different types in this way is called the *Cartesian product* operator. Given two sets, A and B, the *Cartesian product* of A and B, denoted $A \times B$, is defined as follows.

$$A \times B = \{a : A;\ b : B \bullet (a, b)\}$$

This defines the set of all *ordered pairs* (a, b), such that $a \in A$ and $b \in B$. Essentially, the Cartesian product considers all elements of A and all elements of B, and produces a set containing each possible ordered pair of such elements.

Example 6.11 Consider the following sets.

$$Name = \{jack, jill\}$$
$$Age = \{65, 60\}$$

Here, *Name* has two elements—*jack* and *jill*—and *Age* also has two elements—65 and 60. Thus, *jack* and *jill* can both be paired with both elements of *Age*. As such, the Cartesian product *Name* $\times$ *Age* gives the following.

$$\{(jack, 65), (jack, 60), (jill, 65), (jill, 60)\}$$

Note that the first element of each pair is an element of the first set, and the second element of each pair is an element of the second set. As such, $(jack, jack)$ is not an element of *Name* $\times$ *Age*, and neither is $(65, 60)$. Note also that $(jack, 65)$ and $(65, jack)$ are very different entities, as we are dealing with *ordered* pairs. In fact, these pairs are of different types: the former is an element of *Name* $\times$ *Age*, whereas the latter is an element of *Age* $\times$ *Name*. $\square$

Example 6.12 Consider the following sets.

$$Die = \{1, 2, 3, 4, 5, 6\}$$
$$Coin = \{head, tail\}$$

Here,

$$\begin{aligned}
Die \times Coin = \{&(1, head), (2, head), (3, head), (4, head), (5, head), (6, head), \\
&(1, tail), (2, tail), (3, tail), (4, tail), (5, tail), (6, tail)\}
\end{aligned}$$

Furthermore,

$$Coin \times Die = \{(head, 1), (head, 2), (head, 3), (head, 4), (head, 5), (head, 6),$$
$$(tail, 1), (tail, 2), (tail, 3), (tail, 4), (tail, 5), (tail, 6)\}$$

□

Exercise 6.14 Consider the following sets.

$$A = \{0, 1\}$$
$$B = \{yes, no\}$$

Which Cartesian products do the following pairs belong to?

1. $(0, yes)$

2. $(1, 0)$

3. (no, yes)

4. $(no, 1)$

5. $(no, \{yes\})$

6. $(no, \{(1, yes)\})$

□

Exercise 6.15 Consider the sets $A = \{0, 1\}$ and $B = \{yes, no\}$. List the elements of the following sets.

1. $A \times A$

2. $A \times B$

3. $B \times A$

4. $B \times B$

5. $A \times \mathbb{P}\, B$

6. $A \times \emptyset$

□

Exercise 6.16 Consider the sets $A = \{0, 1\}$ and $B = \{yes, no\}$. For each of the following sets, exhibit an example element.

1. $A \times B$

2. $A \times \mathbb{P}\, B$

3. $A \times \mathbb{P}\, \emptyset$

4. $(\mathbb{P}\, A) \times B$

5. $(\mathbb{P}\, A) \times (\mathbb{P}\, B)$

6. $(\mathbb{P}\ A) \times (\mathbb{P}\ \emptyset)$

7. $\mathbb{P}\,(A \times B)$

8. $\mathbb{P}\,(A \times \mathbb{P}\ B)$

9. $\mathbb{P}\,(A \times \mathbb{P}\ \emptyset)$

□

Note that the size of a Cartesian product is determined by the size of the contributing sets. For example, in the above exercises, both A and B had two elements; as such, the size of $A \times B$ was four: there were four possible pairs. This relationship is captured by the following law.

Law 6.1 For any sets X and Y,

$$\#\,(X \times Y) = \#\,X \times \#\,Y$$

□

Exercise 6.17 How many elements do the following sets have?

1. $\{a, b, c\} \times \{0, 1\}$

2. $\{0, 1\} \times \{a, b, c\}$

3. $\{a, b, c\} \times \emptyset$

4. $\{a, b, c\} \times \mathbb{P}\,\{0, 1\}$

5. $\{a, b, c\} \times \mathbb{P}\,\emptyset$

6. $\mathbb{P}\,(\{0, 1\} \times \{a, b, c\})$

7. $\mathbb{P}\,(\{a, b, c\} \times \emptyset)$

8. $\mathbb{P}\,(\mathbb{P}\,\{0, 1\} \times \mathbb{P}\,\{a, b, c\})$

□

The notion of Cartesian products—as, indeed, we saw in the definition of them—allow us to define sets of more complicated structures via set comprehension than would otherwise be the case. For example, the set of all pairs of natural numbers less than 3 may be specified via set comprehension in the following fashion.

$$low\_pairs = \{m, n : \mathbb{N} \mid m < 3 \wedge n < 3 \bullet (m, n)\}$$

This generates the following set.

$$\{(0, 0), (0, 1), (0, 2), (1, 0), (1, 1), (1, 2), (2, 0), (2, 1), (2, 2)\}$$

Here, the term part of the set comprehension is unnecessary, as the characteristic tuple of this set comprehension is (m, n). Whenever there is no term component in a set comprehension in which the declaration consists of two elements, the characteristic tuple is a pair of the appropriate type. In this case, it is a pair of type $\mathbb{N} \times \mathbb{N}$.

As a second example, the following set comprehension has the characteristic tuple (n, a), where $n \in Name$ and $a \in Age$.

$$\{n : Name;\ a : Age \mid n = Norman \lor a = 31\}$$

The order of declarations here is important—the characteristic tuple of the following set comprehension is not (n, a), but (a, n).

$$\{a : Age;\ n : Name \mid n = Norman \lor a = 31\}$$

The above declarations identify a convention for declarations that we shall continue with throughout the remainder of this book: declarations of entities of different types must be separated by semicolons. When dealing with declarations of entities of the same type, however, the use of commas is permissible (as seen, for example, in the definition of low_pairs). In fact, the use of commas in such cases is simply a shorthand notation; we could equally well have defined low_pairs in the following manner.

$$low\_pairs = \{m : \mathbb{N};\ n : \mathbb{N} \mid m < 3 \land n < 3 \bullet (m, n)\}$$

Exercise 6.18 Determine the characteristic tuples of the following set comprehensions.

1. $\{n : \mathbb{N} \mid p\}$
2. $\{m, n : \mathbb{N} \mid p\}$
3. $\{m : \mathbb{N} \times \mathbb{N};\ n : \mathbb{N} \mid p\}$

□

Exercise 6.19 Assuming the sets $Girls$ and $Boys$, defined by

$$Girls = \{emily, rachel, anna\}$$
$$Boys = \{michael, john\}$$

list the elements of the following sets.

1. $\{b : Boys;\ g : Girls\}$
2. $\{g : Girls;\ b : Boys\}$
3. $\{b, c : Boys\}$
4. $\{b, c : Boys \mid b \neq c\}$
5. $\{b : Boys \bullet (b, b)\}$
6. $\{b : Boys;\ g : Girls \bullet b\}$
7. $\{b : Boys;\ g : Girls \mid b = john\}$
8. $\{b : Boys;\ g : Girls \mid b = john \bullet g\}$

□

As Cartesian products are themselves sets (sets of pairs, in fact), all of the set theory operators that we have seen thus far are applicable to Cartesian products. So, for example,

$$\{(jack, 65), (jill, 60)\} \subseteq \{(jack, 65), (jack, 60), (jill, 65), (jill, 60)\}$$

is a legitimate statement, and, what is more, it is true.

Exercise 6.20 Which of the following applications of $\setminus$ are legal?

1. $(\mathbb{N} \times \mathbb{N}) \setminus \{3, 4\}$
2. $(\mathbb{N} \times \mathbb{N}) \setminus (3, 4)$
3. $(\mathbb{N} \times \mathbb{N}) \setminus \{(3, 4)\}$
4. $(\mathbb{P}(\mathbb{N} \times \mathbb{N})) \setminus \{(3, 4)\}$
5. $(\mathbb{P}(\mathbb{N} \times \mathbb{N})) \setminus \{\{(3, 4)\}\}$

$\square$

Exercise 6.21 Determine the results of the following applications of $\cap$, $\cup$, and $\times$.

1. $(\{0, 1, 2\} \cap \{1, 2, 3\}) \times \{2, 3\}$
2. $(\{0, 1, 2\} \cup \{1, 2, 3\}) \times \{2, 3\}$
3. $(\{0, 1, 2\} \setminus \{1, 2, 3\}) \times \{2, 3\}$

$\square$

So far we have restricted our use of Cartesian products to the generation of pairs. However, the use of Cartesian products is more general than this—we can use them to define triples, quadruples, quintuples, in fact *tuples* of any (finite) size. Furthermore, we can consider pairs of pairs, triples of pairs, pairs of triples, and so on.

Example 6.13 Consider a database that stores names, ages, and email addresses; these are represented by the types *Name*, *Age*, and *Email* respectively. The Cartesian product

$Name \times Age \times Email$

generates tuples of size 3 (or *triples*). Assuming that *Name*, *Age*, and *Email* are defined as follows

$$Name = \{jack, jill\}$$
$$Age = \{60, 65\}$$
$$Email = \{jack@the\_hill.com, jill@the\_hill.com\}$$

a typical element of this Cartesian product is $(jack, 65, jack@the\_hill.com)$. $\square$

Just as order is important in Cartesian products, so are parentheses. The Cartesian product *Name* $\times$ *Age* $\times$ *Email* produces triples; the Cartesian product *Name* $\times$ (*Age* $\times$ *Email*) produces *pairs*, in which the first component of each pair is a member of *Name*, and the second component of each pair is itself a pair of type *Age* $\times$ *Email*. As an example, $(jack, (65, jack@the\_hill.com))$ is an element *Name* $\times$ (*Age* $\times$ *Email*).

Exercise 6.22 Assuming the sets *Name*, *Age*, and *Email* given above, list the elements of the following sets.

1. *Name* × *Age* × *Email*
2. *Name* × (*Age* × *Email*)
3. (*Name* × *Age*) × *Email*

□

Exercise 6.23 Assuming the sets *Name*, *Age*, and *Email* given above, list the elements of the following sets.

1. $\{a, b, c : Age\}$
2. $\{n : Name \bullet (n, n, n)\}$
3. $\{m, n : Name \mid m = jack \bullet (m, (m, n))\}$
4. $\{m, n : Name \bullet ((m, n), (m, n))\}$
5. $\{a : Age \times Email\}$
6. $\{a : Age;\ n : Name;\ e : Email \mid a = 65 \wedge n = jack \bullet (n, a, e)\}$

□

To this point, we have considered how to build up new sets from existing ones via Cartesian products. Sometimes, however, it may be desirable to work the other way; that is, when presented with a set of pairs, we might wish to, for example, consider only the first component of each pair. We may access such information via *component selection*.

Example 6.14 Suppose that (*jack*, 65, *jack@the_hill.com*) is an element of the Cartesian product *Name* × *Age* × *Email*, and we wish to know Jack's age, i.e., the second element of the triple. To obtain a component of a tuple, we indicate the position with which we are concerned via a dot. So, for example, the component selection (*jack*, 65, *jack@the_hill.com*).2 indicates that we are concerned with the second component of this triple. As such,

$$(jack, 65, jack@the\_hill.com).2 = 65$$

Furthermore, we also have the following.

$$(jack, 65, jack@the\_hill.com).1 = jack$$
$$(jack, 65, jack@the\_hill.com).3 = jack@the\_hill.com$$

□

Example 6.15 Consider the following set comprehension.

$$\{x : Name \times Age \times Email \bullet x.1\}$$

Here, the declaration part of the set comprehension generates all triples of the appropriate type, while the term part of the set comprehension extracts the first component of each triple. As such, this set is equivalent to $\{jack, jill\}$. □

Exercise 6.24 Calculate the following.

1. (*emily, goldfish*).1

2. (*emily, goldfish*).2

3. (*duncan,* (*gerbil, red*)).1

4. (*duncan,* (*gerbil, red*)).2

5. ((*emily, goldfish*), (*duncan, gerbil*)).1

6. ((*emily, goldfish*), (*duncan, gerbil*)).2

7. ((*emily, goldfish*), (*duncan, gerbil*)).1.2

□

Exercise 6.25 Assuming the sets *Girls* and *Boys*, such that

$$Girls = \{emily, rachel, anna\}$$
$$Boys = \{michael, john\}$$

list the elements of the following sets.

1. $\{x : Boys \times Girls \bullet x.1\}$

2. $\{x : Girls \times Boys \bullet x.1\}$

3. $\{x : Girls \times Boys \mid x.2 = michael\}$

4. $\{x : Girls \times Boys \mid x.2 = michael \bullet x.1\}$

5. $\{x : Girls \times Boys \mid x.2 = michael \bullet x.2\}$

□

6.7 Axiomatic definitions

We have, to this point, seen a number of ways of defining sets: sometimes it is sufficient to introduce them by name, as an abstract entity; at other times we have defined sets by extension, by listing all of their members; at other times still, set comprehension will suffice. The constructs that we shall meet in later chapters—relations, functions, and sequences—can also be defined via set comprehension, for, as we shall see, these structures will be based upon the notion of sets. Often, however, particularly when dealing with more complex structures—as one often must when providing a mathematical formulation of a complex system—the use of set comprehension is a particularly cumbersome means of introducing an entity into such a description. Furthermore, in some cases, such declarations are impossible using set comprehension.

To this end, we introduce the notion of an *axiomatic definition*. The form of an axiomatic definition is similar to that of a set comprehension in that it has both a declaration part and a predicate part (but no term part, however). Indeed, we may use a set comprehension in the predicate part of an axiomatic definition (but not vice versa).

The structure of an axiomatic definition is given below.

$$\begin{array}{|l}
declaration \\
\hline
predicate
\end{array}$$

The declaration provides the name and type of the entity that is being declared, while the predicate states the properties (if any) that it must satisfy.

Example 6.16 The following axiomatic definition introduces a set of natural numbers, X, such that X has at least three members.

$$\begin{array}{|l}
X : \mathbb{P}\,\mathbb{N} \\
\hline
\#\,X \geq 3
\end{array}$$

$\square$

The purpose of an axiomatic definition is to introduce an entity into a mathematical specification; this will prove to be important in Chapter 14, in which we describe and reason about some case studies.

Note how much more flexible an axiomatic definition can be than a set comprehension: the latter states exactly what it takes for an element to be a member of the set in question. Our axiomatic definition, on the other hand, simply states that X is a set of natural numbers and there are at least three elements in X—the actual values of these elements are left undefined.

Of course, axiomatic definitions do not have to be so loose: we can specify exactly what we want the make-up of X to be; the following, more strict, definition of X does exactly this.

$$\begin{array}{|l}
X : \mathbb{P}\,\mathbb{N} \\
\hline
X = \{17, 19, 23\}
\end{array}$$

Taking things to the other extreme, we can do away with the predicate part altogether, and simply introduce X, a set of natural numbers.

$$\begin{array}{|l}
X : \mathbb{P}\,\mathbb{N}
\end{array}$$

Here, as was the case with set comprehension, the default predicate is *true*.

Exercise 6.26 For each of the following axiomatic definitions, give the value of the entity being declared.

1.

$$\begin{array}{|l}
x : \mathbb{N} \\
\hline
x < 1
\end{array}$$

2.

$$\begin{array}{|l}
x : \mathbb{P}\,\mathbb{N} \\
\hline
\#\,x = 0
\end{array}$$

3.

$$
\begin{array}{|l}
x : \mathbb{N} \times \mathbb{N} \\
\hline
(x.1 = x.2) \wedge (x.1 = 7 \text{ div } 2)
\end{array}
$$

□

Exercise 6.27 Use axiomatic definitions to define the following.

1. x, of type $\mathbb{N}$, with the value 42.

2. X, of type $\mathbb{P}\,\mathbb{N}$, such that X contains at least one element and it does not contain 0.

3. X, of type $\mathbb{N} \times \mathbb{N}$, such that for every pair in X, the first component does not equal the second component.

□

6.8   Further exercises

Exercise 6.28 Assuming the type *Dog* and appropriate predicates, define the following using set comprehension.

1. The set of smelly dogs.

2. The set of well-trained dogs.

3. The set of small dogs.

4. The set of smelly, well-trained dogs.

5. The set of smelly, well-trained, small dogs.

□

Exercise 6.29 Assuming the types *Dog* and *Person*, and appropriate predicates and operations, define the following using set comprehension.

1. The set of ages of smelly, well-trained dogs

2. The set of owners of smelly, well-trained dogs.

3. The set of addresses of owners of smelly, well-trained dogs.

□

Exercise 6.30 Assuming the types *Dog*, *Person* and *Age*, and appropriate predicates and operations, define the following using set comprehension.

1. The set of all pairs of type *Dog* × *Person*.

2. The set of all pairs of type *Dog* × *Person*, such that a pair (d, p) appears in the set if, and only if, d is small and p owns d.

3. The set of all pairs of type $Dog \times Person$, such that a pair (d, p) appears in the set if, and only if, d is small, p is small, and p owns d.

4. The set of all pairs of type $Dog \times Age$, such that a pair (d, a) appears in the set if, and only if, d is smelly and a is the age of d's owner.

5. The set of all pairs of type $(Dog \times Age) \times (Person \times Age)$, such that a pair $((d, a_1), (p, a_2))$ appears in the set if, and only if, a_1 is d's age, a_2 is p's age, and p owns d.

□

Exercise 6.31 Assume the following set.

$$D = \{(fido, ken), (spot, barry), (barker, nigel), (lassie, david)\}$$

Given this set, define the following sets in terms of D using set comprehension and component selection.

1. $\{fido, spot, barker, lassie\}$

2. $\{ken, barry, nigel, david\}$

3. $\{((fido, ken), ken), ((spot, barry), barry),$
   $\quad\quad ((barker, nigel), nigel), ((lassie, david), david)\}$

4. $\{(ken, fido), (barry, spot), (nigel, barker), (david, lassie)\}$

□

Exercise 6.32 List the elements of the following sets.

1. $\{n : \mathbb{N} \mid n > 1 \wedge n < 4\}$

2. $\{n : \mathbb{N} \mid n > 1 \wedge n < 4 \bullet n^2\}$

3. $\{n : \mathbb{N} \mid n > 1 \wedge n < 4 \bullet (n, n)\}$

4. $\{m, n : \mathbb{N} \mid m + n = 4\}$

5. $\{m, n : \mathbb{N} \mid m + n = 4 \bullet n\}$

□

Exercise 6.33 Define the following using set comprehension.

1. The set of natural numbers between 0 and 100 (inclusive).

2. The set of even integers.

3. The set of countries that are members of the United Nations.

□

Exercise 6.34

1. Define, using set comprehension, the set $A = \{1, 2\}$.

2. Define, using set comprehension, the set $B = \{2, 3\}$.

3. Define, using set comprehension, the set $id = \{(1, 1), (2, 2), (3, 3), (4, 4)\}$.

4. Define, using set comprehension, the set $double = \{(1, 2), (2, 4), (3, 6), (4, 8)\}$.

□

Exercise 6.35 Is $\times$ associative? □

Exercise 6.36 Is $\times$ commutative? □

Exercise 6.37 Is $\times$ idempotent? □

Exercise 6.38 Prove that

$$A \times (B \cup C) = (A \times B) \cup (A \times C)$$

□

Exercise 6.39 Prove that

$$(A \cap B) \times C = (A \times C) \cap (B \times C)$$

□

Exercise 6.40 Assume the set C, such that

$$C = \{(jon, no, 14), (dave, yes, 64), (jim, yes, 29)\}$$

Calculate the following.

1. $\{c : C \mid c.2 = yes\}$

2. $\{c : C \mid c.2 = yes \land c.3 = 14\}$

3. $\{c : C \mid c.2 = no \bullet c.3\}$

□

Exercise 6.41 Which of the following sets are equivalent to the empty set?

1. $\{n : \mathbb{N} \mid n < 0\}$

2. $\{n : \mathbb{N} \mid n \neq n\}$

3. $\{n : \mathbb{N} \mid n = n + 1\}$

4. $\{n : \mathbb{N} \mid n \leq n - 1\}$

□

Exercise 6.42 Assuming the set *People* = {*becky, laura, jane*}, use axiomatic definitions to define the following subsets of *People*.

1. The set A, containing only *becky* and *laura*.

2. The set B, such that if B is not empty it must include *jane*.

3. The set C, such that if *jane* appears in the set, then so must *laura*.

4. The non-empty set of pairs, D.

5. The set of pairs E, which contains (*laura, jane*) as an element.

6. The set of pairs F, such that only *becky* may appear as a first component in any pair.

□

6.9 Solutions

Solution 6.1 1, 4, and 6 are legal applications of ∩. □

Solution 6.2 2 and 4 are legal applications of ∩. □

Solution 6.3 2, 4, 5, 7, and 9 are legal applications of ∈ or ⊆. □

Solution 6.4 2 and 3 are defined; 1 and 4 are undefined. □

Solution 6.5

1. $A = \{0, 1, 2, 3, 4, 5, 6\}$

2. $B = \{0, 3, 6, 9\}$

3. $C = \{2, 3\}$

4. $D = \emptyset$

□

Solution 6.6

1. $\{c : Car \mid c$ is red$\}$

2. $\{p : People \mid p$ is a vegetarian$\}$

3. $\{c : Car \mid c$ has a sun roof$\}$

4. $\{p : People \mid p$ is rich$\}$

□

Solution 6.7

1. $A = \{0, 1, 4, 9, 16, 25, 36\}$

2. $B = \{0, 3, 6, 9\}$

   3. $C = \{1\}$

   4. $D = \emptyset$

□

Solution 6.8

   1. $\{c : Car \mid c$ is red $\bullet$ $registration(c)\}$

   2. $\{p : People \mid p$ is a vegetarian $\bullet$ $age(p)\}$

   3. $\{c : Car \mid c$ has a sun roof $\bullet$ $owner(c)\}$

   4. $\{p : People \mid p$ is rich $\bullet$ $address(p)\}$

□

Solution 6.9

   1. $A = \{0, 1, 4, 9, 16, 25, 36, 49, 64, \ldots$

   2. $B = \{0, 1, 2, 3, 4, 5, 6, 7, 8, \ldots$

   3. $C = \{0, 0, 1, 1, 2, 2, 3, 3, 4, \ldots$

   4. $D = \{1, 2, 3, 4, 5, 6, 7, 8, 9, \ldots$

□

Solution 6.10

   1. $\{c : Car \bullet registration(c)\}$

   2. $\{p : People \bullet age(p)\}$

   3. $\{c : Car \bullet owner(c)\}$

   4. $\{p : People \bullet address(p)\}$

□

Solution 6.11

   1. n

   2. m

□

Solution 6.12

   1. $\{0, 1, 2, 3, 4, 5\}$

   2. $\{1\}$

   3. $\{4, 5\}$

4. $\{0, 1, 2, 3\}$

□

Solution 6.13

1. $\{7, 8, 9, 10\}$
2. $\{2, 4, 6, 8, 10\}$
3. $\{3, 6, 9\}$

□

Solution 6.14

1. $A \times B$
2. $A \times A$
3. $B \times B$
4. $B \times A$
5. $B \times \mathbb{P} \, B$
6. $B \times \mathbb{P} \, (A \times B)$

□

Solution 6.15

1. $\{(0, 0), (0, 1), (1, 0), (1, 1)\}$
2. $\{(0, yes), (0, no), (1, yes), (1, no)\}$
3. $\{(yes, 0), (yes, 1), (no, 0), (no, 1)\}$
4. $\{(yes, yes), (yes, no), (no, yes), (no, no)\}$
5. $\{(0, \emptyset), (0, \{yes\}), (0, \{no\}), (0, \{yes, no\}),$
   $(1, \emptyset), (1, \{yes\}), (1, \{no\}), (1, \{yes, no\})\}$
6. $\emptyset$

□

Solution 6.16

1. $(0, yes)$
2. $(0, \{yes\})$
3. $(0, \emptyset)$
4. $(\{0\}, yes)$
5. $(\{0\}, \{yes\})$

6. $(\{0\}, \emptyset)$

7. $\{(0, yes)\}$

8. $\{(0, \{yes\})\}$

9. $\{(0, \emptyset)\}$

□

Solution 6.17

1. $3 \times 2 = 6$

2. $2 \times 3 = 6$

3. $3 \times 0 = 0$

4. $3 \times 2^2 = 12$

5. $3 \times 2^0 = 3$

6. $2^6 = 64$

7. $2^0 = 1$

8. $2^{2^2 \times 2^3} = 2^{32}$

□

Solution 6.18

1. n

2. (m, n)

3. (m, n)

□

Solution 6.19

1. $\{(michael, emily), (michael, rachel), (michael, anna),$
   $(john, emily), (john, rachel), (john, anna)\}$

2. $\{(emily, michael), (emily, john), (rachel, michael),$
   $(rachel, john), (anna, michael), (anna, john)\}$

3. $\{(michael, michael), (michael, john), (john, michael), (john, john)\}$

4. $\{(michael, john), (john, michael)\}$

5. $\{(michael, michael), (john, john)\}$

6. $\{michael, john\}$

7. $\{(john, emily), (john, rachel), (john, anna)\}$

8. $\{emily, rachel, anna\}$

$\square$

Solution 6.20 3 and 5 are legal applications of $\backslash$, whereas the others are not. $\square$

Solution 6.21

1. $\{(1,2),(1,3),(2,2),(2,3)\}$

2. $\{(0,2),(0,3),(1,2),(1,3),(2,2),(2,3),(3,2),(3,3)\}$

3. $\{(0,2),(0,3)\}$

$\square$

Solution 6.22

1. $\{(jack, 60, jack@the\_hill.com), (jack, 60, jill@the\_hill.com),$
   $(jack, 65, jack@the\_hill.com), (jack, 65, jill@the\_hill.com),$
   $(jill, 60, jack@the\_hill.com), (jill, 60, jill@the\_hill.com),$
   $(jill, 65, jack@the\_hill.com), (jill, 65, jill@the\_hill.com)\}$

2. $\{(jack, (60, jack@the\_hill.com)), (jack, (60, jill@the\_hill.com)),$
   $(jack, (65, jack@the\_hill.com)), (jack, (65, jill@the\_hill.com)),$
   $(jill, (60, jack@the\_hill.com)), (jill, (60, jill@the\_hill.com)),$
   $(jill, (65, jack@the\_hill.com)), (jill, (65, jill@the\_hill.com))\}$

3. $\{((jack, 60), jack@the\_hill.com), ((jack, 60), jill@the\_hill.com),$
   $((jack, 65), jack@the\_hill.com), ((jack, 65), jill@the\_hill.com),$
   $((jill, 60), jack@the\_hill.com), ((jill, 60), jill@the\_hill.com),$
   $((jill, 65), jack@the\_hill.com), ((jill, 65), jill@the\_hill.com)\}$

$\square$

Solution 6.23

1. $\{(60, 60, 60), (60, 60, 65), (60, 65, 60), (60, 65, 65),$
   $(65, 60, 60), (65, 60, 65), (65, 65, 60), (65, 65, 65)\}$

2. $\{(jack, jack, jack), (jill, jill, jill)\}$

3. $\{(jack, (jack, jack)), (jack, (jack, jill))\}$

4. $\{((jack, jack), (jack, jack)), ((jack, jill), (jack, jill)),$
   $((jill, jack), (jill, jack)), ((jill, jill), (jill, jill))\}$

5. $\{(60, jack@the\_hill.com), (60, jill@the\_hill.com),$
   $\qquad (65, jack@the\_hill.com), (65, jill@the\_hill.com)\}$

6. $\{(jack, 65, jack@the\_hill.com), (jack, 65, jill@the\_hill.com)\}$

□

Solution 6.24

1. *emily*

2. *goldfish*

3. *duncan*

4. $(gerbil, red)$

5. $(emily, goldfish)$

6. $(duncan, gerbil)$

7. *goldfish*

□

Solution 6.25

1. $\{michael, john\}$

2. $\{emily, rachel, anna\}$

3. $\{(emily, michael), (rachel, michael), (anna, michael)\}$

4. $\{emily, rachel, anna\}$

5. $\{michael\}$

□

Solution 6.26

1. $x = 0$

2. $x = \emptyset$

3. $x = (3, 3)$

□

Solution 6.27

1.

$$\begin{array}{|l|} \hline x : \mathbb{N} \\ \hline x = 42 \\ \hline \end{array}$$

2.

$$
\begin{array}{|l}
X : \mathbb{P}\,\mathbb{N} \\
\hline
X \neq \emptyset \wedge 0 \notin X
\end{array}
$$

3.

$$
\begin{array}{|l}
X : \mathbb{N} \times \mathbb{N} \\
\hline
X \subseteq \{x, y : \mathbb{N} \mid x \neq y\}
\end{array}
$$

□

Solution 6.28

1. $\{d : Dog \mid d \text{ is smelly}\}$
2. $\{d : Dog \mid d \text{ is well-trained}\}$
3. $\{d : Dog \mid d \text{ is small}\}$
4. $\{d : Dog \mid d \text{ is smelly} \wedge d \text{ is well-trained}\}$
5. $\{d : Dog \mid d \text{ is smelly} \wedge d \text{ is well-trained} \wedge d \text{ is small}\}$

□

Solution 6.29

1. $\{d : Dog \mid d \text{ is smelly} \wedge d \text{ is well-trained} \bullet age(d)\}$
2. $\{d : Dog;\ p : Person \mid d \text{ is smelly} \wedge d \text{ is well-trained} \wedge p \text{ owns } d \bullet p\}$
3. $\{d : Dog;\ p : Person \mid d \text{ is smelly} \wedge d \text{ is well-trained} \wedge p \text{ owns } d \bullet address(p)\}$

□

Solution 6.30

1. $\{d : Dog;\ p : Person\}$
2. $\{d : Dog;\ p : Person \mid d \text{ is small} \wedge p \text{ owns } d\}$
3. $\{d : Dog;\ p : Person \mid d \text{ is small} \wedge p \text{ is small} \wedge p \text{ owns } d\}$
4. $\{d : Dog;\ p : Person \mid d \text{ is smelly} \wedge p \text{ owns } d \bullet (d, age(p))\}$
5. $\{d : Dog;\ p : Person \mid p \text{ owns } d \bullet ((d, age(d)), (p, age(p)))\}$

□

Solution 6.31

1. $\{d : D \bullet d.1\}$
2. $\{d : D \bullet d.2\}$

3. $\{d : D \bullet (d, d.2)\}$

4. $\{d : D \bullet (d.2, d.1)\}$

□

Solution 6.32

1. $\{2, 3\}$

2. $\{4, 9\}$

3. $\{(2, 2), (3, 3)\}$

4. $\{(0, 4), (1, 3), (2, 2), (3, 1), (4, 0)\}$

5. $\{0, 1, 2, 3, 4\}$

□

Solution 6.33

1. $\{n : \mathbb{N} \mid n \leq 100\}$

2. $\{n : \mathbb{Z} \mid n \bmod 2 = 0\}$

3. $\{c : Country \mid c \text{ is a member of the UN}\}$

□

Solution 6.34

1. $A = \{n : \mathbb{N} \mid n > 0 \wedge n < 3\}$

2. $B = \{n : \mathbb{N} \mid n > 1 \wedge n < 4\}$

3. $id = \{n : \mathbb{N} \mid n > 0 \wedge n < 5 \bullet (n, n)\}$

4. $double = \{n : \mathbb{N} \mid n > 0 \wedge n < 5 \bullet (n, n \times 2)\}$

□

Solution 6.35 No. Consider the sets $A = \{1, 2\}$, $B = \{a\}$, and $C = \{jon\}$. Here,

$$\begin{aligned} A \times (B \times C) &= \{a : A; \; d : B \times C \bullet (a, d)\} \\ &= \{a : A; \; b : B; \; c : C \bullet (a, (b, c))\} \\ &= \{(1, (a, jon)), (2, (a, jon))\} \end{aligned}$$

and

$$\begin{aligned} (A \times B) \times C &= \{d : A \times B; \; c : C \bullet (d, c)\} \\ &= \{a : A; \; b : B; \; c : C \bullet ((a, b), c)\} \\ &= \{((1, a), jon), ((2, a), jon)\} \end{aligned}$$

Here, the sets $A \times (B \times C)$ and $(A \times B) \times C$ are not equal. As such, $\times$ is not associative. □

Solution 6.36 No. Consider the sets $A = \{1, 2\}$ and $B = \{a\}$. Here,

$$A \times B = \{a : A; \ b : B \bullet (a, b)\}$$
$$= \{(1, a), (2, a)\}$$

and

$$B \times A = \{b : B; \ a : A \bullet (b, a)\}$$
$$= \{(a, 1), (a, 2)\}$$

Here, the sets $A \times B$ and $B \times A$ are not equal. As such, $\times$ is not commutative. $\square$

Solution 6.37 No. Consider the set $A = \{1, 2\}$. Here,

$$A \times A = \{a : A; \ b : A \bullet (a, b)\}$$
$$= \{(1, 1), (1, 2), (2, 1), (2, 2)\}$$

Here, the sets A and $A \times A$ are not equal. As such, $\times$ is not idempotent. $\square$

Solution 6.38 Assume some pair (x, y) of the appropriate type. Now,

$$(x, y) \in A \times (B \cup C)$$
$$\Leftrightarrow x \in A \wedge y \in B \cup C$$
$$\Leftrightarrow x \in A \wedge (y \in B \vee y \in C)$$
$$\Leftrightarrow (x \in A \wedge y \in B) \vee (x \in A \wedge y \in C)$$
$$\Leftrightarrow (x, y) \in A \times B \vee (x, y) \in A \times C$$
$$\Leftrightarrow (x, y) \in (A \times B) \cup (A \times C)$$

$\square$

Solution 6.39 Assume some pair (x, y) of the appropriate type. Now,

$$(x, y) \in (A \cap B) \times C$$
$$\Leftrightarrow (x \in A \cap B) \wedge y \in C$$
$$\Leftrightarrow x \in A \wedge x \in B \wedge y \in C$$
$$\Leftrightarrow x \in A \wedge x \in B \wedge y \in C \wedge y \in C$$
$$\Leftrightarrow x \in A \wedge y \in C \wedge x \in B \wedge y \in C$$
$$\Leftrightarrow (x, y) \in A \times C \wedge (x, y) \in B \times C$$
$$\Leftrightarrow (x, y) \in (A \times C) \cap (B \times C)$$

$\square$

Solution 6.40

1. $\{(dave, yes, 64), (jim, yes, 29)\}$

2. $\emptyset$

3. $\{14\}$

□

Solution 6.41 They all are. □

Solution 6.42

1.

$$
\begin{array}{|l}
\hline
A : \mathbb{P}\ People \\
\hline
A = \{becky, laura\}
\end{array}
$$

2.

$$
\begin{array}{|l}
\hline
B : \mathbb{P}\ People \\
\hline
B \neq \emptyset \Rightarrow jane \in B
\end{array}
$$

3.

$$
\begin{array}{|l}
\hline
C : \mathbb{P}\ People \\
\hline
jane \in C \Rightarrow laura \in C
\end{array}
$$

4.

$$
\begin{array}{|l}
\hline
D : People \times People \\
\hline
D \neq \emptyset
\end{array}
$$

5.

$$
\begin{array}{|l}
\hline
E : People \times People \\
\hline
(laura, jane) \in E
\end{array}
$$

6.

$$
\begin{array}{|l}
\hline
F : People \times People \\
\hline
F = \emptyset \\
\lor \\
F \subseteq \{p : People \times People \mid p.1 = becky\}
\end{array}
$$

□

Chapter 7

Predicate Logic

7.1 The need for quantification

Truth tables are a perfectly satisfactory means of determining the truth or falsity of propositions that consist of a relatively small number of atomic propositions. However, as we have seen, the process of constructing truth tables can become rather cumbersome when one wishes to determine the truth or falsity of propositions consisting of four or more atomic propositions. As an example, the truth table associated with the proposition $(p \lor \lnot q) \land (r \lor s)$ consists of 16 rows, while the truth table associated with $(p \lor \lnot q) \land (r \lor s) \land \lnot t$ consists of 32 rows. (Recall that for a proposition consisting of n atomic propositions, the associated truth table contains 2^n rows.)

Furthermore, assume that we wanted to formulate the statement "45 is a prime number, or 46 is a prime number, or 47 is a prime number, or 48 is a prime number, etc." using propositional logic.

Assuming the existence of some relation $prime$, such that $prime\,(x)$ is true if, and only if, $x \in \mathbb{N}$ is a prime number, we might express this statement via propositional logic by

$$prime\,(45) \lor prime\,(46) \lor prime\,(47) \lor prime\,(48) \lor \ldots$$

It would take either a very large truth table (infinitely large, in fact) to determine the truth or falsity of such a statement. Here, the drawback is not associated with truth tables but with propositional logic itself: it is impossible to write such a proposition formally in terms of propositional logic, let alone reason about it (even though we know that the statement is true). As such, we need a means of representing such statements more succinctly; *predicate logic* provides us with this means.

7.2 Universal quantification

As we saw in Chapter 4, $\bigcap$ is a generalised form of $\cap$. In a similar vein, our first predicate logic operator—*universal quantification*—which is written $\forall$, is a generalised form of conjunction ($\land$). Universal quantification allows us to capture statements of the form "for all" or "for every". For example, the statement "every natural number is greater than or equal to zero" can be written formally as follows.

$$\forall\, n : \mathbb{N} \bullet n \geq 0$$

A quantified statement of this form consists of three parts: the quantifier (in this case ∀); the quantification, which describes the variable(s) and the type(s) of variable(s) with which the statement is concerned; and the predicate, which is normally some statement about the quantified variable(s). Thus every predicate logic statement (for the moment, at least) can be considered to be of the following form.

> quantifier quantification • predicate

Of course, the above statement is logically equivalent to (and a lot more succinct than) the conjunction

$$(0 \geq 0) \wedge (1 \geq 0) \wedge (2 \geq 0) \wedge (3 \geq 0) \wedge \ldots$$

As was the case with our declarations in set theory, we are not restricted to declaring one variable. For example, the following predicate states that, given two natural numbers, m and n, it is always the case that $m < n$ or $m = n$ or $m > n$.

> $\forall m, n : \mathbb{N} \bullet m < n \vee m = n \vee m > n$

Note that our conventions for declarations are consistent with those of Chapter 6; the above could equally well have been written as follows.

> $\forall m : \mathbb{N}; \ n : \mathbb{N} \bullet m < n \vee m = n \vee m > n$

Furthermore, the following is also valid.

> $\forall m : \mathbb{N} \bullet \forall n : \mathbb{N} \bullet m < n \vee m = n \vee m > n$

The following is also valid.

> $\forall n : \mathbb{N} \bullet \forall m : \mathbb{N} \bullet m < n \vee m = n \vee m > n$

In addition, we are not restricted to quantifying over variables of one type. The following universally quantified statement expresses a possible relationship between small dogs and small people.

> $\forall p : People ; \ d : Dog \bullet p$ is small $\wedge \ p$ owns $d \Rightarrow d$ is small

Exercise 7.1 State, in natural language, what the following statements represent.

1. $\forall d : Dog \bullet d$ is smelly

2. $\forall d : Dog \bullet (d$ is well-trained $\vee \ d$ is small)

3. $(\forall d : Dog \bullet d$ is well-trained$) \vee (\forall d : Dog \bullet d$ is small$)$

4. $\forall d : Dog \bullet (d$ is small $\Rightarrow d$ is well-trained$)$

5. $(\forall d : Dog \bullet d$ is small$) \Rightarrow (\forall d : Dog \bullet d$ is well-trained$)$

□

Exercise 7.2 Consider statements 4 and 5 of Exercise 7.1. Are these statements logically equivalent? Justify your answer. □

Exercise 7.3 Assume the existence of the set *People*. Using this set, together with predicates "likes Jaffa cakes" and "is a vegetarian", formulate the following statements using predicate logic.

1. "Everybody likes Jaffa cakes."

2. "All vegetarians don't like Jaffa cakes."

3. "Everybody either likes Jaffa cakes or is a vegetarian."

4. "Either everybody likes Jaffa cakes or everybody is a vegetarian."

5. "Everybody likes Jaffa cakes and is a vegetarian."

6. "Everybody likes Jaffa cakes and everybody is a vegetarian."

□

Exercise 7.4 Consider statements 3 and 4 of Exercise 7.3. Are these statements logically equivalent? Justify your answer. □

Note that, in the above exercise, although statement 3 is not equivalent to statement 4, statement 4 does imply statement 3. This can be stated generally as follows.

Law 7.1

$$((\forall x : X \bullet p) \vee (\forall x : X \bullet q)) \Rightarrow (\forall x : X \bullet p \vee q)$$

□

Example 7.1 The following statement holds for a given set of people.

$$((\forall p : Person \bullet p \text{ is tall}) \vee (\forall p : Person \bullet p \text{ is strong})) \Rightarrow$$
$$(\forall p : Person \bullet p \text{ is tall} \vee p \text{ is strong})$$

□

Exercise 7.5 Consider statements 5 and 6 of Exercise 7.3. Are these statements logically equivalent? Justify your answer. □

The above equivalence can be stated generally as follows.

Law 7.2

$$(\forall x : X \bullet p \wedge q) \Leftrightarrow ((\forall x : X \bullet p) \wedge (\forall x : X \bullet q))$$

□

Example 7.2 The following statement holds for a given set of people.

$$((\forall p : Person \bullet p \text{ is tall}) \wedge (\forall p : Person \bullet p \text{ is strong})) \Leftrightarrow$$
$$(\forall p : Person \bullet p \text{ is tall} \wedge p \text{ is strong})$$

$\square$

This equivalence holds due to the fact that the universal quantifier behaves like a 'big conjunction'. Recall that when we combine propositions via conjunction, the order in which the propositions are conjoined is of no importance. As such, universal quantification distributes through conjunction: it doesn't matter in what order we apply our 'and's.

Exercise 7.6 Assume the set *People* and the set of pairs *likes* $\in$ *People* $\times$ *People*, which are defined as follows.

$$People = \{jim, jon, rick\}$$
$$likes = \{(jim, jon), (jon, jim), (rick, jon), (jon, rick)\}$$

Determine the truth or falsity of the following statements.

1. $\forall l : likes \bullet l.1 = rick$

2. $\forall l : likes \bullet \neg (l.1 = rick)$

3. $\forall p, q : People \bullet (p, q) \in likes \Rightarrow (q, p) \in likes$

4. $\forall l : likes \bullet l.1 = rick \Rightarrow l.2 = jon$

$\square$

7.3   Existential quantification

Our second quantifier is the *existential quantifier*, which is written $\exists$. Whereas universal quantification is used to assert that a certain property holds of *every* element of a set, existential quantification is used to assert that such a property holds of *some* (or *at least one*) elements of a set. Furthermore, just as $\forall$ can be viewed as a generalised form of $\wedge$, $\exists$ can be viewed as a generalised form of $\vee$.

Example 7.3 The statement "some natural numbers are divisible by 3" can be written as

$$\exists n : \mathbb{N} \bullet n \bmod 3 = 0$$

$\square$

Note that the form of existentially quantified statements is similar to that of universally quantified statements, i.e., it is

quantifier quantification $\bullet$ predicate

Exercise 7.7 State, in natural language, what the following statements represent.

1. $\exists\, d : Dog \bullet d$ is smelly

2. $\exists\, d : Dog \bullet (d$ is well-trained $\vee\ d$ is small$)$

3. $\exists\, d : Dog \bullet (d$ is small $\Rightarrow d$ is well-trained$)$

4. $(\exists\, d : Dog \bullet d$ is small$) \Rightarrow (\exists\, d : Dog \bullet d$ is well-trained$)$

□

Exercise 7.8 Assume the existence of the set *People*. Using this set, together with the predicates "likes Jaffa cakes" and "is a vegetarian", formulate the following statements using predicate logic.

1. "Some people like Jaffa cakes."

2. "Some vegetarians don't like Jaffa cakes."

3. "Some people either like Jaffa cakes or are vegetarian."

4. "Either some people like Jaffa cakes or some people are vegetarian."

5. "Some people like Jaffa cakes and are vegetarian."

6. "Some people like Jaffa cakes and some people are vegetarian."

□

Exercise 7.9 Consider statements 3 and 4 of Exercise 7.8. Are these statements logically equivalent? Justify your answer. □

The above equivalence can be state generally as follows.

Law 7.3

$$(\exists\, x : X \bullet p \vee q) \Leftrightarrow ((\exists\, x : X \bullet p) \vee (\exists\, x : X \bullet q))$$

□

Example 7.4 The following statement holds for a given set of cars.

$$(\exists\, c : Car \bullet fast\,(c) \vee small\,(c)) \Leftrightarrow ((\exists\, c : Car \bullet fast\,(c)) \vee (\exists\, c : Car \bullet small\,(c)))$$

□

The justification for this law is similar to that for Law 7.2. Recall that the existential quantifier behaves like a 'big disjunction'. Recall also that, just as is the case with conjunction, when we combine propositions via disjunction, the order in which the propositions are disjoined is of no importance. As such, existential quantification distributes through disjunction.

Exercise 7.10 Consider statements 5 and 6 of Exercise 7.8. Are these statements logically equivalent? Justify your answer. □

Note that statement 5 implies statement 6. This is stated generally as follows.

Law 7.4

$$(\exists\, x : X \bullet p \wedge q) \Rightarrow ((\exists\, x : X \bullet p) \wedge (\exists\, x : X \bullet q))$$

☐

Example 7.5 The following statement holds for a given set of cars.

$$(\exists\, c : Car \bullet fast\,(c) \wedge small\,(c)) \Rightarrow ((\exists\, c : Car \bullet fast\,(c)) \wedge (\exists\, c : Car \bullet small\,(c)))$$

☐

Exercise 7.11 Assume the set *People* and the set of pairs *likes* $\in$ *People* $\times$ *People*, which are defined as follows.

$$People = \{jim, jon, rick\}$$
$$likes = \{(jim, jon), (jon, jim), (rick, jon), (jon, rick)\}$$

Determine the truth or falsity of the following statements.

1. $\exists\, l : likes \bullet l.1 = rick$
2. $\exists\, l : likes \bullet \neg\,(l.1 = rick)$
3. $\exists\, p, q : People \bullet (p, q) \in likes \Rightarrow (q, p) \in likes$
4. $\exists\, l : likes \bullet l.1 = rick \Rightarrow l.2 = jon$

☐

Exercise 7.12 In general, is the proposition

$$(\forall\, x : X \bullet p) \Rightarrow (\exists\, x : X \bullet p)$$

true? ☐

Exercise 7.13 In general, is the proposition

$$(\exists\, x : X \bullet p) \Rightarrow (\forall\, x : X \bullet p)$$

true? ☐

Exercise 7.14

1. Exhibit a predicate p that makes the following statement true.

$$(\exists\, n : \mathbb{N} \bullet p) \Rightarrow (\forall\, n : \mathbb{N} \bullet p)$$

2. Exhibit a predicate p that makes the above statement false.

☐

7.4   Satisfaction and validity

The predicate $n > 3$ can be considered neither true nor false unless we know the value associated with n: if this value is less than 4, then the statement is equivalent to *false*; otherwise it is equivalent to *true*.

We say that a predicate p is *valid* if, and only if, it is true for all possible values of the appropriate type. That is, if a predicate p is associated with a variable x of type X, then p is valid if, and only if, $\forall\, x : X \bullet p$ is true.

Example 7.6 The predicate $n \geq 0$ is valid, as $\forall\, n : \mathbb{N} \bullet n \geq 0$ is equivalent to *true*. $\square$

We say that a predicate p is *satisfiable* if, and only if, it is true for some values of the appropriate type. That is, if a predicate p is associated with a variable x of type X, then p is satisfiable if, and only if, $\exists\, x : X \bullet p$ is true.

Example 7.7 The predicate $n > 0$ is satisfiable, as $\exists\, n : \mathbb{N} \bullet n > 0$ is equivalent to *true*. $\square$

Finally, we say that a predicate p is *unsatisfiable* if, and only if, it is false for all possible values of the appropriate type. That is, if a predicate p is associated with a variable x of type X, then p is unsatisfiable if, and only if, $\exists\, x : X \bullet p$ is false.

Example 7.8 The predicate $n \neq n$ is unsatisfiable, as $\exists\, n : \mathbb{N} \bullet n \neq n$ is equivalent to *false*. $\square$

Note the analogies between valid, satisfiable and unsatisfiable predicates, and tautologies, contingencies and contradictions: valid predicates and tautologies are always true; satisfiable predicates and contingencies are sometimes true and sometimes false; and unsatisfiable predicates and contradictions are never true.

Exercise 7.15 Assume that the set *People* is given by $\{alex, anne, henry\}$. For each of the following predicates, state whether they are valid, satisfiable, or unsatisfiable. You may assume that x and y are of type *People*.

1. $x = alex$

2. $x = alex \wedge x = anne$

3. $x = alex \wedge y = anne$

4. $x \in People$

5. $x \in \emptyset$

$\square$

7.5   The negation of quantifiers

As we have already seen, we are free to combine quantified statements via our propositional logic operators, $\wedge$, $\vee$, $\Rightarrow$, and $\Leftrightarrow$. We may also apply our negation operator, $\neg$, to quantified expressions; this is the subject of this section.

The statement "somebody likes Brian" may be expressed via predicate logic as

$\exists\, p : Person \bullet p$ likes Brian

If we wished to negate this expression, we would write

$\neg\ \exists\,p : Person \bullet p$ likes Brian

which, in natural language, may be expressed as "nobody likes Brian".

Logically, to say "nobody likes Brian" is of, course, equivalent to saying "everybody doesn't like Brian". The negation of quantifiers behaves exactly in this fashion: just as, in natural language, "nobody likes Brian" and "everybody doesn't like Brian" are equivalent, so, in predicate logic,

$\neg\ (\exists\,p : Person \bullet p$ likes Brian$)$

and

$\forall\,p : Person \bullet \neg\ (p$ likes Brian$)$

are equivalent.

As a further example, the statement "everybody likes Brian" may be expressed via predicate logic as

$\forall\,p : Person \bullet p$ likes Brian

If we were to negate this expression, we would write

$\neg\ \forall\,p : Person \bullet p$ likes Brian

which, in natural language, may be expressed as "not everybody likes Brian".

Logically, to say "not everybody likes Brian" is of, course, equivalent to saying "there is somebody who doesn't like Brian". Just as, in natural language, "not everybody likes Brian" and "there is somebody who doesn't like Brian" are equivalent, so, in predicate logic,

$\neg\ (\forall\,p : Person \bullet p$ likes Brian$)$

and

$\exists\,p : Person \bullet \neg\ (p$ likes Brian$)$

are equivalent. These equivalences can be stated generally as follows.

Law 7.5

$\neg\ (\exists\,x : X \bullet p) \Leftrightarrow \forall\,x : X \bullet \neg\ p$

$\neg\ (\forall\,x : X \bullet p) \Leftrightarrow \exists\,x : X \bullet \neg\ p$

$\square$

Effectively, when negation is applied to a quantified expression, it 'flips' quantifiers as it moves inwards (i.e., negation turns all universal quantifiers to existential quantifiers and vice versa, and negates all predicates). This behaviour becomes more apparent with expressions which feature more than one quantifier. For example, consider the statement

$\forall\,x : X \bullet \exists\,y : Y \bullet \forall\,z : Z \bullet p\,(x, y, z)$

Here, the effect of the negation of this expression can be determined as follows.

$$\neg \, \forall \, x : X \bullet \exists \, y : Y \bullet \forall \, z : Z \bullet p\,(x, y, z)$$
$$\Leftrightarrow \exists \, x : X \bullet \neg \, \exists \, y : Y \bullet \forall \, z : Z \bullet p\,(x, y, z)$$
$$\Leftrightarrow \exists \, x : X \bullet \forall \, y : Y \bullet \neg \, \forall \, z : Z \bullet p\,(x, y, z)$$
$$\Leftrightarrow \exists \, x : X \bullet \forall \, y : Y \bullet \exists \, z : Z \bullet \neg \, p\,(x, y, z)$$

As the negation is 'pulled' through the expression, it converts every quantifier that appears in its path.

Exercise 7.16   Provide the negation of the following statements.

1. "Everyone likes everyone."

2. "Everyone likes someone."

3. "Someone likes everyone."

4. "Someone likes someone."

☐

Exercise 7.17 Assuming the type *Person*, and the predicate "likes", such that x likes y if, and only if, person x likes person y, write the statements of Solution 7.16 in terms of predicate logic.
☐

Exercise 7.18 Give the negation of each of the following predicates without using $\neg$.

1. $\forall \, x : \mathbb{N} \bullet x = x$

2. $\forall \, x : \mathbb{N} \bullet \exists \, y : \mathbb{N} \bullet x \geq y$

3. $\forall \, x : \mathbb{N} \bullet \forall \, y : \mathbb{N} \bullet x > y \vee x = y \vee x < y$

☐

7.6   Free and bound variables

The predicate $n > 5$ states something about a natural number n: that it is greater than 5. Unfortunately, we do not know the value of n: n is simply a placeholder, or *variable*, the value of which cannot be determined by this predicate alone. As such—as we have already seen—we cannot determine the truth of falsity of the predicate $n > 5$.

Consider, on the other hand, the statement $\exists \, n : \mathbb{N} \bullet n > 5$, which—as we have seen previously—is equivalent to

$$(0 > 5) \vee (1 > 5) \vee (2 > 5) \vee (3 > 5) \vee (4 > 5) \vee (5 > 5) \vee (6 > 5) \ldots$$

This predicate is, of course, true: provided that there is at least one natural number that is greater than 5, the statement is satisfied. Thus, this quantified statement has a very different meaning to the unquantified predicated $n > 5$.

In the statement $\exists\, n : \mathbb{N} \bullet n > 5$ we say that the variable n is *bound*. This is because we know exactly what role it is playing: it is an existentially quantified variable. In the statement $n > 5$, on the other hand, as we can say nothing more about n than the fact that it is an integer. In such circumstances we say that n is *free*: we have in no sense determined—or even restricted—its value.

Example 7.9 Assume the set *People*, given by

$$People = \{robin, richard, robert\}$$

together with the following expression.

$$(\forall\, y : People \bullet y \; likes \; robin) \lor x \; likes \; richard$$

In this statement, we have two variables: y and x. The former is universally quantified and, as such, is bound, whereas the latter is free. □

In general, given a statement $\forall\, x : X \bullet p$, we say the occurrence of x in the declaration is *binding*, any occurrences of x in p are *bound*, and any occurrences of any variables other than x (and which are not bound by another quantifier) are *free*. This is illustrated below.

$$(\forall\ \overset{binding}{\overbrace{x}}\ : \mathbb{N} \bullet\ \overset{bound}{\overbrace{x}} \geq 0 \land \overset{free}{\overbrace{y}} \geq \overset{bound}{\overbrace{x}}\,) \land \overset{free}{\overbrace{x}} = 3$$

Note that in this expression the *scope* of the universal quantifier reaches only to the right parenthesis: the rightmost occurrence of x is outside the scope of this quantifier, and, as such, is free.

$$(\forall\, x : \mathbb{N} \bullet \underbrace{x \geq 0 \land y \geq x}_{scope}) \land x = 3$$

Exercise 7.19 Determine the scope of the universal and existential quantifiers of the following predicate.

$$(\forall\, x : X \bullet p \land (\exists\, y : Y \bullet q) \land r)$$

□

Exercise 7.20 Which variables are free and which are bound in the following expressions?

1. $x \geq 0$
2. $\forall\, x : \mathbb{N} \bullet x \geq 0$
3. $\forall\, x : \mathbb{N} \bullet x \geq x$
4. $\forall\, x : \mathbb{N} \bullet x \geq y$

5. $y \geq 0 \wedge \forall x : \mathbb{N} \bullet x \geq y$

6. $x \geq 0 \wedge \forall x : \mathbb{N} \bullet x \geq y$

7. $\forall x : \mathbb{N} \bullet \exists y : \mathbb{N} \bullet x \geq y$

8. $\forall x : \mathbb{N} \bullet (y = 0 \wedge \exists y : \mathbb{N} \bullet x \geq y)$

$\square$

Exercise 7.21 To which of the predicate logic expressions of Exercise 7.20 may we attribute truth values? $\square$

Having discussed the notion of free and bound variables, we are now in a position to formalise the rather subtle distinction that exists between predicates and propositions. At its simplest this difference is captured by predicates being logical statements which contain free variables (or 'logical holes'), while propositions are logical statements which contain no such holes.

Example 7.10 The expression $n > 7$ is a predicate. On the other hand, $\exists n : \mathbb{N} \bullet n > 7$ and $8 > 7$ are both propositions. $\square$

Exercise 7.22 Which of the following are predicates and which are propositions, assuming that *richard* and *duncan* are both elements of *Person*?

1. *richard* is taller than *duncan*

2. *richard* is taller than x

3. $\exists x : X \bullet$ *richard* is taller than x

$\square$

7.7 Substitution

Consider again the statement $\exists n : \mathbb{N} \bullet n > 5$. In the previous section, we stated that, in this expression, the variable n was *bound* by the existential quantifier: we could determine the role it plays in the overall statement.

In the unquantified predicate $n > 5$, on the other hand, we stated that the variable n was *free*: it may assume any integer value. If that value was to be less than 6, then the truth value of the statement would be *false*; otherwise it would be *true*. For example, if n were to take the value 3, then the statement would be *false*; on the other hand, if it were to be replaced by the term $3 + 4$, then it would be equivalent to *true*.

The fact that the variable n is free in this sense allows us to conclude that we may replace the *name* n with any *term* t, provided that t is of the same type as n. This process is called *substitution*: the expression $p\,[t/n]$ denotes the fact that the term t is being substituted for the variable name n in the predicate p. For example, we may denote the substitution of 3 for n in the predicate $n > 5$ by

$n > 5\,[3/n]$

This gives the result

$$3 > 5$$

Furthermore, we may denote the substitution of $3 + 4$ for n in $n > 5$ by

$$n > 5 \, [3 + 4/n]$$

This gives the result

$$3 + 4 > 5$$

Note that in both cases, the resulting expressions are no longer *predicates*: they are *propositions*. (Recall that a predicate is a logical expression which features at least one free variable, whereas a proposition is a logical expression which has no free variables.)

Note that it is only *free* variables that we may substitute for; substituting for bound variables has no effect. As an example,

$$((\exists\, n : \mathbb{N} \bullet n > 5) \, [3/n]) \Leftrightarrow \exists\, n : \mathbb{N} \bullet n > 5$$

This is exactly as we would desire: if this substitution were to be valid, the result would be $\exists\, n : \mathbb{N} \bullet 3 > 5$, which, logically, has a very different meaning to $\exists\, n : \mathbb{N} \bullet n > 5$.

Exercise 7.23 Which variables may be substituted for in the following expressions?

1. $4 + 3 = 7 - x$
2. $\exists\, x : \mathbb{N} \bullet 4 + 3 = 7 - x$
3. $\exists\, x : \mathbb{N} \bullet 4 + 3 = 7 - y$
4. $(\exists\, x : \mathbb{N} \bullet 4 + 3 = 7 - x \wedge x = y)$
5. $(\exists\, x : \mathbb{N} \bullet 4 + 3 = 7 - x) \wedge x = y$

□

Exercise 7.24 Determine the effects of the following substitutions.

1. x is happy $[nigel/x]$
2. $(x$ is happy $\wedge y$ is sad$)\,[nigel/x]$
3. $(x$ is happy $\vee x$ is sad$)\,[nigel/x]$
4. x is happy $\vee (x$ is sad $[nigel/x])$
5. $(\forall\, x : People \bullet x$ is happy$)\,[nigel/x]$
6. $(\forall\, x : People \bullet x$ is happy $\vee x$ is sad$)\,[nigel/x]$
7. $(\forall\, x : People \bullet x$ is happy$) \vee x$ is sad $[nigel/x]$

□

Exercise 7.25 Determine the effects of the following substitutions.

1. $(\forall\, x : \mathbb{N} \bullet x \geq 0)\,[3/x]$

2. $(\forall\, x : \mathbb{N} \bullet x \geq y)\,[3/y]$

3. $(y > 0 \land \forall\, x : \mathbb{N} \bullet x \geq y)\,[3/y]$

4. $(x > 0 \land \forall\, x : \mathbb{N} \bullet x \geq y)\,[3/x]$

□

Thus far we have only considered the substitution of one variable in an expression, yet there may of course be more than one free variable in an expression; under such circumstances, we may wish to perform more than one substitution.

Consider the predicate "x is happy $\land$ y is sad". Here, x and y are both free variables. We may substitute *nigel* for x as follows.

$(x$ is happy $\land$ y is sad$)\,[nigel/x] \Leftrightarrow$
    $nigel$ is happy $\land$ y is sad

We are now in a position to substitute *ken* for y:

$(nigel$ is happy $\land$ y is sad$)\,[ken/y]$
    $nigel$ is happy $\land$ ken is sad

Rather than perform these substitutions in two discrete stages, we may denote this overall substitution as

$(x$ is happy $\land$ y is sad$)\,[nigel/x]\,[ken/y]$

This indicates that the two substitutions—*nigel* for x and *ken* for y—are to take place *sequentially*. First, *nigel* is substituted for x, then *ken* is substituted for y. This has exactly the same effect as the substitution described above.

If on, the other hand, we wish the two substitutions to take place *simultaneously*, we would write

$(x$ is happy $\land$ y is sad$)\,[nigel/x, ken/y]$

Here, the two substitutions take place *at the same time*, and, again, giving the same result.

Example 7.11 To illustrate the difference between sequential and simultaneous substitution consider the effects of the following substitutions:

$(x > 3 \land y > 7)\,[y/x, 8/y]$
$\Leftrightarrow y > 3 \land 8 > 7$

$(x > 3 \land y > 7)\,[y/x]\,[8/y]$
$\Leftrightarrow y > 3 \land y > 7\,[8/y]$
$\Leftrightarrow 8 > 3 \land 8 > 7$
$\Leftrightarrow true$

In the latter, the sequential nature of the substitutions means that there are two occurrences of y to substitute 8 for, whereas in the former, the substitution of 8 for y happens when there is only one occurrence of y to substitute for. □

Exercise 7.26 Determine the effects of the following substitutions.

1. $x^2 = 49\,[y + 3/x]\,[4/y]$
2. $x^2 = 49\,[y + 3/x, 4/y]$
3. $x^2 > y\,[y + 3/x]\,[4/y]$
4. $x^2 > y\,[y + 3/x, 4/y]$

□

Substitution must be carried out with care. In particular, we must take care when substituting variables, or terms containing variables, for other variables. Take, as an example, the following statement.

$$\exists\,n : \mathbb{N} \bullet (n > 5 \wedge m < 5)$$

Here, the truth of this statement is dependent upon the value of m: if m is less than 5, then the statement is true, otherwise it is false.

If we were to substitute n for m in this statement, the result would be the following.

$$\exists\,n : \mathbb{N} \bullet (n > 5 \wedge n < 5)$$

Here, the truth value of the statement is no longer dependent upon the value of m; following this substitution, the value of the statement is equivalent to *false*.

What has occurred is the phenomenon of *variable capture*: we have substituted a term containing a variable name into an environment in which that variable name is already bound. Previously, m was a free variable in the statement: substituting n for m meant that this free variable was replaced by a bound one. This is obviously not desirable.

As such, whenever a term containing a variable name or whenever a term which is a variable name is being substituted into a quantified expression, if that variable name is the same as that of a bound variable in the quantified expression, then we must *rename the bound variable*. This change will have no effect on the truth or falsity of the expression. To convince oneself of this, one needs only to consider the expressions $\forall\,x : \mathbb{N} \bullet x = x$ and $\forall\,y : \mathbb{N} \bullet y = y$. Here, different variable names are used to express the predicate "all natural numbers are equal to themselves", yet, logically, the two expressions are equivalent.

Example 7.12 The predicate $\exists\,n : \mathbb{N} \bullet (n > 5 \wedge m < 5)$ is equivalent to

$$\exists\,l : \mathbb{N} \bullet (l > 5 \wedge m < 5)$$

As such, the effect of substituting n for m in our original expression can be achieved as follows.

$$\exists\,n : \mathbb{N} \bullet (n > 5 \wedge m < 5)\,[n/m]$$
$$\Leftrightarrow \exists\,l : \mathbb{N} \bullet (l > 5 \wedge n < 5)$$

This is, of course, the result that we desire. □

Exercise 7.27 Which of the following substitutions may result in variable capture?

1. $\forall\, x : \mathbb{N} \bullet x \geq y\,[z/y]$

2. $\forall\, x : \mathbb{N} \bullet x \geq y\,[x/y]$

3. $\forall\, x : \mathbb{N} \bullet x \geq y\,[x+y/y]$

4. $\forall\, x, y : \mathbb{N} \bullet x \geq y\,[x/y, y/x]$

5. $\forall\, x : \mathbb{N} \bullet x \geq y\,[z/y]\,[x/z]$

6. $\forall\, x : \mathbb{N} \bullet x \geq y\,[z/x]\,[x/z]$

7. $(y > 0 \wedge \forall\, x : \mathbb{N} \bullet x \geq y)\,[x+3/y]$

□

Exercise 7.28 Determine the results of the substitutions of the previous exercise. □

7.8   Restriction

In the previous chapter, we saw how the set of squares of prime numbers might be defined via set comprehension. This could be accomplished in the following way.

$$square\_primes = \{n : \mathbb{N} \mid prime\,(n) \bullet n^2\}$$

Here, the predicate part of this set comprehension behaves like a 'filter'—checking through the natural numbers, keeping those which satisfy the predicate, and throwing away the rest.

Filters can also play a role in predicate logic. As an example, the statement "all natural numbers which are prime numbers and are greater than 2 are odd" may be written as

$$\forall\, n : \mathbb{N} \bullet (n \text{ is prime} \wedge n > 2) \Rightarrow n \text{ is odd}$$

Using a filter, we may write the statement as

$$\forall\, n : \mathbb{N} \mid n \text{ is prime} \wedge n > 2 \bullet n \text{ is odd}$$

Here, the set over which the predicate ranges is *restricted* (or filtered) by the condition

$$n \text{ is prime} \wedge n > 2$$

The form of such restricted predicates is similar to that of set comprehensions, i.e.,

quantifier quantification | restriction • predicate

Again, as was the case with set comprehension, the lack of a restriction is entirely equivalent to having the restriction *true*. So, for example, the following statements are equivalent.

$$\forall\, p : People \mid true \bullet p \text{ likes spam} \Leftrightarrow$$
$$\forall\, p : People \bullet p \text{ likes spam}$$

Of course, such a restriction can also be used in existential quantifications. As an example, the statement "there has been a female British Prime Minister" may be written as

$$\exists\, p : People \mid p \text{ is female} \bullet p \text{ was a British Prime Minister}$$

Any statement written in this form may be rewritten without the filter. When doing this, it is important to remember the differences between the quantifiers.

Recall our expression

$$\forall\, n : \mathbb{N} \mid n \text{ is prime} \wedge n > 2 \bullet n \text{ is odd}$$

We argued that this was equivalent to

$$\forall\, n : \mathbb{N} \bullet (n \text{ is prime} \wedge n > 2) \Rightarrow n \text{ is odd}$$

That is, it represents the predicate "given any natural number, n, if n is prime and is greater than 2, then it is an odd number". Here, an expression of the form

universal quantification | restriction • predicate

has been transformed into one of the form

universal quantification • restriction ⇒ predicate

Now, if we were to apply such a transformation to

$$\exists\, p : People \mid p \text{ is female} \bullet p \text{ was a British Prime Minister}$$

we would obtain

$$\exists\, p : People \bullet p \text{ is female} \Rightarrow p \text{ was a British Prime Minister}$$

The first statement was true for exactly one person. This second, transformed statement, however, is true of *every male* (recall that *false* ⇒ *true* is logically equivalent to *true*), as well as one female. Thus, the two statements are not equivalent. The problem lies with the fact that restriction behaves differently in existentially quantified statements than it does in universally quantified ones: in the latter it is effectively behaving as an implication; in the former it is effectively behaving as a conjunction. Thus, the correct elimination of the restriction gives the result

$$\exists\, p : People \bullet p \text{ is female} \wedge p \text{ was a British Prime Minister}$$

This is clearly logically equivalent to the original statement.

Example 7.13 The statement

$$\exists\, n : \mathbb{N} \mid odd\,(n) \bullet n = 5$$

is equivalent to

$$\exists\, n : \mathbb{N} \bullet odd\,(n) \wedge n = 5$$

□

Example 7.14 The statement

$$\forall\, n : \mathbb{N} \mid odd\,(n) \bullet n = 5$$

is equivalent to

$$\forall\, n : \mathbb{N} \bullet odd\,(n) \Rightarrow n = 5$$

□

These equivalences are stated generally as follows.

Law 7.6

$$(\forall\, x : X \mid p \bullet t) \Leftrightarrow (\forall\, x : X \bullet p \Rightarrow t)$$

□

Law 7.7

$$(\exists\, x : X \mid p \bullet t) \Leftrightarrow (\exists\, x : X \bullet p \wedge t)$$

□

Exercise 7.29 Represent the following statements in the form

quantifier quantification | restriction • predicate

You may assume the type *People*, as well as the predicates "likes spam", "likes cheese" and "likes bananas".

1. "Everyone who likes cheese likes spam."
2. "There is a person who likes cheese that likes spam."
3. "Everyone who likes cheese also likes spam and likes bananas."
4. "Everyone who likes cheese and spam also likes bananas."
5. "There is a person who likes cheese and spam that likes bananas."
6. "There is a person who likes cheese that likes spam and bananas."

□

Exercise 7.30 Rewrite the predicates of Solution 7.29 in the form

quantifier quantification • predicate

□

Exercise 7.31 Determine the truth values of the following.

1. $\exists\, n : \mathbb{N} \mid p(n) \bullet true \Rightarrow \neg\, p(n)$

2. $\forall\, n : \mathbb{N} \mid p(n) \bullet false \Rightarrow \neg\, p(n)$

3. $\exists\, n : \mathbb{N} \mid n \in \emptyset \bullet n$ is prime

4. $\forall\, n : \mathbb{N} \mid n \in \emptyset \bullet n$ is prime

□

7.9   Uniqueness

Our existential quantifier, $\exists$, allows us to represent statements such as "there is at least one x, such that ..." Sometimes we may wish to be more specific and consider statements of the form "there is exactly one x, such that ..."

As an example, the statement $\exists\, x : \mathbb{N} \bullet x + 1 = 1$ is equivalent to *true*. Furthermore, it is the case that the predicate $x + 1 = 1$ is true for exactly one natural number: 0. The existence of an operator which might allow us to state that "there is exactly one natural number, x, such that $x + 1 = 1$" or "there is exactly one superpower" is desirable. Fortunately, we may define such an operator: $\exists_1$.

The structure of expressions which are quantified using this new operator are exactly the same as for our other quantifiers. As an example, the statement "there is exactly one natural number, x, such that $x + 1 = 1$", can be written via our $\exists_1$ operator as

$$\exists_1\, x : \mathbb{N} \bullet x + 1 = 1$$

Without this operator, we could only represent this statement via predicate logic as follows.

$$\exists\, x : \mathbb{N} \bullet (x + 1 = 1 \land$$
$$\forall\, y : \mathbb{N} \mid x \neq y \bullet y + 1 \neq 1)$$

This states that there is a natural number, x, for which this property holds, and, for all other natural numbers, y, such that $y \neq x$, this property does not hold. Of course, this is exactly what was captured in our original expression, albeit more succinctly.

Example 7.15 The predicate "there is exactly one superpower" may be written as

$$\exists_1\, c : Country \bullet c \text{ is a superpower}$$

□

Again, if we did not have our $\exists_1$ operator, we would be forced to represent the above predicate in a far less succinct manner, i.e.,

$$\exists\, c : Country \bullet (c \text{ is a superpower} \land$$
$$\forall\, d : Country \mid c \neq d \bullet \neg\, d \text{ is a superpower})$$

Exercise 7.32 Which of the following statements are true, and which are false?

1. $\exists_1 \, n : \mathbb{N} \bullet n$ is a divisor of $7 \wedge n \neq 1$

2. $\exists_1 \, n : \mathbb{N} \bullet \forall m : \mathbb{N} \bullet n \leq m$

3. $\exists_1 \, n : \mathbb{N} \bullet \forall m : \mathbb{N} \bullet n \geq m$

4. $\forall n : \mathbb{N} \bullet \exists_1 \, m : \mathbb{N} \bullet n \geq m$

□

Exercise 7.33 Rewrite the expressions of Exercise 7.32 in terms of $\exists$ and $\forall$. □

The statement $\exists_1 \, x : \mathbb{N} \bullet x + 1 = 1$ asserts that there is exactly one natural number that is exactly one less than 1; of course, we know that this value is 0. In addition, the statement $\exists_1 \, c : Country \bullet c$ is a superpower asserts that there is exactly one superpower; again it is possible to determine exactly which element of the underlying type this is.

Assuming that the statement $\exists_1 \, x : X \bullet p$ is equivalent to *true*, it is sometimes useful to pinpoint exactly which element of X it is that the predicate p holds for. Furthermore, it might be convenient for us to perform some operation on this value. The μ (pronounced "mu") operator allows us to do exactly this: the statement $(\mu \, x : X \mid p)$ is read as "the unique x from the set X such that p holds of x". For example, the μ expression

$$(\mu \, x : \mathbb{N} \mid x + 1 = 1)$$

is associated with the value 0, and

$$(\mu \, c : Country \mid c \text{ is a superpower})$$

gives the value *USA*.

When a unique element of the relevant type does possess the relevant property, then we are free to construct statements such as that given above. On the other hand, μ expressions that are not associated with a unique element generate an undefined value. As an example, the μ expressions

$$(\mu \, n : \mathbb{N} \mid n \text{ is even})$$

and

$$(\mu \, c : Country \mid population \, (c) > 10,000,000)$$

are both undefined.

Just as we could incorporate terms into set comprehensions, so we can apply terms in μ expressions. Expressions of the form

$$(\mu \, x : X \mid p \bullet t)$$

return the result of applying the term t to the unique element of the set X which satisfies the predicate p. For example,

$$(\mu \, n : \mathbb{N} \mid x + 1 = 1 \bullet x + 1)$$

gives the value 1, while

$$(\mu\, c : Country \mid c \text{ is a superpower} \bullet capital\,(c))$$

gives the value *washington*.

Again, if there is no unique element which possesses the relevant property, then the result is undefined. As such,

$$(\mu\, n : \mathbb{N} \mid n \text{ is even} \bullet n^2)$$

and

$$(\mu\, c : Country \mid population\,(c) > 10,000,000 \bullet capital\,(c))$$

are both undefined.

Furthermore, it is important to realise that the resulting term also has an associated type. As such, while the proposition

$$(\mu\, c : Country \mid c \text{ is a superpower} \bullet capital\,(c)) = paris$$

is equivalent to *false*, the proposition

$$(\mu\, c : Country \mid c \text{ is a superpower} \bullet capital\,(c)) = 1$$

is undefined: the type of the left-hand side of the equation is *City*, while the type of the right-hand side of the equation is $\mathbb{N}$.

Exercise 7.34 Assuming the existence of appropriate types, predicates and functions, write the following in terms of μ expressions.

1. The tallest mountain in the world.
2. The height of the tallest mountain in the world.
3. The oldest person in the world.
4. The nationality of the oldest person in the world.
5. The smallest natural number.

□

Exercise 7.35 For each of the statements below, state whether they are true, false, or undefined.

1. $(\mu\, n : \mathbb{N} \mid \forall m : \mathbb{N} \bullet n \leq m) = 0$
2. $(\mu\, n : \mathbb{N} \mid \forall m : \mathbb{N} \bullet n \geq m) \neq 0$
3. $(\mu\, n : \mathbb{N} \mid n^2 = 1) \neq 0$
4. $(\mu\, n : \mathbb{N} \mid n^2 = n) = 0$
5. $(\mu\, n : \mathbb{N} \mid n^2 = n \land n \neq 1) = 0$

□

7.10   Equational reasoning

In Chapter 3, we saw a number of ways of reasoning about the validity, or otherwise, of propositions. We shall reprise two of the methods in this chapter, namely equational reasoning and natural deduction.

We have already met a number of laws of predicate logic; for example, Laws 7.2 and 7.3 stated that the universal quantifier distributes over conjunction and that the existential quantifier distributes over disjunction. Furthermore, Law 7.5 described the effect of negation on quantified expressions. In addition, we have seen how quantified expressions of the form $\forall x : X \mid p \bullet t$ and $\exists x : X \mid p \bullet t$ are equivalent to expressions of the form $\forall x : X \bullet p \Rightarrow t$ and $\exists x : X \bullet p \wedge t$ respectively. Other, relatively straightforward, laws are given below.

First, if a predicate holds for all elements of a set, then it holds for some of them.

Law 7.8

$$\forall x : X \bullet p \Rightarrow \exists x : X \bullet p$$

□

Next, if a predicate holds for exactly one element of a set, then it holds for *at least* one of them.

Law 7.9

$$\exists_1 x : X \bullet p \Rightarrow \exists x : X \bullet p$$

□

If p holds for all elements of X and t is a term of type X, then p holds for t.

Law 7.10

$$(\forall x : X \bullet p(x) \wedge t \in X) \Rightarrow p(t)$$

□

Example 7.16 The predicate

$$prime(x) \wedge x > 2 \Rightarrow x \text{ is odd}$$

holds for all natural numbers. The term $3 + 4$ is of type $\mathbb{N}$ and, as such,

$$prime(3 + 4) \wedge 3 + 4 > 2 \Rightarrow 3 + 4 \text{ is odd}$$

holds. □

Our next law states that if t is a term of type X and p holds of t, then we can conclude that p holds for some elements of X.

Law 7.11

$$(t \in X \wedge p(t)) \Rightarrow (\exists x : X \bullet p(x))$$

□

Example 7.17 The proposition $7 \in \mathbb{N}$ is true, and, furthermore, 7 is prime. As such,

$$\exists\, n : \mathbb{N} \bullet n \text{ is prime}$$

holds. □

The quantification of any variables with regards to propositions which are always true or always false has absolutely no effect. This notion is captured by the following laws.

Law 7.12

$$(\forall\, x : X \bullet true) \Leftrightarrow true$$

□

Law 7.13

$$(\forall\, x : X \bullet false) \Leftrightarrow false$$

□

Law 7.14

$$(\exists\, x : X \bullet true) \Leftrightarrow true$$

□

Law 7.15

$$(\exists\, x : X \bullet false) \Leftrightarrow false$$

□

Such laws, together with those of Chapter 3, allow us to apply equational reasoning to predicate logic statements. Just as was the case in Chapter 3, equivalences such as those described above can allow us to determine the truth or falsity of certain expressions, or test whether a given logical statement is a theorem.

Example 7.18 We may establish that the following is a theorem of predicate logic.

$$(\neg\, \forall\, x : X \bullet p \Rightarrow q) \Leftrightarrow (\exists\, x : X \bullet p \wedge \neg\, q)$$

A proof of this is given below.

$$
\begin{aligned}
&\quad \neg\, \forall\, x : X \bullet p \Rightarrow q \\
&\Leftrightarrow \exists\, x : X \bullet \neg\, (p \Rightarrow q) \quad \text{[Law 7.5]} \\
&\Leftrightarrow \exists\, x : X \bullet \neg\, (\neg\, p \vee q) \quad \text{[Law 3.18]} \\
&\Leftrightarrow \exists\, x : X \bullet p \wedge \neg\, q \qquad \text{[Law 3.9]}
\end{aligned}
$$

□

Example 7.19 We may establish the truth value of $\neg \, (\forall \, x : X \mid false \bullet p)$ as follows.

$$
\begin{aligned}
&\neg \, (\forall \, x : X \mid false \bullet p) \\
\Leftrightarrow \; &\neg \, (\forall \, x : X \bullet false \Rightarrow p) && \text{[Law 7.6]} \\
\Leftrightarrow \; &\exists \, x : X \bullet \neg \, (false \Rightarrow p) && \text{[Law 7.5]} \\
\Leftrightarrow \; &\exists \, x : X \bullet \neg \, (true \vee p) && \text{[Law 3.18]} \\
\Leftrightarrow \; &\exists \, x : X \bullet false \wedge \neg \, p && \text{[Law 3.9]} \\
\Leftrightarrow \; &\exists \, x : X \bullet false && \text{[Law 3.5]} \\
\Leftrightarrow \; &false && \text{[Law 7.15]}
\end{aligned}
$$

□

Exercise 7.36 Prove, via equational reasoning, that each of the following are theorems of predicate logic.

1. $((\exists \, x : X \bullet p) \Rightarrow (\exists \, x : X \bullet q)) \Rightarrow (\exists \, x : X \bullet p \Rightarrow q)$

2. $(\exists \, x : X \bullet p \Rightarrow q) \Leftrightarrow ((\forall \, x : X \bullet p) \Rightarrow (\exists \, x : X \bullet q))$

□

The predicate $\exists \, x : X \bullet p\,(x) \wedge y = z$ is concerned with three variables: x, y, and z. Of these, x is bound, and y and z are free. As x does not appear in the predicate $y = z$, we are free to assume that the existential quantification of x has no effect on the truth or falsity of that predicate. As such, we may conclude that the following equivalence holds.

$$
\begin{aligned}
&\exists \, x : X \bullet (p\,(x) \wedge y = z) \Leftrightarrow \\
&\quad (\exists \, x : X \bullet p\,(x)) \wedge y = z
\end{aligned}
$$

This principle forms the basis of the following laws.

Law 7.16 Provided that x does not appear free in q, the following holds.

$$(\forall \, x : X \bullet p \wedge q) \Leftrightarrow ((\forall \, x : X \bullet p) \wedge q)$$

□

Example 7.20 The following equivalence holds.

$$(\forall \, n : \mathbb{N} \bullet n > 3 \wedge l = 10) \Leftrightarrow ((\forall \, n : \mathbb{N} \bullet n > 3) \wedge l = 10)$$

□

Law 7.17 Provided that x does not appear free in q, the following holds.

$$(\forall \, x : X \bullet p \vee q) \Leftrightarrow ((\forall \, x : X \bullet p) \vee q)$$

□

Example 7.21 The following equivalence holds.

$$(\forall\, n : \mathbb{N} \bullet n > 3 \vee l = 10) \Leftrightarrow ((\forall\, n : \mathbb{N} \bullet n > 3) \vee l = 10)$$

□

Law 7.18 Provided that x does not appear free in q, the following holds.

$$(\exists\, x : X \bullet p \wedge q) \Leftrightarrow ((\exists\, x : X \bullet p) \wedge q)$$

□

Example 7.22 The following equivalence holds.

$$(\exists\, p : Person \bullet p \text{ likes spam} \wedge l = 10) \Leftrightarrow ((\exists\, p : Person \bullet p \text{ likes spam}) \wedge l = 10)$$

□

Law 7.19 Provided that x does not appear free in q, the following holds.

$$(\exists\, x : X \bullet p \vee q) \Leftrightarrow ((\exists\, x : X \bullet p) \vee q)$$

□

Example 7.23 The following equivalence holds.

$$(\exists\, p : Person \bullet p \text{ likes spam} \vee l = 10) \Leftrightarrow ((\exists\, p : Person \bullet p \text{ likes spam}) \vee l = 10)$$

□

Exercise 7.37 Prove, via equational reasoning, that each of the following are tautologies. You may assume that x does not appear free in q in each case.

1. $(\forall\, x : X \bullet p \Rightarrow q) \Leftrightarrow ((\exists\, x : X \bullet p) \Rightarrow q)$
2. $(\exists\, x : X \bullet p \Rightarrow q) \Leftrightarrow ((\forall\, x : X \bullet p) \Rightarrow q)$

□

7.11   Natural deduction

As was the case with equational reasoning, we can also extend natural deduction to predicate logic. Furthermore, just as each propositional logic operator had associated introduction and elimination operators, so do both of our predicate logic operators. We consider the natural deduction rules for universal quantification first.

$$\frac{\forall\, x : X \bullet p \quad t \in X}{p\,[t/x]} \; [\forall\,\text{elim}]$$

The motivation behind this rule is similar to that for $\wedge$ elimination. Essentially, this rule states that if p holds for all values of X, then—provided that t is of type X—p holds for t.

Example 7.24 All natural numbers are greater than or equal to 0. $6+3$ is a natural number and, as such, we may conclude that $6+3$ is greater than 0.

$$\frac{\forall\, n : \mathbb{N} \bullet n \geq 0 \quad 6 + 3 \in \mathbb{N}}{6 + 3 \geq 0} \ [\forall\,\text{elim}]$$

□

With regards to the more general form of universal quantification, we need to ensure that the term satisfies the predicate. As such, the other form of this proof rule is given by

$$\frac{\forall\, x : X \mid p \bullet q \quad p\,[t/x] \quad t \in X}{q\,[t/x]} \ [\forall\,\text{elim}]$$

Example 7.25 All natural numbers which are prime and greater than 2 are odd. 7 is prime and greater than 2 and, as such, we may conclude that it is odd.

$$\frac{\forall\, n : \mathbb{N} \mid prime\,(n) \wedge n > 2 \bullet odd\,(n) \quad prime\,(7) \wedge 7 > 2 \quad 7 \in \mathbb{N}}{odd\,(7)} \ [\forall\,\text{elim}]$$

□

We now consider the introduction rule for universal quantification.

Here, the motivation is that if we can prove that a predicate p holds for *any* value x, such that there are no special assumptions or conditions involved in the choice of x, then we can conclude that it holds for *every* value of the relevant type.

$$\lceil x \in X \rceil_1$$
$$\vdots$$
$$\frac{p}{\forall\, x : X \bullet p} \ [\forall\,\text{intro}_1] \left(\begin{array}{l} \text{Provided that } x \text{ does not appear} \\ \text{free in the assumptions of } p \end{array} \right)]$$

Example 7.26

$$\frac{\dfrac{\lceil n \in \mathbb{N} \rceil_1}{n \geq 0} \ [\text{mathematics}]}{\forall\, n : \mathbb{N} \bullet n \geq 0} \ [\forall\,\text{intro}_1]$$

□

With regards to restricted predicates of the form $\forall\, x : X \mid p \bullet q$, the variable x must also satisfy the constraint p.

$$\lceil x \in X \rceil_1 \quad \lceil p \rceil_1$$
$$\vdots$$
$$\frac{q}{\forall\, x : X \mid p \bullet q} \ [\forall\,\text{intro}_1] \left(\begin{array}{l} \text{Provided that } x \text{ does not appear} \\ \text{free in the assumptions of } q \end{array} \right)]$$

Example 7.27

$$\frac{\dfrac{\lceil n \in \mathbb{N} \rceil_1 \quad \lceil prime\,(n) \wedge n > 2 \rceil_1}{odd\,(n)}\text{[mathematics]}}{\forall\, n : \mathbb{N} \mid prime\,(n) \wedge n > 2 \bullet odd\,(n)}\ [\forall\,\text{intro}_1]$$

□

Exercise 7.38 Prove the following using the rules of natural deduction seen thus far.

$$(\forall\, x : X \bullet p \wedge \forall\, x : X \bullet q) \Rightarrow (\forall\, x : X \bullet p \wedge q)$$

□

Exercise 7.39 Prove the following using the rules of natural deduction seen thus far.

$$(\forall\, x : X \bullet p \wedge q) \Rightarrow (\forall\, x : X \bullet p \wedge \forall\, x : X \bullet q)$$

□

Having considered the introduction and elimination rules for universal quantification, we now consider the corresponding rules for existential quantification.

The motivation behind the elimination rule for $\exists$ is similar to that for $\vee$ elimination. If we have assumed some $x \in X$ such that p holds for x, then, if we are able to derive a further predicate, q in which x does not appear free, then we may conclude q.

$$\frac{\exists\, x : X \bullet p \qquad \begin{array}{c}\lceil x \in X \wedge p \rceil_1 \\ \vdots \\ q \end{array}}{q}\ [\exists\,\text{elim}_i] \left(\begin{array}{l}\text{Provided that } x \text{ does not appear free} \\ \text{in the assumptions, and that } x \text{ does} \\ \text{not appear free in } q\end{array}\right)]$$

In the case of predicates with restrictions, the restriction is included in the assumption; this law is given below.

$$\frac{\exists\, x : X \mid p \bullet q \qquad \begin{array}{c}\lceil x \in X \wedge p \wedge q \rceil_1 \\ \vdots \\ r \end{array}}{r}\ [\exists\,\text{elim}_i] \left(\begin{array}{l}\text{Provided that } x \text{ does not} \\ \text{appear free in the} \\ \text{assumptions, and that } x \\ \text{does not appear free in } r\end{array}\right)]$$

We now turn to the introduction rule for existential quantification. The basis of this rule is that if p is true for any term t of the set X, then we may conclude that $\exists\, x : X \bullet p$ holds.

$$\frac{t \in X \quad p\,[t/x]}{\exists\, x : X \bullet p}\ [\exists\,\text{intro}]$$

Example 7.28 7 is a natural number, which is prime. As such, $\exists\, n : \mathbb{N} \bullet prime\,(n)$ holds.

$$\frac{7 \in \mathbb{N} \quad prime\,(7)}{\exists\, n : \mathbb{N} \bullet prime\,(n)}\ [\exists\,\text{intro}]$$

□

Of course, if we are dealing with the restricted form of existential quantification, then the constraint also has to be true of t.

$$\frac{t \in X \quad p\,[t/x] \quad q\,[t/x]}{\exists\, x : X \mid p \bullet q}\ [\exists\,\text{intro}]$$

Example 7.29 7 is a natural number, which is prime, is greater than 2, and odd. As such

$$\exists\, n : \mathbb{N} \mid prime\,(n) \wedge n > 2 \bullet odd\,(n)$$

holds.

$$\frac{7 \in \mathbb{N} \quad prime\,(7) \wedge 7 > 2 \quad odd\,(7)}{\exists\, n : \mathbb{N} \mid prime\,(n) \wedge n > 2 \bullet odd\,(n)}\ [\exists\,\text{intro}]$$

□

Example 7.30 We may establish that

$$\exists\, x : X \bullet p \wedge q \Rightarrow (\exists\, x : X \bullet p \wedge \exists\, x : X \bullet q)$$

is a theorem of predicate logic in the following way.

Exercise 7.40 Prove the following using the rules of natural deduction.

$$((\exists\, x : X \bullet p) \vee (\exists\, x : X \bullet q)) \Rightarrow (\exists\, x : X \bullet p \vee q)$$

□

Exercise 7.41 Prove the following using the rules of natural deduction.

$$(\exists\, x : X \bullet \exists\, y : Y \bullet p) \Rightarrow (\exists\, y : Y \bullet \exists\, x : X \bullet p)$$

□

7.12 The one-point rule

Consider the following predicate.

$$\exists\, p : Person \bullet p \ likes \ emily \wedge p = rick$$

This predicate not only asserts that there is a person, p, that likes Emily, but also that the name of that person is Rick. As such, it is safe to conclude that $rick \in Person$ and $rick \ likes \ emily$ are both true.

This is the motivation behind the *one-point rule*. If we have established that $\exists\, x : X \bullet p$ holds and, within p, the value of x is determined to be equal to some term, t, then we can conclude that both $p\,[t/x]$ and $t \in X$ hold. This, of course, has the effect of eliminating both the existential quantifier and the bound variable.

Law 7.20 Provided that x does not appear free in t,

$$(\exists\, x : X \bullet p \wedge x = t) \Leftrightarrow (t \in X \wedge p\,[t/x])$$

holds. □

Example 7.31

$$\exists\, n : \mathbb{N} \bullet n = n^2 \wedge n = 0 \Leftrightarrow$$
$$0^2 = 0 \wedge 0 \in \mathbb{N}$$

□

In the above example, $n = 0$ is the equation which determines n, not $n = n^2$. Without the requirement that the bound variable does not appear in the term which is being substituted, we might end up with

$$\exists\, n : \mathbb{N} \bullet n = n^2 \wedge n = 0 \Leftrightarrow$$
$$n^2 = 0 \wedge n^2 \in \mathbb{N}$$

This, of course, fails to remove the original bound variable name from the new expression, which is the purpose of applying the one-point rule.

Exercise 7.42 To which of the following statements may we apply the one-point rule?

1. $\exists\, n : \mathbb{N} \bullet n \geq 3 \wedge n = 5$

2. $\exists\, n : \mathbb{N} \bullet n \geq 3 \wedge n = m$

3. $\exists\, n : \mathbb{N} \bullet n \geq 3 \wedge n = m + n$

4. $\exists\, n : \mathbb{N} \bullet n \geq 3 \vee n = 5$

□

Exercise 7.43 Apply the one-point rule to each of the following statements.

1. $\exists\, p : Person \bullet p$ owns a red car $\land\ p = becky$

2. $\exists\, p : Person \bullet p$ owns a red car $\land\ p\ likes\ rick \land\ p = becky$

3. $\exists\, p : Person \bullet p$ owns a red car $\land\ p = becky\ \land$
    $\exists\, q : Person \bullet \neg\, q$ owns a red car $\land\ p\ likes\ q$

4. $\exists\, p : Person \bullet p$ owns a red car $\land\ p = becky\ \land$
    $\exists\, q : Person \bullet \neg\, q$ owns a red car $\land\ p\ likes\ q \land\ q = rick$

$\square$

The natural deduction form of the one-point rule is given below.

$$\frac{\exists\, x : X \bullet p \land x = t}{p\,[t/x] \land t \in X} \quad \text{[one-point rule]}$$

Example 7.32

$$\frac{\exists\, p : Person \bullet p \text{ owns a red car} \land p = becky}{becky \text{ owns a red car} \land becky \in Person} \quad \text{[one-point rule]}$$

$\square$

Note that, as the one-point rule is an equivalence, the following is also a valid inference rule.

$$\frac{p\,[t/x] \land t \in X}{\exists\, x : X \bullet p \land x = t} \quad \text{[one-point rule]}$$

Example 7.33

$$\frac{becky \text{ owns a red car} \land becky \in Person}{\exists\, p : Person \bullet p \text{ owns a red car} \land p = becky} \quad \text{[one-point rule]}$$

$\square$

7.13   Further exercises

Exercise 7.44 Consider the set *People*, such that $\{jon, steve, ali\} \subseteq People$, together with the predicates *happy*, *in_pain* and *angry*, all of which are associated with the set *People*.
    Represent the following using predicate logic.

1. "Ali is happy."

2. "Jon is happy if, and only if, Ali is angry and Steve is in pain."

3. "If Ali is happy, then Jon is not happy."

4. "If someone is not happy or someone is in pain, then Jon is happy."

5. "If no-one is in pain, then Jon is not happy."

6. "If everyone is happy, then Jon is in pain."

□

Exercise 7.45 Symbolise the following predicates about dogs.

1. "Some dogs have been well trained."

2. "Any dog is attractive if it is neat and well trained."

3. "Some dogs are gentle and have been well trained."

4. "All gentle dogs have been well trained."

5. "Some dogs are gentle only if they have been groomed by every trainer."

□

Exercise 7.46 Assume the predicates *plane*, *train* and *boat*, such that *plane* (*c*, *d*) denotes the fact that one can travel from country *c* to country *d* by plane, *train* (*c*, *d*) denotes the fact that one can travel from country *c* to country *d* by train, and *boat* (*c*, *d*) denotes the fact that one can travel from country *c* to country *d* by boat.

Express the following using predicate logic.

1. One can travel from France to Singapore by plane.

2. One can travel from France to Germany by train and by plane.

3. One can travel from France to the United Kingdom by air, by train, and by boat.

4. There is at least one country that can be reached by train from the United Kingdom.

5. Any country that can be reached by plane from France can also be reached by plane from the United Kingdom.

□

Exercise 7.47 Assuming the predicates of the previous exercise, give the natural language equivalents of each of the following.

1. $\exists\, c : Country \bullet train\,(germany, c) \wedge \neg\, train\,(ireland, c)$

2. $\neg\, \exists\, c : Country \bullet boat\,(switzerland, c)$

3. $\forall\, c, d : Country \bullet plane\,(c, d) \Rightarrow plane\,(d, c)$

4. $\forall\, c, d : Country \bullet plane\,(c, d) \Rightarrow boat\,(c, d)$

5. $\forall\, c : Country \bullet \neg\, plane\,(c, c)$

□

Exercise 7.48

1. Give a predicate, p, in which x occurs, such that $\forall x : \mathbb{N} \bullet p$ and $\exists x : \mathbb{N} \bullet p$ are both equivalent to *false*.

2. Give a predicate, p, in which x occurs, such that $\forall x : \mathbb{N} \bullet p$ and $\exists x : \mathbb{N} \bullet p$ are both equivalent to *true*.

3. Give a predicate, p, in which x occurs, such that $\forall x : \mathbb{N} \bullet p$ is equivalent to *false* and $\exists x : \mathbb{N} \bullet p$ is equivalent to *true*.

$\square$

Exercise 7.49 Assume $p = x - y > y$. What are the truth values of the following statements?

1. $\forall x : \mathbb{N} \bullet \exists y : \mathbb{N} \bullet p$

2. $\exists x : \mathbb{N} \bullet \forall y : \mathbb{N} \bullet p$

3. $\forall y : \mathbb{N} \bullet \exists x : \mathbb{N} \bullet p$

4. $\exists y : \mathbb{N} \bullet \forall x : \mathbb{N} \bullet p$

$\square$

Exercise 7.50 Assume the following set, such that $Capital = Country \times City$.

$$Capital = \{(france, paris), (USA, washington)\}$$

Assuming that $c \in Country$ and $d \in City$, which of the following predicates are valid, which are satisfiable, and which are unsatisfiable?

1. $(c, d) \in Capital$

2. $(USA, d) \in Capital$

3. $(USA, d) \in Capital \land d = paris$

4. $(USA, d) \in Capital \land d = washington$

5. $(USA, d) \in Capital \lor (france, d) \in Capital$

$\square$

Exercise 7.51 For each of the following expressions, state which variables are free.

1. $\forall x : X \bullet (x = y \land \exists y : X \bullet y = z)$

2. $(\forall x : X \bullet x = y) \land (\exists y : X \bullet y = z)$

3. $\forall x : X \bullet \exists y : X \bullet (x = y \land y = z)$

$\square$

Exercise 7.52 Give the results of each of the following substitutions.

1. $\forall\, x : X \bullet (x = y \wedge \exists\, y : X \bullet y = z)\,[w/x]$

2. $\forall\, x : X \bullet (x = y \wedge \exists\, y : X \bullet y = z)\,[w/y]$

3. $\forall\, x : X \bullet (x = y \wedge \exists\, y : X \bullet y = z)\,[w/z]$

4. $\forall\, x : X \bullet (x = y \wedge \exists\, y : X \bullet y = z)\,[x/z]$

5. $\forall\, x : X \bullet (x = y \wedge \exists\, y : X \bullet y = z)\,[y/z]\,[w/y]$

6. $\forall\, x : X \bullet (x = y \wedge \exists\, y : X \bullet y = z)\,[y/z, w/y]$

$\square$

Exercise 7.53 Determine whether the following propositions are true, false, or undefined.

1. $(\mu\, n : \mathbb{N} \mid n - 1 < n) \geq 4$

2. $(\mu\, n : \mathbb{N} \mid n - 1 = 3) > 4$

3. $(\mu\, n : \mathbb{N} \mid n - 1 = 3 \bullet n^2) > 4$

$\square$

7.14   Solutions

Solution 7.1

1. All dogs are smelly.

2. All dogs are either well-trained or small.

3. Either all dogs are well-trained or all dogs are small.

4. For any dog, it is the case that if it is small then it is well-trained.

5. If all dogs are small, then all dogs are well-trained.

$\square$

Solution 7.2 No. Assume that *Dog* consists of only two elements: *fido* and *lassie*, such that *fido* is not small and is well-trained, and *lassie* is small and is not well-trained. In this case, statement 4 is equivalent to

> (*fido* is small $\Rightarrow$ *fido* is well-trained) $\wedge$
>   (*lassie* is small $\Rightarrow$ *lassie* is well-trained)

which is equivalent to *false*. On the other hand, the second statement is equivalent to

> (*fido* is small $\wedge$ *lassie* is small) $\Rightarrow$
>   (*fido* is well-trained $\wedge$ *lassie* is well-trained)

which is equivalent to *true*. $\square$

Solution 7.3

1. $\forall\, x : People \bullet x$ likes Jaffa cakes

2. $\forall\, x : People \bullet (x$ is a vegetarian $\Rightarrow \neg\,(x$ likes Jaffa cakes$))$

3. $\forall\, x : People \bullet (x$ likes Jaffa cakes $\lor x$ is a vegetarian$)$

4. $(\forall\, x : People \bullet x$ likes Jaffa cakes$)$
   $\lor$
   $(\forall\, x : People \bullet x$ is a vegetarian$)$

5. $\forall\, x : People \bullet (x$ likes Jaffa cakes $\land x$ is a vegetarian$)$

6. $(\forall\, x : People \bullet x$ likes Jaffa cakes$)$
   $\land$
   $(\forall\, x : People \bullet x$ is a vegetarian$)$

$\square$

Solution 7.4 No. Consider the set *People*, containing only *jim* and *jon*. Assume that *jim* is a vegetarian and doesn't like Jaffa cakes, whereas *jon* is not a vegetarian and does like Jaffa cakes. Here, statement 3 is equivalent to *true* and statement 4 is equivalent to *false*. $\square$

Solution 7.5 Yes. $\forall$ is a generalised form of $\land$ and, as such, the former distributes through the latter. $\square$

Solution 7.6 1 and 2 are false, and 3 and 4 are true. $\square$

Solution 7.7

1. "Some dogs are smelly."

2. "Some dogs are either well-trained or small."

3. "For some dogs, it is the case that if they are small then they are well-trained."

4. "If some dogs are small, then some dogs are well-trained."

$\square$

Solution 7.8

1. $\exists\, x : People \bullet x$ likes Jaffa cakes

2. $\exists\, x : People \bullet x$ is a vegetarian $\Rightarrow \neg\,(x$ likes Jaffa cakes$)$

3. $\exists\, x : People \bullet (x$ likes Jaffa cakes $\lor x$ is a vegetarian$)$

4. $(\exists\, x : People \bullet x$ likes Jaffa cakes$)$
   $\lor$
   $(\exists\, x : People \bullet x$ is a vegetarian$)$

5. $\exists\, x : People \bullet (x$ likes Jaffa cakes $\land x$ is a vegetarian$)$

6. $(\exists\, x : People \bullet x$ likes Jaffa cakes$)$

   $\wedge$

   $(\exists\, x : People \bullet x$ is a vegetarian$)$

□

Solution 7.9 Yes. $\exists$ is a generalised form of $\vee$ and, as such, the former distributes through the latter. □

Solution 7.10 No. Consider the set *People*, containing only *jim* and *jon*. Assume that *jim* is a vegetarian but doesn't like Jaffa cakes, and that *jon* is not a vegetarian but does like Jaffa cakes. Here, statement 5 is equivalent to *false* and statement 6 is equivalent to *true*. □

Solution 7.11 They are all true. □

Solution 7.12 Yes. If it is the case that p holds for *all* elements of the set X, then we may conclude that it holds for some of them. □

Solution 7.13 No. Just because p holds for *some* elements of the set X, it is not necessarily the case that it holds for all of them. □

Solution 7.14

1. $p \Leftrightarrow (n = n)$
2. $p \Leftrightarrow (n > 3)$

□

Solution 7.15 4 is valid; 1 and 3 are satisfiable; 2 and 5 are unsatisfiable. □

Solution 7.16

1. "Someone doesn't like someone."
2. "Someone doesn't like everyone."
3. "Everyone doesn't like someone."
4. "Everyone doesn't like everyone."

□

Solution 7.17

1. $\exists\, p, q : Person \bullet \neg\, p$ likes q
2. $\exists\, p : Person \bullet \forall\, q : Person \bullet \neg\, p$ likes q
3. $\forall\, p : Person \bullet \exists\, q : Person \bullet \neg\, p$ likes q
4. $\forall\, p, q : Person \bullet \neg\, p$ likes q

□

Solution 7.18

1. $\exists\, x : \mathbb{N} \bullet x \neq x$

2. $\exists\, x : \mathbb{N} \bullet \forall\, y : \mathbb{N} \bullet x < y$

3. $\exists\, x : \mathbb{N} \bullet \exists\, y : \mathbb{N} \bullet x \leq y \wedge x \neq y \wedge x \geq y$

$\square$

Solution 7.19

$$(\forall\, x : X \bullet \overbrace{p \wedge (\exists\, y : Y \bullet \underbrace{q}_{scope\ of\ y}\ ) \wedge r}^{scope\ of\ x})$$

$\square$

Solution 7.20

1. x is free.

2. x is bound.

3. Both occurrences of x are bound.

4. x is bound and y is free.

5. x is bound and both occurrences of y are free.

6. The first occurrence of x and the sole occurrence of y are both free, and the second occurrence of x is bound.

7. x and y are both bound.

8. The first occurrence of y is free, and the second occurrence of y and the sole occurrence of x are both bound.

$\square$

Solution 7.21 We may attribute truth values to all of those that feature no free variables, i.e., 2, 3, and 7. $\square$

Solution 7.22 1 and 3 are propositions, whereas 2 is a predicate. $\square$

Solution 7.23

1. x may be substituted for in this expression.

2. There are no variables which may be substituted for in this expression.

3. y may be substituted for in this expression.

4. y may be substituted for in this expression, but x may not.

5. y and the second occurrence of x may be substituted for in this expression, but the first occurrence of x may not.

$\square$

Solution 7.24

1. *nigel* is happy

2. *nigel* is happy $\wedge$ y is sad

3. *nigel* is happy $\vee$ *nigel* is sad

4. x is happy $\vee$ *nigel* is sad

5. $\forall\, x : People \bullet x$ is happy

6. $\forall\, x : People \bullet x$ is happy $\vee$ x is sad

7. $(\forall\, x : People \bullet x$ is happy$) \vee$ *nigel* is sad

$\square$

Solution 7.25

1. $\forall\, x : \mathbb{N} \bullet x \geq 0$

2. $\forall\, x : \mathbb{N} \bullet x \geq 3$

3. $3 > 0 \wedge \forall\, x : \mathbb{N} \bullet x \geq 3$

4. $3 > 0 \wedge \forall\, x : \mathbb{N} \bullet x \geq y$

$\square$

Solution 7.26

1.
$$x^2 = 49\,[y + 3/x]\,[4/y]$$
$$\Leftrightarrow (y + 3)^2 = 49\,[4/y]$$
$$\Leftrightarrow (4 + 3)^2 = 49$$
$$\Leftrightarrow true$$

2.
$$x^2 = 49\,[y + 3/x, 4/y]$$
$$\Leftrightarrow (y + 3)^2 = 49$$

3.
$$x^2 > y\,[y + 3/x]\,[4/y]$$
$$\Leftrightarrow (y + 3)^2 > y\,[4/y]$$
$$\Leftrightarrow (4 + 3)^2 > 4$$
$$\Leftrightarrow true$$

4.
$$x^2 > y\,[y + 3/x, 4/y]$$
$$\Leftrightarrow (y + 3)^2 > 4$$

$\square$

Solution 7.27 2, 3, 5, and 7 may result in variable capture. □

Solution 7.28

1. $\forall x : \mathbb{N} \bullet x \geq z$

2. $\forall z : \mathbb{N} \bullet z \geq x$

3. $\forall z : \mathbb{N} \bullet z \geq x + y$

4. $\forall x, y : \mathbb{N} \bullet x \geq y$

5. $\forall w : \mathbb{N} \bullet w \geq x$

6. $\forall x : \mathbb{N} \bullet x \geq y$

7. $x + 3 > 0 \wedge \forall z : \mathbb{N} \bullet z \geq x + 3$

□

Solution 7.29

1. $\forall p : People \mid p$ likes cheese $\bullet$ p likes spam

2. $\exists p : People \mid p$ likes cheese $\bullet$ p likes spam

3. $\forall p : People \mid p$ likes cheese $\bullet$ p likes spam $\wedge$ p likes bananas

4. $\forall p : People \mid p$ likes cheese $\wedge$ p likes spam $\bullet$ p likes bananas

5. $\exists p : People \mid p$ likes cheese $\wedge$ p likes spam $\bullet$ p likes bananas

6. $\exists p : People \mid p$ likes cheese $\bullet$ p likes spam $\wedge$ p likes bananas

□

Solution 7.30

1. $\forall p : People \bullet p$ likes cheese $\Rightarrow p$ likes spam

2. $\exists p : People \bullet p$ likes cheese $\wedge p$ likes spam

3. $\forall p : People \bullet p$ likes cheese $\Rightarrow (p$ likes spam $\wedge p$ likes bananas$)$

4. $\forall p : People \bullet (p$ likes cheese $\wedge p$ likes spam$) \Rightarrow p$ likes bananas

5. $\exists p : People \bullet p$ likes cheese $\wedge p$ likes spam $\wedge p$ likes bananas

6. $\exists p : People \bullet p$ likes cheese $\wedge p$ likes spam $\wedge p$ likes bananas

□

Solution 7.31

1. $\exists n : \mathbb{N} \mid p(n) \bullet true \Rightarrow \neg\, p(n)$
   $\Leftrightarrow \exists n : \mathbb{N} \bullet p(n) \wedge (true \Rightarrow \neg\, p(n))$

If $p(n)$ is equivalent to *true*, then the second conjunct is *false*; if $p(n)$ is equivalent to *false*, then the first conjunct is *false*. As such, the statement is equivalent to *false*.

2. $\forall\, n : \mathbb{N} \mid p\,(n) \bullet false \Rightarrow \neg\, p\,(n)$
   $\Leftrightarrow \forall\, n : \mathbb{N} \bullet p\,(n) \Rightarrow (false \Rightarrow \neg\, p\,(n))$

The consequent is always equivalent to *true*. As such, the statement is equivalent to *true*.

3. $\exists\, n : \mathbb{N} \mid n \in \emptyset \bullet n$ is prime
   $\Leftrightarrow \exists\, n : \mathbb{N} \bullet n \in \emptyset \wedge n$ is prime
   $\Leftrightarrow \exists\, n : \mathbb{N} \bullet false \wedge n$ is prime
   $\Leftrightarrow false$

4. $\forall\, n : \mathbb{N} \mid n \in \emptyset \bullet n$ is prime
   $\Leftrightarrow \forall\, n : \mathbb{N} \bullet n \in \emptyset \Rightarrow n$ is prime
   $\Leftrightarrow \forall\, n : \mathbb{N} \bullet false \Rightarrow n$ is prime
   $\Leftrightarrow true$

☐

Solution 7.32 1, 2 and 4 are true, and 3 is false. ☐

Solution 7.33

1. $\exists\, n : \mathbb{N} \bullet (n$ is a divisor of $7 \wedge n \neq 1 \wedge$
   $\forall\, m : \mathbb{N} \mid m \neq n \bullet \neg\, m$ is a divisor of $7 \vee m = 1)$

2. $\exists\, n : \mathbb{N} \bullet (\forall\, m : \mathbb{N} \bullet n \leq m \wedge$
   $\forall\, l : \mathbb{N} \mid l \neq n \bullet \exists\, m : \mathbb{N} \bullet l > m)$

3. $\exists\, n : \mathbb{N} \bullet (\forall\, m : \mathbb{N} \bullet n \geq m \wedge$
   $\forall\, l : \mathbb{N} \mid l \neq n \bullet \exists\, m : \mathbb{N} \bullet l < m)$

4. $\forall\, n : \mathbb{N} \bullet (\exists\, m : \mathbb{N} \bullet (n \geq m \wedge \forall\, l : \mathbb{N} \mid l \neq m \bullet n < l))$

☐

Solution 7.34

1. $(\mu\, m : Mountain \mid m$ is the tallest in the world$)$

2. $(\mu\, m : Mountain \mid m$ is the tallest in the world $\bullet height(m))$

3. $(\mu\, p : People \mid p$ is the oldest in the world$)$

4. $(\mu\, p : People \mid p$ is the oldest in the world $\bullet nationality(p))$

5. $(\mu\, n : \mathbb{N} \mid (\forall\, m : \mathbb{N} \bullet n \leq m))$

☐

Solution 7.35 1, 3 and 5 are true, and 2 and 4 are undefined. ☐

Solution 7.36

1. $(\exists\, x : X \bullet p) \Rightarrow (\exists\, x : X \bullet q)$
$\Leftrightarrow \neg\,(\exists\, x : X \bullet p) \vee (\exists\, x : X \bullet q)$ [Law 3.18]
$\Leftrightarrow (\forall\, x : X \bullet \neg\, p) \vee (\exists\, x : X \bullet q)$ [Law 7.5]
$\Rightarrow (\exists\, x : X \bullet \neg\, p) \vee (\exists\, x : X \bullet q)$ [Law 7.8]
$\Leftrightarrow \exists\, x : X \bullet \neg\, p \vee q$ [Law 7.3]
$\Leftrightarrow \exists\, x : X \bullet p \Rightarrow q$ [Law 3.18]

2. $\exists\, x : X \bullet p \Rightarrow q$
$\Leftrightarrow \exists\, x : X \bullet \neg\, p \vee q$ [Law 3.18]
$\Leftrightarrow (\exists\, x : X \bullet \neg\, p) \vee (\exists\, x : X \bullet q)$ [Law 7.3]
$\Leftrightarrow \neg\,(\forall\, x : X \bullet p) \vee (\exists\, x : X \bullet q)$ [Law 7.5]
$\Leftrightarrow (\forall\, x : X \bullet p) \Rightarrow (\exists\, x : X \bullet q)$ [Law 3.18]

□

Solution 7.37

1. $\forall\, x : X \bullet p \Rightarrow q$
$\Leftrightarrow \forall\, x : X \bullet \neg\, p \vee q$ [Law 3.18]
$\Leftrightarrow (\forall\, x : X \bullet \neg\, p) \vee q$ [Law 7.17]
$\Leftrightarrow (\neg\, \exists\, x : X \bullet p) \vee q$ [Law 7.5]
$\Leftrightarrow (\exists\, x : X \bullet p) \Rightarrow q$ [Law 3.18]

2. $\exists\, x : X \bullet p \Rightarrow q$
$\Leftrightarrow \exists\, x : X \bullet \neg\, p \vee q$ [Law 3.18]
$\Leftrightarrow (\exists\, x : X \bullet \neg\, p) \vee q$ [Law 7.19]
$\Leftrightarrow (\neg\, \forall\, x : X \bullet p) \vee q$ [Law 7.5]
$\Leftrightarrow (\forall\, x : X \bullet p) \Rightarrow q$ [Law 3.18]

□

Solution 7.38

$$\dfrac{\dfrac{\lceil \forall\, x : X \bullet p \wedge \forall\, x : X \bullet q \rceil_1}{\dfrac{\forall\, x : X \bullet p}{p}\,[\forall\,\mathrm{e}]}\,[\wedge\,\mathrm{e1}] \qquad \dfrac{\lceil \forall\, x : X \bullet p \wedge \forall\, x : X \bullet q \rceil_1}{\dfrac{\forall\, x : X \bullet q}{q}\,[\forall\,\mathrm{e}]}\,[\wedge\,\mathrm{e2}]}{\dfrac{\dfrac{p \wedge q}{\dfrac{\forall\, x : X \bullet p \wedge q}{(\forall\, x : X \bullet p \wedge \forall\, x : X \bullet q) \Rightarrow \forall\, x : X \bullet p \wedge q}\,[\Rightarrow\,\mathrm{i}_1]}\,\lceil x \in X \rceil_2\;[\forall\,\mathrm{i}_2]}{}}\,[\wedge\,\mathrm{i1}]$$

□

Solution 7.39

$$\cfrac{\cfrac{\cfrac{\lceil x \in X \rceil_2 \quad \lceil \forall x : X \bullet p \wedge q \rceil_1}{p \wedge q} \, [\forall \, \text{elim}]}{\cfrac{p}{\forall x : X \bullet p} \, [\wedge \, \text{elim1}]}{[\forall \, \text{intro}_2]} \wedge \cfrac{\cfrac{\cfrac{\lceil x \in X \rceil_3 \quad \lceil \forall x : X \bullet p \wedge q \rceil_1}{p \wedge q} \, [\forall \, \text{elim}]}{\cfrac{q}{\forall x : X \bullet q} \, [\wedge \, \text{elim}_2]}{[\forall \, \text{intro}_3]}}{\cfrac{(\forall x : X \bullet p) \wedge (\forall x : X \bullet q)}{(\forall x : X \bullet p \wedge q) \Rightarrow (\forall x : X \bullet p \wedge \forall x : X \bullet q)} \, [\Rightarrow \, \text{intro}_1]} \, [\wedge \, \text{intro}]$$

□

Solution 7.40

$$\cfrac{\lceil \exists x : X \bullet p \vee \exists x : X \bullet q \rceil_1 \quad \cfrac{\cfrac{\cfrac{\lceil \exists x : X \bullet p \rceil_2}{p} \, [\exists \, \text{elim}]}{p \vee q} \, [\vee \, \text{intro1}]}{\exists x : X \bullet p \vee q} \, [\exists \, \text{intro}] \quad \cfrac{\cfrac{\cfrac{\lceil \exists x : X \bullet q \rceil_2}{q} \, [\exists \, \text{elim}]}{p \vee q} \, [\vee \, \text{intro2}]}{\exists x : X \bullet p \vee q} \, [\exists \, \text{intro}]}{\cfrac{\exists x : X \bullet p \vee q}{((\exists x : X \bullet p) \vee (\exists x : X \bullet q)) \Rightarrow (\exists x : X \bullet p \vee q)} \, [\Rightarrow \, \text{intro}_1]} \, [\vee \, \text{elim}_2]$$

□

Solution 7.41 We start at the root of the tree with the statement that we wish to prove:

$$\exists x : X \bullet \exists y : Y \bullet p \Rightarrow \exists y : Y \bullet \exists x : X \bullet p$$

Applying ⇒ introduction at this stage gives us

$$\cfrac{\exists y : Y \bullet \exists x : X \bullet p}{\exists x : X \bullet \exists y : Y \bullet p \Rightarrow \exists y : Y \bullet \exists x : X \bullet p} \, [\Rightarrow \, \text{intro}_1]$$

As such, we are left with the task of establishing that $\exists y : Y \bullet \exists x : X \bullet p$ follows from $\exists x : X \bullet \exists y : Y \bullet p$. We can do this by means of $\exists$ elimination. As such, the next step in our proof is given by

$$\cfrac{\lceil \exists x : X \bullet \exists y : Y \bullet p \rceil_1 \quad \exists y : Y \bullet \exists x : X \bullet p}{\exists y : Y \bullet \exists x : X \bullet p} \, [\exists \, \text{elim}_2]$$

We can complete our tree in the following fashion.

$$\cfrac{\lceil \exists y : Y \bullet p \rceil_2 \quad \cfrac{\cfrac{\lceil y \in Y \wedge p \rceil_3}{y \in Y} \, [\wedge \, -\text{e1}] \quad \cfrac{\cfrac{\lceil x \in X \wedge p \rceil_2}{x \in X} \, [\wedge \, -\text{e1}] \quad \cfrac{\lceil x \in X \wedge p \rceil_3}{p} \, [\wedge \, -\text{e2}]}{\exists x : X \bullet p} \, [\exists \, -\text{i}]}{\exists y : Y \bullet \exists x : X \bullet p} \, [\exists \, -\text{i}]}{\exists y : Y \bullet \exists x : X \bullet p} \, [\exists \, -\text{e}_3]$$

As there are no undischarged assumptions, we may conclude that

$$\exists x : X \bullet \exists y : Y \bullet p \Rightarrow \exists y : Y \bullet \exists x : X \bullet p$$

is a theorem of predicate logic. □

Solution 7.42 We may apply the one-point rule only to statements 1 and 2. In 3, the term $m + n$ has the original bound variable appearing free. In 4, the value of n isn't determined exactly. □

Solution 7.43

1. *becky* ∈ *Person* ∧ *becky* owns a red car

2. *becky* ∈ *Person* ∧ *becky* owns a red car ∧ *becky likes rick*

3. *becky* ∈ *Person* ∧ *becky* owns a red car ∧
   ∃ *q* : *Person* • ¬ *q* owns a red car ∧ *becky likes q*

4. *becky* ∈ *Person* ∧ *becky* owns a red car ∧
   rick ∈ *Person* ∧ ¬ *rick* owns a red car ∧ *becky likes rick*

□

Solution 7.44

1. *happy* (*ali*)

2. *happy* (*jon*) ⇔ *angry* (*ali*) ∧ *in_pain* (*steve*)

3. *happy* (*ali*) ⇒ ¬ *happy* (*jon*)

4. ((∃ *p* : *People* • ¬ *happy* (*p*)) ∨ (∃ *q* : *People* • *in_pain* (*q*))) ⇒ *happy* (*jon*)

5. (¬ ∃ *p* : *People* • *in_pain* (*p*)) ⇒ ¬ *happy* (*jon*)

6. (∀ *p* : *People* • *happy* (*p*)) ⇒ *in_pain* (*jon*)

□

Solution 7.45

1. ∃ *d* : *Dog* • *well_trained* (*d*)

2. ∀ *d* : *Dog* • *neat* (*d*) ∧ *well_trained* (*d*) ⇒ *attractive*(*d*)

3. ∃ *d* : *Dog* • *gentle* (*d*) ∧ *well_trained* (*d*)

4. ∀ *d* : *Dog* • *gentle* (*d*) ⇒ *well_trained* (*d*)

5. ∃ *d* : *Dog* • *gentle* (*d*) ⇒ ∀ *t* : *Trainer* • *groomed* (*d*, *t*)

□

Solution 7.46

1. *plane* (*france*, *singapore*)

2. *train* (*france*, *germany*) ∧ *plane* (*france*, *germany*)

3. *train* (*france*, *UK*) ∧ *plane* (*france*, *UK*) ∧ *boat* (*france*, *UK*)

4. $\exists\, c : Country \bullet train\,(UK, c)$

5. $\forall\, c : Country \bullet plane\,(france, c) \Rightarrow plane\,(UK, c)$

□

Solution 7.47

1. "There is at least one country which can be reached by train from Germany and cannot be reached by train from Ireland."

2. "There are no countries which may be reached by boat from Switzerland."

3. "If one can travel from one country to another by plane, then one can also travel in the reverse direction by plane."

4. "If one can travel from one country to another by plane, then one can make the same journey by boat."

5. "One cannot fly from one country to itself."

□

Solution 7.48

1. $x \neq x$

2. $x = x$

3. $x = 0$

□

Solution 7.49 1, 2 and 4 are false, and 3 is true. □

Solution 7.50 1, 2 and 4 are satisfiable; 3 is unsatisfiable; 5 is valid. □

Solution 7.51

1. The first occurrence of y is free, as is z.

2. The first occurrence of y is free, as is z.

3. z is free.

□

Solution 7.52

1. $\forall\, x : X \bullet (x = y \wedge \exists\, y : X \bullet y = z)$

2. $\forall\, x : X \bullet (x = w \wedge \exists\, y : X \bullet y = z)$

3. $\forall\, x : X \bullet (x = y \wedge \exists\, y : X \bullet y = w)$

4. $\forall\, v : X \bullet (v = y \wedge \exists\, y : X \bullet y = x)$

5. $\forall\, x : X \bullet (x = w \wedge \exists\, v : X \bullet v = w)$

6. $\forall\, x : X \bullet (x = w \wedge \exists\, v : X \bullet v = y)$

$\square$

Solution 7.53 1 is undefined, 2 is false, and 3 is true. $\square$

Chapter 8

Relations

In Chapter 6 we encountered Cartesian products. Given two sets X and Y, the Cartesian product $X \times Y$ was defined to be the set of all pairs of the form (x, y), such that $x \in X$ and $y \in Y$. Then, in Chapter 7, we encountered predicates. There, we could reason about some predicate p holding of some values u and v. For example, $p(u, v)$ would indicate that the predicate p holds for the pair (u, v). Furthermore, if u is an element of X and v is an element of Y, we may conclude that $(u, v) \in X \times Y$ and the set containing all such pairs that satisfy p is a subset of $X \times Y$. The consideration of such subsets of Cartesian products, or *relations*, is the subject of this chapter. Readers who require further reading on this topic are referred to [DG00], which is a text that concentrates solely on set theory, relations (the subject of this chapter), and functions (the subject of the next chapter).

8.1   Binary relations

In Chapter 6 we saw how Cartesian products allowed us to define sets of pairs. The Cartesian product $A \times B$ is the set of all pairs of the form (a, b) such that $a \in A$ and $b \in B$. In addition, we saw that all of our standard set theory operators could be applied to Cartesian products. So, for example,

$$\{(jack, 65), (jill, 60)\} \subset \{jack, jill\} \times \{60, 65\}$$

Here, the set of pairs $\{(jack, 65), (jill, 60)\}$ associates Jack and Jill with their ages, whereas the Cartesian product simply produces all possible pairs of the appropriate type. The set of pairs $\{(jack, 65), (jill, 60)\}$ is termed a *relation*—some sort of a relationship holds between the values Jack and 65, and the values Jill and 60. More generally, any subset R of a Cartesian product $A \times B$ is a relation.

Example 8.1 Consider the sets $A = \{a, b, c\}$ and $B = \{0, 1\}$. The Cartesian product $A \times B$ is given by

$$A \times B = \{(a, 0), (a, 1), (b, 0), (b, 1), (c, 0), (c, 1)\}$$

All of the subsets of this set (including the empty set and the set itself) are relations of this type. $\square$

If some pair (a, b) appears in a relation R, then we may say that "the relation R holds between a and b". This may be written in any one of a number of ways. For example, we may write this as $(a, b) \in R$, as $R(a, b)$, as $a \mapsto b \in R$, or, in some circumstances, as $a \, R \, b$. The pair (a, b) when written $a \mapsto b$ is referred to as a *maplet* of the relation.

Example 8.2 We might define the relation *less_than* on the natural numbers as follows.

$$less\_than = \{m, n : \mathbb{N} \mid m < n \bullet (m, n)\}$$

It follows that we may write $3 \mapsto 5 \in less\_than$, $less\_than(3, 5)$, or $(3, 5) \in less\_than$. It may also be possible to write $3 \, less\_than \, 5$. $\square$

We let $A \leftrightarrow B$ denote the set of all relations between the sets A and B. Here,

$$A \leftrightarrow B = \mathbb{P}(A \times B)$$

As such, $less\_than \in \mathbb{N} \leftrightarrow \mathbb{N}$.

Example 8.3 The set of relations $\{jack, jill\} \leftrightarrow \{60, 65\}$ is as follows.

$\{\emptyset, \{jack \mapsto 60\}, \{jack \mapsto 65\}, \{jill \mapsto 60\}, \{jill \mapsto 65\},$
$\{jack \mapsto 60, jack \mapsto 65\}, \{jack \mapsto 60, jill \mapsto 60\}, \{jack \mapsto 60, jill \mapsto 65\},$
$\{jill \mapsto 60, jack \mapsto 65\}, \{jill \mapsto 60, jill \mapsto 65\}, \{jack \mapsto 65, jill \mapsto 65\},$
$\{jack \mapsto 60, jack \mapsto 65, jill \mapsto 60\}, \{jack \mapsto 60, jack \mapsto 65, jill \mapsto 65\},$
$\{jack \mapsto 60, jill \mapsto 60, jill \mapsto 65\}, \{jack \mapsto 65, jill \mapsto 60, jill \mapsto 65\},$
$\{jack \mapsto 60, jack \mapsto 65, jill \mapsto 60, jill \mapsto 65\}\}$

$\square$

Exercise 8.1 Give the set of relations $A \leftrightarrow B$, where $A = \{0, 1\}$ and $B = \{x, y\}$. $\square$

Exercise 8.2 Recall the notion of axiomatic definitions from Chapter 6. There, we could define a natural number, n, such that $n < 100$ by

$$\begin{array}{|l}
n : \mathbb{N} \\
\hline
n < 100
\end{array}$$

Define, using axiomatic definitions, the following relations.

1. The relation *equals* on the natural numbers, where $(m, n) \in equals$ if, and only if, $m = n$.

2. The relation *looks_like* on the set *People*. Here, $(p, q) \in looks\_like$ holds if, and only if, person p looks like person q.

3. The relation *likes* on the sets *People* and *Food*. Here, $(p, f) \in likes$ holds if, and only if, person p likes food f.

4. The relation *double_of_prime*, such that $(m, n) \in double\_of\_prime$ holds if, and only if, n is a prime number and m is its double.

$\square$

8.2 Reasoning about relations

Essentially, relations are nothing more than sets of pairs. As such, the technique which we employed to establish whether two sets are equal can also be employed to establish whether two relations are equal.

We shall say that two relations R and S are equal when they contain precisely the same elements; this is true exactly when $(x, y) \in R \Leftrightarrow (x, y) \in S$ holds for any pair (x, y) of the appropriate type. Equivalences presented in this chapter, together with the laws presented in previous chapters, may be used to establish such truths.

Exercise 8.3 Prove that

$$\{m : \mathbb{N} \bullet (m, m)\} = \{m, n : \mathbb{N} \mid m = n\}$$

□

8.3 Domain and range

Given a relation $R \in A \leftrightarrow B$, the set A is referred to as the *source* of R, and the set B is referred to as the *target* of R. Thus, the source and target of $R \in \{jack, jill\} \leftrightarrow \{60, 65\}$ are $\{jack, jill\}$ and $\{60, 65\}$ respectively, while the source and target of $S \in \mathbb{N} \leftrightarrow \mathbb{N}$ are both $\mathbb{N}$. The terminology is evocative of the maplet notation. For example, if R is defined by

$$R = \{jack \mapsto 65, jill \mapsto 60\}$$

then those elements of the source appear on the side which the maplets are *pointing from*, whereas those elements of the target appear on the side which the maplets are *pointing to*.

One means of representing relations graphically is via 'potato diagrams'; such diagrams further emphasise the distinction between source and target. In such diagrams, all elements of the source and target are illustrated, together with an indication of which elements of the source are related to which elements of the target.

Example 8.4 Assume the following set and relation.

$$House\_mates = \{karen, kerry, matt\}$$
$$likes = \{karen \mapsto matt, matt \mapsto kerry\}$$

The potato diagram for *likes* is given below.

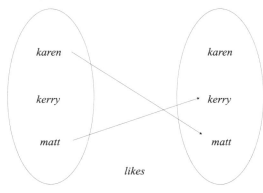

Here, the source of *likes* appears on the left-hand side and the target of *likes* appears on the right-hand side. In addition, arrows indicate the maplets contained in the relation: Karen likes Matt and Matt likes Kerry. It is important to note that the names listed on each side describe the source and target of the relation; *all* elements of these sets must appear in the diagram, not just those that are mapped to or mapped from. □

We say that the *domain* of a relation R consists of those elements of the source which appear in the relation, i.e., those elements which are mapped *from*. If $R \in A \leftrightarrow B$, then this is defined by

$$\text{dom } R = \{a : A;\ b : B \mid a \mapsto b \in R \bullet a\}$$

The dot notation provides us with an alternative definition of dom, which is given as follows.

$$\text{dom } R = \{r : R \bullet r.1\}$$

Example 8.5 For the case of

$$likes = \{karen \mapsto matt, matt \mapsto kerry\}$$

we have

$$\text{dom } likes = \{karen, matt\}$$

□

The *range* of a relation R consists of those elements of the target which appear in the relation, i.e., those elements which are mapped *to*. If $R \in A \leftrightarrow B$, then this is defined by

$$\text{ran } R = \{a : A;\ b : B \mid a \mapsto b \in R \bullet b\}$$

Again, the dot notation provides us with an alternative definition, which is given as follows.

$$\text{ran } R = \{r : R \bullet r.2\}$$

Example 8.6 For the case of

$$likes = \{karen \mapsto matt, matt \mapsto kerry\}$$

we have

$$\text{ran } likes = \{kerry, matt\}$$

□

Exercise 8.4 Give the domain and range of the following relations.

1. $\{(1, 2), (2, 3), (3, 4)\}$
2. $\{n : \mathbb{N} \mid n \leq 10 \bullet (n, 2n)\}$
3. $\{n : \mathbb{N} \bullet (n, n)\}$

4. $\{n : \mathbb{N} \bullet (n, n+1)\}$

5. $\{n : \mathbb{N} \mid n \neq n \bullet (n, n)\}$

$\square$

Exercise 8.5 Assuming that R and S are both elements of $X \leftrightarrow Y$, prove that

$$\text{dom}\,(R \cup S) = (\text{dom}\ R) \cup (\text{dom}\ S)$$

holds. $\square$

Exercise 8.6 Assuming that R and S are both elements of $X \leftrightarrow Y$, prove that

$$\text{ran}\,(R \cup S) = (\text{ran}\ R) \cup (\text{ran}\ S)$$

holds. $\square$

Exercise 8.7 Assuming that R and S are both elements of $X \leftrightarrow Y$, prove that

$$\text{ran}\,(R \cap \emptyset) = \text{ran}\,(S \cap \emptyset)$$

holds. $\square$

8.4   Relational inverse

The *relational inverse* of a relation $R \in A \leftrightarrow B$, is simply its mirror image. For example, if the relation R is given by

$$\{(a1, b1), (a2, b2)\}$$

then its inverse—which we may denote by either $R^{\sim}$ or R^{-1}—is given by

$$\{(b1, a1), (b2, a2)\}$$

The domain of R is the range of its inverse, while the range of R is the domain of its inverse. Furthermore, if $R \in A \leftrightarrow B$, then $R^{\sim} \in B \leftrightarrow A$.

We may define the inverse of a relation R formally as follows.

$$R^{\sim} = \{a : A;\ b : B \mid a \mapsto b \in R \bullet b \mapsto a\}$$

Example 8.7 Recall the relation *likes*, which was given by

$$likes = \{karen \mapsto matt, matt \mapsto kerry\}$$

The inverse of this relation, *likes* $^{\sim}$, is given by

$$likes^{\sim} = \{kerry \mapsto matt, matt \mapsto karen\}$$

$\square$

In potato diagram form, the relation of the above example is represented as follows.

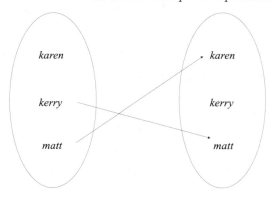

Law 8.1 For any relation R,

$$\text{dom } R^{\sim} = \text{ran } R$$
$$\text{ran } R^{\sim} = \text{dom } R$$

$\square$

Example 8.8 Consider the sets *People* and *Channels*, together with the relation $tv \in People \leftrightarrow Channels$, all of which are defined as follows.

$$People = \{becky, emily, john, michael\}$$
$$Channels = \{bbc1, bbc2, itv, c4, c5\}$$
$$tv = \{becky \mapsto bbc1, emily \mapsto itv, becky \mapsto itv, john \mapsto c5\}$$

Here, $tv^{\sim} \in Channels \leftrightarrow People$, and

$$tv^{\sim} = \{bbc1 \mapsto becky, itv \mapsto emily, itv \mapsto becky, c5 \mapsto john\}$$

$\square$

Exercise 8.8 Give the relational inverse of each of the following.

1. $\emptyset$
2. $\{(1,1),(2,2)\}$
3. $\{(1,2),(2,3),(3,4)\}$

$\square$

Exercise 8.9 Consider the relations of Exercise 8.8. Assuming that each of these relations is of type $\{1,2,3,4\} \times \{1,2,3,4\}$, illustrate the relational inverse of each via potato diagrams. $\square$

Exercise 8.10 Assume the sets *Person*, *Mode* and *Animal*, which are given by

$$Person = \{andy, dave, jim, jon, rick\}$$
$$Mode = \{cycles, drives, flies, walks\}$$
$$Animal = \{cat, dog, gerbil, goldfish, parrot\}$$

together with the following relations.

$$owns = \{(andy, dog), (rick, parrot), (jim, goldfish), (jon, cat)\}$$
$$travels = \{\, (andy, walks), (jim, cycles), (jim, drives),$$
$$(jon, cycles), (jon, drives), (dave, cycles)\}$$

Calculate each of the following.

1. *owns* $^\sim$

2. *travels* $^\sim$

3. dom (*owns* $^\sim$)

4. ran (*owns* $^\sim$)

5. dom (*travels* $^\sim$)

6. ran (*travels* $^\sim$)

7. ran (*travels*$^\sim$) $\cap$ ran (*owns*$^\sim$)

8. ran (*travels*$^\sim$) $\cup$ ran (*owns*$^\sim$)

9. ran (*travels*$^\sim$) $\setminus$ ran (*owns*$^\sim$)

$\square$

Exercise 8.11 Assuming that R is an element of $X \leftrightarrow Y$, prove that

$$\text{dom } R^\sim = \text{ran } R$$

holds. $\square$

Exercise 8.12 Assuming that R is an element of $X \leftrightarrow Y$, prove that

$$R^{\sim\,\sim} = R$$

holds. $\square$

8.5   Operations on relations

As we have seen, given a relation $R \in X \leftrightarrow Y$ the domain of R is a subset of its source (in this case, the set X), and the range of R is a subset of its target (in this case, the set Y). As an example, the relation *travels* of Exercise 8.10 was of type *Person* $\leftrightarrow$ *Mode*, and was defined by

$$travels = \{\, (andy, walks), (jim, cycles), (jim, drives),$$
$$(jon, cycles), (jon, drives), (dave, cycles)\}$$

Here, the target of *travels* is given by the set $\{cycles, drives, flies, walks\}$, while the range is given by $\{cycles, drives, walks\}$. In this case, the range is a strict subset of the target. In addition, the domain of *travels*—the set $\{andy, dave, jim, jon\}$—is a strict subset of the source—the set $\{andy, dave, jim, jon, rick\}$.

It may be—for whatever reason—that we wish to consider only the modes of transport used by Jon and Jim. The *domain restriction* operator allows us to do exactly this. Given a relation R, together with some subset of the source $A \subseteq X$, the domain restriction $A \lhd R$ denotes the domain restriction of R to A: only those maplets originating from elements of A are considered; all others are thrown away. As such,

$$\{jim, jon\} \lhd travels = \{(jim, cycles), (jim, drives), (jon, cycles), (jon, drives)\}$$

Note that the domain restriction symbol is pointing to the left, indicating that it is the *domain* that we are restricting, rather than the *range*. Note also that whenever we write $A \lhd R$, A must be a set and R must be a relation.

We define domain restriction formally as follows (assuming that R is of type $X \leftrightarrow Y$).

$$A \lhd R = \{x : X; \ y : Y \mid x \in A \wedge x \mapsto y \in R \bullet (x, y)\}$$

An alternative definition, using the dot notation, is given below.

$$A \lhd R = \{z : R \mid z.1 \in A\}$$

Exercise 8.13 Assuming the relation *travels*, which is defined by

$$travels = \{(andy, walks), (jim, cycles), (jim, drives),$$
$$(jon, cycles), (jon, drives), (dave, cycles)\}$$

calculate the following.

1. $\emptyset \lhd travels$

2. $Person \lhd travels$

3. $\{andy\} \lhd travels$

4. $\{andy, dave\} \lhd travels$

5. $\{andy, dave, rick\} \lhd travels$

$\square$

Exercise 8.14 Assuming that R is an element of $X \leftrightarrow Y$ and that $A \subseteq X$, prove that

$$\operatorname{dom}(A \lhd R) = A \cap \operatorname{dom} R$$

holds. $\square$

The *domain co-restriction* operator is the complement of the domain restriction operator. Given a relation R, together with some subset of the source $A \subseteq X$, the domain co-restriction $A \ntriangleleft R$ denotes the domain co-restriction of R to A: those maplets originating from elements of A are thrown away; all others are kept. As such,

$$\{jim, jon\} \ntriangleleft travels = \{(andy, walks), (dave, cycles)\}$$

Again, the domain co-restriction symbol is pointing to the left, indicating that it is the domain that we are restricting. Also, in this case, the symbol has a 'minus' sign inside the triangle, indicating that it is the 'negative' co-restriction operator, rather than the 'positive' restriction operator.

We define domain co-restriction formally as follows (again, assuming that $R \in X \leftrightarrow Y$).

$$A \triangleleft\!\!\!- R = \{x : X; \ y : Y \mid x \notin A \wedge x \mapsto y \in R \bullet (x, y)\}$$

An alternative definition, using the dot notation, is given below.

$$A \triangleleft\!\!\!- R = \{z : R \mid z.1 \notin A\}$$

Exercise 8.15 Again, assuming the relation *travels*, calculate the following.

1. $\emptyset \triangleleft\!\!\!- travels$

2. $Person \triangleleft\!\!\!- travels$

3. $\{andy\} \triangleleft\!\!\!- travels$

4. $\{andy, dave\} \triangleleft\!\!\!- travels$

5. $\{andy, dave, rick\} \triangleleft\!\!\!- travels$

$\square$

Exercise 8.16 Assuming that R is an element of $X \leftrightarrow Y$ and $A \subseteq X$, prove that

$$\text{dom}\,(A \triangleleft\!\!\!- R) = \text{dom}\,R \setminus A$$

holds. $\square$

Continuing with our *travels* relation, we may only wish to consider those people in *travels* who cycle. The *range restriction* operator allows us to do this. Given a relation R, together with some subset of the target $B \subseteq Y$, $R \triangleright B$ denotes the range restriction of R to B: only those maplets pointing at elements of B are considered; all others are thrown away. Here,

$$travels \triangleright \{cycles\} = \{(jim, cycles), (jon, cycles), (dave, cycles)\}$$

Note that the range restriction symbol points to the right, indicating that it is the *range* that we are restricting. Note also that whenever we write $R \triangleright B$, R must be a relation and B must be a set.

We define range restriction formally as follows (again, assuming that R is of type $X \leftrightarrow Y$).

$$R \triangleright B = \{x : X; \ y : Y \mid y \in B \wedge x \mapsto y \in R \bullet (x, y)\}$$

An alternative definition, using the dot notation, is given below.

$$R \triangleright B = \{z : R \mid z.2 \in B\}$$

Exercise 8.17 Calculate the following.

1. $travels \rhd \emptyset$

2. $travels \rhd Mode$

3. $travels \rhd \{walks\}$

4. $travels \rhd \{walks, cycles\}$

□

Exercise 8.18 Assuming that R is an element of $X \leftrightarrow Y$ and $A \subseteq X$, prove that

$$(A \lhd R)^{\sim} = (R^{\sim} \rhd A)$$

holds. □

Finally, the *range co-restriction* operator is the complement of the range restriction operator. Given a relation R, together with some subset of the target $B \subseteq Y$, $R \rhd\!\!\!- B$ denotes the range co-restriction of R to B: those maplets pointing at elements of B are thrown away; all others are kept. Here,

$$travels \rhd\!\!\!- \{cycles\} = \{(andy, walks), (jim, drives), (jon, drives)\}$$

Again, the range co-restriction symbol points to the right, indicating that it is the *range* that we are restricting. Also, as was the case with domain co-restriction, this symbol has a 'minus sign' inside it, indicating that it is 'negative' co-restriction, rather than 'positive' restriction.

We define range co-restriction formally as follows (again, assuming that R is of type $X \leftrightarrow Y$).

$$R \rhd\!\!\!- B = \{x : X; \ y : Y \mid y \notin B \land x \mapsto y \in R \bullet (x, y)\}$$

An alternative definition, using the dot notation, is given below.

$$R \rhd\!\!\!- B = \{z : R \mid z.2 \notin B\}$$

Exercise 8.19 Calculate the following.

1. $travels \rhd\!\!\!- \emptyset$

2. $travels \rhd\!\!\!- Mode$

3. $travels \rhd\!\!\!- \{walks\}$

4. $travels \rhd\!\!\!- \{walks, cycles\}$

□

Exercise 8.20 Assuming that R is an element of $X \leftrightarrow Y$ and $A \subseteq Y$, prove that

$$(R \rhd\!\!\!- A)^{\sim} = (A \lhd R^{\sim})$$

holds. □

Having considered how we may restrict a relation to a certain subset of the domain or the range, it is only natural to wish to consider the *effect* of a particular relation on a subset of the domain. The *relational image* operator allows us to do exactly this. Given a relation R, together with some subset A of the domain, the relational image, $R (\!| \ A \ |\!)$, returns a set containing those elements of the *range* which are mapped to in R by those elements of A.

Example 8.9 The domain restriction $\{andy, dave\} \lhd travels$ gives the result

$$\{(andy, walks), (dave, cycles)\}$$

Correspondingly, we have

$$travels (\!| \ \{andy, dave\} \ |\!) = \{walks, cycles\}$$

☐

If we consider the potato diagram of *travels*, which is given below, the effect of the relational image operator becomes clear.

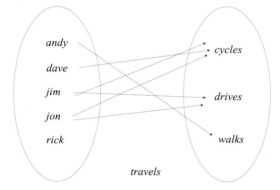

Here, if we consider only those arrows originating from *andy* or from *dave* and remove all others, then we are left with the following diagram.

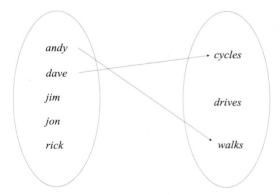

Thus, a domain restriction returns a subset of the relation (a set of pairs, which is itself a relation), whereas a relational image returns a subset of the range (a set of elements).

Example 8.10 The relation *showing* is defined as follows.

$$showing = \{bbc1 \mapsto news, bbc2 \mapsto documentary,$$
$$itv \mapsto game\_show, c4 \mapsto american\_comedy,$$
$$c5 \mapsto adult\_thriller\}$$

Here,

$$showing (\!| \{bbc1, bbc2\} |\!) = \{news, documentary\}$$

□

A formal definition of relational image (assuming that $R \in X \leftrightarrow Y$) is given as follows.

$$R (\!| A |\!) = \{x : X; \; y : Y \mid x \mapsto y \in R \land x \in A \bullet y\}$$

An alternative definition, using the dot notation, is given below.

$$R (\!| A |\!) = \{x : R \mid x.1 \in A \bullet x.2\}$$

Exercise 8.21 Recall that the relation *travels* was defined as

$$travels = \{(andy, walks), (jim, cycles), (jim, drives),$$
$$(jon, cycles), (jon, drives), (dave, cycles)\}$$

Calculate the following.

1. $travels (\!| \emptyset |\!)$
2. $travels (\!| \{rick\} |\!)$
3. $travels (\!| \{jon\} |\!)$
4. $travels (\!| \{jon, jim\} |\!)$
5. $travels (\!| Person |\!)$

□

Exercise 8.22 Assuming that R is an element of $X \leftrightarrow Y$ and $A \subseteq X$, prove that

$$R (\!| A |\!) = \operatorname{ran}(A \lhd R)$$

holds. □

8.6 Relational composition

In Section 8.4 we considered the relational inverse operator. There, given a relation R, its relational inverse, $R^{\sim}$, was effectively its mirror image. Then, in Section 8.5, we considered a number of operations on relations: range and domain restrictions and co-restrictions, as well as relational image. All of these operations are concerned with individual relations; the next operation which we consider allows us to combine two relations to produce a third relation which is defined in terms of the first two.

Given two relations R and S, we denote the *relational composition* of R and S by $R \, \overset{\circ}{,} \, S$.[1] Essentially, this operator allows us to 'glue' the two relations together to produce a new one. This new relation describes the overall effect of R, followed by S.

Example 8.11 Consider the following relations.

$$likes = \{karen \mapsto matt, matt \mapsto kerry\}$$
$$hobbies = \{karen \mapsto reading, karen \mapsto squash,$$
$$kerry \mapsto squash, kerry \mapsto tv,$$
$$matt \mapsto reading, matt \mapsto tv\}$$

Recall that the potato diagram for *likes* is given by

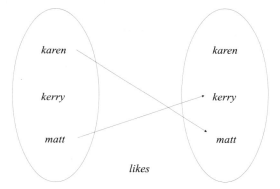

likes

Furthermore, the potato diagram for *hobbies* is given below.

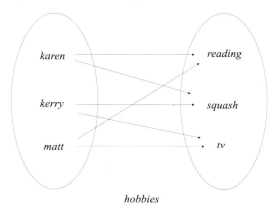

hobbies

The relational composition of these two relations effectively joins them together: the target of the first relation is combined with the source of the second relation; of course, the target of the first relation and the source of the second relation must be the same type for such a composition to

---

[1] In many discrete mathematics texts, the composition of relations $R \, \overset{\circ}{,} \, S$ is written $S \circ R$. We choose to use the notation $R \, \overset{\circ}{,} \, S$ as it is more evocative of the result of combining two relations using this operation: the effect of the composition is equivalent to R *followed* by S.

work. It is as if we draw the two potato diagrams side-by-side, and then pull them together so that the target of the first potato and the source of the second potato become one.

The combining of the relations *likes* and *hobbies* in this fashion gives the following.

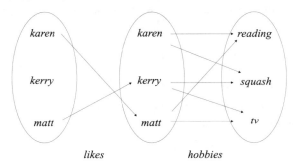

Here, we can see that the person who Matt likes counts playing squash and watching television among their hobbies: this is precisely the type of information that we wish such a relational composition to provide. Furthermore, the person that Karen likes counts reading and watching television among their hobbies.

Finally, as every relation has only one source and only one target, we require the result of a relational composition—which is itself a relation—to conform to these rules. As such, the 'middle potato' is hidden.

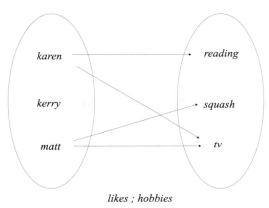

likes ; hobbies

Here, because *likes* ∈ *People* ↔ *People* and *hobbies* ∈ *People* ↔ *Hobby*, we may conclude that the relational composition *likes* ⨾ *hobbies* is of type *People* ↔ *Hobby*. This relation is given by

$$likes \, \mathbf{\S} \, hobbies = \{karen \mapsto reading, karen \mapsto tv, matt \mapsto squash, matt \mapsto tv\}$$

□

Note that in the above example, it would not have been possible to define the relational composition *hobbies* ⨾ *likes* because the target of *hobbies* and the source of *likes* are different types.

Formally, the relational composition of two relations $R \in X \leftrightarrow Y$ and $S \in Y \leftrightarrow Z$ is defined as follows.

$$R \, \mathbf{\S} \, S = \{x : X; \ y : Y; \ z : Z \mid x \mapsto y \in R \wedge y \mapsto z \in S \bullet (x, z)\}$$

An alternative definition, using the dot notation, is given below.

$$R \mathbin{\raise0.2ex\hbox{$\scriptstyle\circ$}} S = \{x : R;\ y : S \mid x.2 = y.1 \bullet (x.1, y.2)\}$$

Example 8.12 Recall that the relations *tv* and *showing* are defined as follows.

$$tv = \{becky \mapsto bbc1, emily \mapsto itv, becky \mapsto itv, john \mapsto c5\}$$
$$showing = \{bbc1 \mapsto news, bbc2 \mapsto documentary,$$
$$itv \mapsto game\_show, c4 \mapsto american\_comedy,$$
$$c5 \mapsto adult\_thriller\}$$

As such,

$$tv \mathbin{\raise0.2ex\hbox{$\scriptstyle\circ$}} showing = \{becky \mapsto news, emily \mapsto game\_show,$$
$$becky \mapsto game\_show, john \mapsto adult\_thriller\}$$

□

Exercise 8.23 Assume the following relations, such that $X \neq Y$.

$$A \in X \leftrightarrow X$$
$$B \in X \leftrightarrow Y$$
$$C \in Y \leftrightarrow X$$
$$D \in Y \leftrightarrow Y$$

Which of the following relational compositions are legal?

1. $A \mathbin{\raise0.2ex\hbox{$\scriptstyle\circ$}} A$

2. $B \mathbin{\raise0.2ex\hbox{$\scriptstyle\circ$}} B$

3. $C \mathbin{\raise0.2ex\hbox{$\scriptstyle\circ$}} D$

4. $D \mathbin{\raise0.2ex\hbox{$\scriptstyle\circ$}} C$

5. $D \mathbin{\raise0.2ex\hbox{$\scriptstyle\circ$}} D$

□

Exercise 8.24 Consider the following relations.

$$R = \{1 \mapsto 1, 2 \mapsto 2, 3 \mapsto 3\}$$
$$S = \{1 \mapsto 2, 2 \mapsto 3, 3 \mapsto 4\}$$
$$T = \{1 \mapsto 0, 2 \mapsto 0, 3 \mapsto 0\}$$

Calculate the following.

1. $R \mathbin{\raise0.2ex\hbox{$\scriptstyle\circ$}} R$

2. $R \mathbin{\raise0.2ex\hbox{$\scriptstyle\circ$}} S$

3. $R \mathbin{\raise0.2ex\hbox{$\scriptstyle\circ$}} T$

4. $S \,_9^\circ\, R$

5. $S \,_9^\circ\, S$

6. $S \,_9^\circ\, T$

□

Exercise 8.25 Consider the following relations.

$$pets = \{ dog \mapsto fido, dog \mapsto lassie,$$
$$cat \mapsto bramble, cow \mapsto daisy, goldfish \mapsto fish \}$$
$$noises = \{ fido \mapsto bark, fido \mapsto howl,$$
$$lassie \mapsto bark, bramble \mapsto miaow, daisy \mapsto moo \}$$

Calculate the following.

1. $pets \,_9^\circ\, noises$

2. $pets \,_9^\circ\, noises \,(\!|\; \{ dog \} \;|\!)$

3. $\{ dog, cat \} \lhd (pets \,_9^\circ\, noises)$

4. $(\{ dog, cat \} \lhd pets) \,_9^\circ\, noises$

□

Exercise 8.26 Assuming that R is an element of $X \leftrightarrow Y$ and S is an element of $Y \leftrightarrow Z$, prove that

$$\operatorname{dom} R \,_9^\circ\, S \subseteq \operatorname{dom} R$$

holds. □

Exercise 8.27 Assuming that R is an element of $X \leftrightarrow Y$ and S is an element of $Y \leftrightarrow Z$, prove that

$$X \lhd (R \,_9^\circ\, S) = (X \lhd R) \,_9^\circ\, S$$

holds. □

8.7 Homogeneous and heterogeneous relations

We have seen how relational composition allows us to combine two different relations in order to define a further one. There are, of course, circumstances under which we might wish to apply the same relation a number of times. For example, in Exercise 8.23 we considered the relational compositions $A \,_9^\circ\, A$ and $B \,_9^\circ\, B$. Given that $A \,_9^\circ\, A$ is itself a relation, there is no reason for us not to consider the compositions $A \,_9^\circ\, A \,_9^\circ\, A$ and $A \,_9^\circ\, A \,_9^\circ\, A \,_9^\circ\, A$.

As a further example, consider the relation *square*, which is defined as follows.

$$square = \{ 0 \mapsto 0, 1 \mapsto 1, 2 \mapsto 4, 3 \mapsto 9, \dots \}$$

We might define the relation *fourth*, which raises natural numbers to their fourth root, in terms of *square* in the following way.

$$fourth = square \,\text{\small ⨾}\, square$$
$$= \{0 \mapsto 0, 1 \mapsto 1, 2 \mapsto 16, 3 \mapsto 81, \ldots\}$$

Of course, the reason that we are able to compose *square* in this fashion is that its source and target are of the same type: it is said to be a *homogeneous* relation.

On the other hand, the relation *position*, which is given by

$$position = \{1 \mapsto a, 2 \mapsto b, 3 \mapsto c, \ldots\}$$

is not homogeneous: its source and target are of different types. As such, the result of the composition *position* ⨾ *position* is undefined. We refer to those relations which are not homogeneous as being *heterogeneous*. Thus, homogeneous relations are those that can be composed with themselves, whereas heterogeneous relations are those than can't be composed with themselves.

Exercise 8.28 Which of the following relations are homogeneous, and which are heterogeneous?

$$A = \{1 \mapsto 2, 2 \mapsto 3\}$$
$$B = \{\{1\} \mapsto 2, \{2\} \mapsto 3\}$$
$$C = \{1 \mapsto \{2\}, 2 \mapsto \{3\}\}$$
$$D = \{\{1\} \mapsto \{2\}, \{2\} \mapsto \{3\}\}$$

□

8.8   Properties of relations

There are several interesting properties of homogeneous relations which we may wish to reason about. We consider some of these—namely, reflexivity, transitivity, symmetry, asymmetry, anti-symmetry, and totality—in this section.

8.8.1   Reflexivity

A homogeneous relation R of type $X \leftrightarrow X$ is said to be *reflexive* if, and only if, *every* element of X is related to itself in R. That is to say, R is reflexive if, and only if,

$$\forall x : X \bullet (x, x) \in R$$

Example 8.13 Consider the relation *looks_like* $\in$ *People* $\leftrightarrow$ *People*, such that

$$People = \{jim, john, jack\}$$

The relation *looks_like* is illustrated by the following diagram.

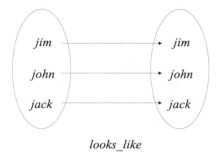

looks_like

Here, every element of the source is related to itself, and, as such, the relation is reflexive. □

Example 8.14 The relation = on the natural numbers is reflexive. Given any natural number $n \in \mathbb{N}$, it is always the case that $n = n$. □

Example 8.15 The relation < on the natural numbers is not reflexive. Given any natural number $n \in \mathbb{N}$, it is never the case that $n < n$. □

Example 8.16 The relation *loves* is not reflexive: it is not the case that every person loves him- or herself. □

Exercise 8.29 Which of the following relations (all of which are of type $\mathbb{N} \leftrightarrow \mathbb{N}$) are reflexive?

1. $\emptyset$

2. $\geq$

3. $\{1 \mapsto 1, 2 \mapsto 2, 3 \mapsto 3\}$

4. $\mathbb{N} \times \mathbb{N}$

□

Exercise 8.30 Which of the following relations (all of which are of type $\{a, b, c\} \leftrightarrow \{a, b, c\}$) are reflexive?

1. $\{a \mapsto a, b \mapsto b\}$

2. $\{a \mapsto a, b \mapsto a, b \mapsto b, c \mapsto c\}$

□

Exercise 8.31 Prove that, for any reflexive relation $R \in X \leftrightarrow X$, it is the case that

$$\{x : X \bullet (x, x)\} \subseteq R$$

holds. □

8.8.2 Transitivity

A homogeneous relation R of type $X \leftrightarrow X$ is said to be *transitive* if, and only if, given pairs (x, y) and (y, z), if (x, y) and (y, z) both appear in R then (x, z) must also appear in R. That is to say, R is transitive if, and only if,

$$\forall\, x, y, z : X \bullet (x, y) \in R \land (y, z) \in R \Rightarrow (x, z) \in R$$

Informally, we can think of a relation being transitive if it has the property that if (x, z) appears in $R \,{}^\circ_9\, R$, then (x, z) must also appear in R.

Example 8.17 Consider the relation *likes* which is defined on $\{jim, john, jack\}$.

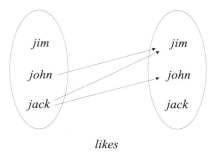

likes

Here, $jack \mapsto john$, $john \mapsto jim$, and $jack \mapsto jim$ are the elements of the relation. As we can map from $jack$ to jim (via $john$) in *likes* ${}^\circ_9$ *likes*, we require that $jack \mapsto jim$ appears in *likes*. It does, and, as there are no other indirect mappings, this relation is transitive. □

Example 8.18 Consider the relation RRN (for reduced rail network).

$\{birmingham \mapsto london, london \mapsto birmingham,$
$\quad birmingham \mapsto manchester, manchester \mapsto birmingham,$
$\quad london \mapsto manchester, manchester \mapsto london,$
$\quad london \mapsto ipswich, ipswich \mapsto london\}$

Here, one can travel from Birmingham to Ipswich via London (i.e., $birmingham \mapsto ipswich$ appears in $RRN \,{}^\circ_9\, RRN$), but one cannot travel from Birmingham to Ipswich directly (i.e., $birmingham \mapsto ipswich$ does not appear in RRN). As such, this relation is not transitive. □

Example 8.19 The relation $=$ on the natural numbers is transitive. Given any natural numbers, $l, m, n \in \mathbb{N}$, it is always the case that if $l = m$ and $m = n$ then $l = n$. □

Example 8.20 The relation $<$ on the natural numbers is transitive. Given any natural numbers, $l, m, n \in \mathbb{N}$, it is always the case that if $l < m$ and $m < n$ then $l < n$. □

Example 8.21 The relation *loves* is not transitive. For example, it does not necessarily follow that if Joe loves Mary and Mary loves Bob, then Joe loves Bob. □

Exercise 8.32 Which of the following relations (all of which are of type $\mathbb{N} \leftrightarrow \mathbb{N}$) are transitive?

1. $\emptyset$

2. $\geq$

3. $\{1 \mapsto 1, 2 \mapsto 2, 3 \mapsto 3\}$

4. $\mathbb{N} \times \mathbb{N}$

$\square$

Exercise 8.33 Which of the following relations (all of which are of type $\{a, b, c\} \leftrightarrow \{a, b, c\}$) are transitive?

1. $\{a \mapsto a, a \mapsto b, a \mapsto c, b \mapsto c\}$

2. $\{a \mapsto b, b \mapsto c\}$

$\square$

Exercise 8.34 Prove that, for any transitive relation $R \in X \leftrightarrow X$, it is the case that $R \, \S \, R \subseteq R$ holds. $\square$

8.8.3   Symmetry

A homogeneous relation R of type $X \leftrightarrow X$ is said to be *symmetric* if, and only if, it is the case that for every pair (x, y) that appears in R, the pair (y, x) also appears in R. That is to say, R is symmetric if, and only if,

$$\forall x, y : X \bullet (x, y) \in R \Rightarrow (y, x) \in R$$

Informally, we can think of a relation being symmetric if it guarantees that its 'mirror image' is contained in the relation, i.e., if x maps to y in R, then y must map to x in R.

Example 8.22 Consider the relation *hates*, which is defined as follows.

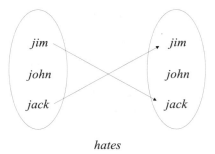

hates

Here, Jack hates Jim and Jim hates Jack. This, together with the fact that there are no other pairs contained in *hates*, means that the relation is symmetric. $\square$

Example 8.23 Consider again the relation RRN, which was defined by

$$RRN = \{birmingham \mapsto london, london \mapsto birmingham,$$
$$birmingham \mapsto manchester, manchester \mapsto birmingham,$$
$$london \mapsto manchester, manchester \mapsto london,$$
$$london \mapsto ipswich, ipswich \mapsto london\}$$

This relation is symmetric as, given any elements a and b, if it is possible to get from a to b in RRN then it is also possible to get from b to a in RRN. □

Example 8.24 Recall that the relation *likes* was given by

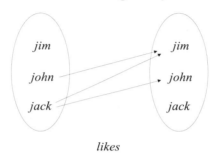

likes

This relation is not symmetric—although John likes Jim, it is not the case that Jim likes John. As such, the relation is not symmetric. □

Example 8.25 The relation = on the natural numbers is symmetric. Given any natural numbers, $m, n \in \mathbb{N}$, it is always the case that if $m = n$ then $n = m$. □

Example 8.26 The relation < on the natural numbers is not symmetric. Given any natural numbers, $m, n \in \mathbb{N}$, it is never the case that if $m < n$ then $n < m$. □

Example 8.27 The relation *loves* is not symmetric—it does not necessarily follow that if Joe loves Mary then Mary loves Joe. □

Exercise 8.35 Which of the following relations (all of which are of type $\mathbb{N} \leftrightarrow \mathbb{N}$) are symmetric?

1. $\emptyset$

2. $\geq$

3. $\{1 \mapsto 1, 2 \mapsto 2, 3 \mapsto 3\}$

4. $\mathbb{N} \times \mathbb{N}$

□

Exercise 8.36 Which of the following relations (all of which are of type $\{a, b, c\} \leftrightarrow \{a, b, c\}$) are symmetric?

1. $\{a \mapsto a, a \mapsto b, b \mapsto a\}$

2. $\{a \mapsto a, b \mapsto b, c \mapsto b\}$

□

Exercise 8.37 Prove that, for any symmetric relation $R \in X \leftrightarrow X$, it is the case that $R^{\sim} = R$.
□

8.8.4 Asymmetry

A homogeneous relation R of type $X \leftrightarrow X$ is said to be *asymmetric* if, and only if, it is the case that for any pair (x, y) appearing in R, (y, x) does not appear in R. That is to say, R is asymmetric if, and only if,

$$\forall x, y : X \bullet (x, y) \in R \Rightarrow (y, x) \notin R$$

Example 8.28 Consider the relation *older_than*.

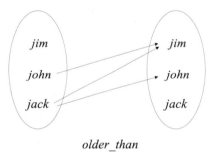

older_than

Here, Jack is older than both John and Jim, and John is older than Jim. This relation is clearly asymmetric. □

Example 8.29 The relation $=$ on the natural numbers is not asymmetric. Given any natural numbers, $m, n \in \mathbb{N}$, it is never the case that if $m = n$ then $n \neq m$. □

Example 8.30 The relation $<$ on the natural numbers is asymmetric. Given any natural numbers, $m, n \in \mathbb{N}$, it is always the case that if $m < n$ holds, then $n < m$ does not hold. □

Example 8.31 The relation *loves* is not asymmetric—it does not necessarily follow that if Joe loves Mary then Mary does not love Joe. □

Exercise 8.38 Which of the following relations (all of which are of type $\mathbb{N} \leftrightarrow \mathbb{N}$) are asymmetric?

1. $\emptyset$

2. $\geq$

3. $\{1 \mapsto 1, 2 \mapsto 2, 3 \mapsto 3\}$

4. $\mathbb{N} \times \mathbb{N}$

□

Exercise 8.39 Which of the following relations (all of which are of type $\{a, b, c\} \leftrightarrow \{a, b, c\}$) are asymmetric?

1. $\{a \mapsto a, a \mapsto b, b \mapsto a\}$

2. $\{a \mapsto b, b \mapsto c\}$

□

8.8.5 Antisymmetry

A homogeneous relation R of type $X \leftrightarrow X$ is said to be *antisymmetric* if, and only if, it is the case that for any pairs (x, y) and (y, x) that appear in R, then $x = y$. That is to say, R is antisymmetric if, and only if,

$$\forall x, y : X \bullet (x, y) \in R \land (y, x) \in R \Rightarrow x = y$$

Example 8.32 Consider again the relation *looks_like*.

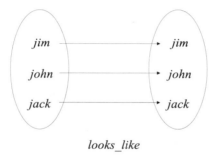

looks_like

This relation is antisymmetric, as every element which is contained in the source maps only to itself. □

Example 8.33 Consider again the relation RRN, which was defined by

$$RRN = \{birmingham \mapsto london, london \mapsto birmingham,$$
$$birmingham \mapsto manchester, manchester \mapsto birmingham,$$
$$london \mapsto manchester, manchester \mapsto london,$$
$$london \mapsto ipswich, ipswich \mapsto london\}$$

This relation is not antisymmetric. As an example, it is possible to get from Ipswich to London and also from London to Ipswich, yet both are clearly distinct elements. □

Example 8.34 The relation $=$ on the natural numbers is antisymmetric. Given any natural numbers, $m, n \in \mathbb{N}$, it is always the case that if $m = n$ and $n = m$ then $m = n$. □

Example 8.35 The relation $<$ on the natural numbers is antisymmetric. There are no natural numbers, $m, n \in \mathbb{N}$, such that $m < n$ and $n < m$. As such, the antecedent of the implication is equivalent to *false*, and so the overall implication is equivalent to *true*. □

Example 8.36 The relation *loves* is not antisymmetric—it does not necessarily follow that if Joe loves Mary and Mary loves Joe then Mary is the same person as Joe. □

Exercise 8.40 Which of the following relations (all of which are of type $\mathbb{N} \leftrightarrow \mathbb{N}$) are antisymmetric?

1. $\emptyset$
2. $\geq$
3. $\{1 \mapsto 1, 2 \mapsto 2, 3 \mapsto 3\}$
4. $\mathbb{N} \times \mathbb{N}$

□

Exercise 8.41 Which of the following relations (all of which are of type $\{a, b, c\} \leftrightarrow \{a, b, c\}$) are antisymmetric?

1. $\{a \mapsto a, a \mapsto b, b \mapsto a\}$
2. $\{a \mapsto b, b \mapsto c\}$

□

8.8.6 Totality

A homogeneous relation R of type $X \leftrightarrow X$ is said to be *total* if, and only if, it is the case that for every distinct $x, y \in X$, either $(x, y) \in R$ or $(y, x) \in R$. That is to say, R is total if, and only if,

$$\forall x, y : X \bullet (x, y) \in R \vee (y, x) \in R \vee x = y$$

Example 8.37 Consider again the relation *older_than*.

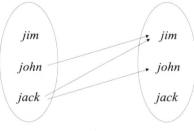

older_than

This relation is total, as, given any pair of elements x and y, it is the case that $x = y$ holds, or it is the case that $(x, y) \in older\_than$ holds, or it is the case that $(y, x) \in older\_than$ holds. □

Example 8.38 The relation RRN, which was given by

$$RRN = \{birmingham \mapsto london, london \mapsto birmingham,$$
$$birmingham \mapsto manchester, manchester \mapsto birmingham,$$
$$london \mapsto manchester, manchester \mapsto london,$$
$$london \mapsto ipswich, ipswich \mapsto london\}$$

is not total as it is not the case that for every pair of distinct elements m and n, $m \mapsto n \in RRN$ or $n \mapsto m \in RRN$. $\square$

Example 8.39 The relation $=$ on the natural numbers is not total. Given any natural numbers, $m, n \in \mathbb{N}$, it is not always the case that $m = n$. $\square$

Example 8.40 The relation $<$ on the natural numbers is total. Given any natural numbers, $m, n \in \mathbb{N}$, it is always the case that $m < n$ or $n < m$ or $m = n$. $\square$

Example 8.41 The relation *loves* is not total. Given any two distinct people, p and q, it is not always the case that p loves q or q loves p. $\square$

Exercise 8.42 Which of the following relations (all of which are of type $\mathbb{N} \leftrightarrow \mathbb{N}$) are total?

1. $\emptyset$

2. $\geq$

3. $\{1 \mapsto 1, 2 \mapsto 2, 3 \mapsto 3\}$

4. $\mathbb{N} \times \mathbb{N}$

$\square$

Exercise 8.43 Which of the following relations (all of which are of type $\{a, b, c\} \leftrightarrow \{a, b, c\}$) are antisymmetric?

1. $\{a \mapsto b, a \mapsto c, b \mapsto a\}$
2. $\{a \mapsto b, b \mapsto c, a \mapsto c, b \mapsto b\}$

$\square$

8.9 Orderings and equivalences

Different combinations of properties give rise to different classes of relations. We look at the three most common such classes of relations in this section: partial orderings, total orderings, and equivalence relations.

8.9.1 Partial orderings

Given a homogeneous relation R, if R is reflexive, transitive and antisymmetric, then we may say that R is a *partial ordering*. If a set X is *partially ordered*, then the elements of the set form some sort of 'hierarchy' (or order), in which some elements are higher than others. The ordering is partial in the sense that not all pairs of elements need to be related by the ordering.

Example 8.42 Consider the relation $\leq$ on the natural numbers. Here, $\leq$ is reflexive, as for every $n \in \mathbb{N}$, $n \leq n$. Furthermore, $\leq$ is transitive, as for every $l, m, n \in \mathbb{N}$, if $l \leq m$ and $m \leq n$ then it follows that $l \leq n$. Finally, $\leq$ is antisymmetric as for every $m, n \in \mathbb{N}$, if $m \leq n$ and $n \leq m$ then it must be the case that $m = n$. As such, we may conclude that $\leq$ is a partial ordering. $\square$

Exercise 8.44 The relation $<$ on the natural numbers is not a partial ordering. Why not? $\square$

Exercise 8.45 Prove that $\subseteq$ forms a partial ordering on the subsets of some set X. $\square$

8.9.2   Total orderings

Given a homogeneous relation R, if R is antisymmetric, transitive and total, then we may say that R is a *total ordering*. The ordering is total in the sense that all pairs of elements are related with respect to the ordering: given any two distinct elements x and y of a *totally ordered* set X, it is always the case that x is higher or lower in the order than y. That is, we may always compare two distinct elements in a totally ordered set, whereas this is not the case for a partially ordered set.

Example 8.43 Consider the relation $<$ on the natural numbers. Here, $<$ is antisymmetric, as for every $m, n \in \mathbb{N}$, if $m < n$ and $n < m$ then $m = n$ (the antecedent is never satisfied, so the implication is satisfied trivially). Furthermore, $<$ is transitive, as for every $l, m, n \in \mathbb{N}$, if $l < m$ and $m < n$ then it follows that $l < n$. Finally, $<$ is total, as for every $m, n \in \mathbb{N}$, it is the case that $m < n$ or $n < m$ or $m = n$. $\square$

Exercise 8.46 The relation $=$ on the natural numbers is not a total ordering. Why not? $\square$

Exercise 8.47 Prove that $\subseteq$ does not form a total ordering on the subsets of some set X. $\square$

8.9.3   Equivalence relations

Given a homogeneous relation R, if R is symmetric, transitive and reflexive, then we may say that R is an *equivalence relation*. If R is an equivalence relation, then if (x, y) appears in R then it follows that x is, in some sense, 'equivalent' to y.

Example 8.44 Consider the relation $=$ on the natural numbers. Here, $=$ is symmetric, as for every $m, n \in \mathbb{N}$, if $m = n$ then $m = n$. Furthermore, $=$ is transitive, as for every $l, m, n \in \mathbb{N}$, if $l = m$ and $m = n$ then it follows that $l = n$. Finally, $=$ is reflexive, as for every $n \in \mathbb{N}$, it is the case that $n = n$. $\square$

Exercise 8.48 The relation $\leq$ on the natural numbers is not an equivalence. Why not? $\square$

Exercise 8.49 Prove that $\subseteq$ does not form an equivalence on the subsets of some set X. $\square$

8.10   Closures

Given a homogeneous relation R which is not reflexive, it is always possible to extend R (if we so desire) by adding maplets to it in such a way that this extended relation is reflexive. In such cases, this extended relation—the smallest relation which both contains R and is reflexive—is called the *reflexive closure* of R. Similarly, we may add maplets to R so that this extended relation is symmetric. In such cases, the smallest relation which both contains R and is symmetric is called the *symmetric closure* of R. Finally, we may do the same for transitivity: the smallest relation which contains R and is transitive is called the *transitive closure* of R. We study reflexive, symmetric, and transitive closures in this section.

8.10.1   Reflexive closures

To form the reflexive closure of a relation S—which we shall denote S^R—we need to find the smallest relation which contains S as a subset and is reflexive.

Consider the identity relation Id, which, for some type X, is the set of all maplets of the form (x, x) such that $x \in X$.

$$Id\,[X] = \{x : X \bullet (x, x)\}$$

As an example,

$$Id\,[\{a, b, c\}] = \{(a, a), (b, b), (c, c)\}$$

Assuming that S is of type $Y \leftrightarrow Y$, the easiest way in which we may form the reflexive closure of S is to combine S—via set union—with $Id\,[Y]$.

$$S^R = S \cup Id\,[Y]$$

This is clearly the smallest reflexive relation containing S. Note that if S is already reflexive, then this definition ensures that $S^R = S$.

Example 8.45 Consider the set $People = \{john, richard, duncan, carsten\}$, together with the following relation, $older\_than$.

$$older\_than = \{john \mapsto richard, john \mapsto duncan, richard \mapsto duncan\}$$

The reflexive closure of this relation, i.e., the smallest relation containing $older\_than$ which is reflexive, is given by the union of $older\_than$ and $Id\,[People]$. By definition, the relation $Id\,[People]$ is given by

$$Id\,[People] = \{john \mapsto john, duncan \mapsto duncan, richard \mapsto richard, carsten \mapsto carsten\}$$

Thus, $older\_than^R$ is given by

$$\begin{aligned}
older\_than^R = \{ & john \mapsto richard, john \mapsto duncan, \\
& richard \mapsto duncan, john \mapsto john, \\
& richard \mapsto richard, duncan \mapsto duncan, carsten \mapsto carsten\}
\end{aligned}$$

□

Exercise 8.50 Assume the set $Pet = \{cat, dog, mouse\}$. Give the reflexive closure of each of the following relations of type $Pet \leftrightarrow Pet$.

1. $\emptyset$

2. $\{cat \mapsto dog, dog \mapsto mouse\}$

3. $\{cat \mapsto dog, mouse \mapsto dog\}$

4. $\{cat \mapsto cat, dog \mapsto dog, mouse \mapsto mouse\}$

□

8.10.2 Transitive closures

To form the transitive closure of a relation R—which we shall denote R^+—we need to find the smallest relation containing R as a subset which is transitive.

Recall the relational composition operator, $R \, \S \, R$. This had the same effect as applying R twice: a pair (x, z) appears in $R \S R$ exactly when there is some y, such that (x, y) appears in R and (y, z) appears in R. An alternative notation for relational composition is R^2, which denotes that the relation R has been applied twice. By extension, the relation R^3 represents the application of R three times: a pair (w, z) appears in $R \S R \S R$ exactly when there are elements x and y, such that (w, x), (x, y) and (y, z) all appear in R. More generally, R^n denotes the application of R n times. This is illustrated below.

$$R^1 = R$$
$$R^2 = R \, \S \, R$$
$$R^3 = R \, \S \, R \, \S \, R$$
$$R^4 = R \, \S \, R \, \S \, R \, \S \, R$$

Given the above, we can define the relation R^+—the transitive closure of R—as follows.

$$R^+ = \bigcup \{n : \mathbb{N} \setminus \{0\} \bullet R^n\}$$

This denotes the set of all pairs that appear in R, or, $R \S R$, or $R \S R \S R$, etc.

Example 8.46 Consider the set *People* = {*john, richard, duncan, carsten*}, together with the following relation, *likes*.

$$likes = \{john \mapsto richard, richard \mapsto duncan, duncan \mapsto carsten\}$$

Here, $likes^2$ is given by

$$likes^2 = \{john \mapsto duncan, richard \mapsto carsten\}$$

Performing the relational composition of $likes^2$ and *likes* gives $likes^3$, which is given by

$$likes^3 = \{john \mapsto carsten\}$$

Applying *likes* again gives $likes^4$, which is given by

$$likes^4 = \emptyset$$

Given the above, we may conclude that $likes^k$, for any $k > 3$, is equivalent to the empty set. As such, any further applications of *likes* will not contribute any new pairs to $likes^+$. Therefore,

$$likes^+ = \{john \mapsto richard, richard \mapsto duncan, duncan \mapsto carsten,$$
$$john \mapsto duncan, richard \mapsto carsten, john \mapsto carsten\}$$

□

Exercise 8.51 Assume the set $Pet = \{cat, dog, mouse\}$. Give the transitive closure of each of the following relations of type $Pet \leftrightarrow Pet$.

1. $\emptyset$

2. $\{cat \mapsto dog, dog \mapsto mouse\}$

3. $\{cat \mapsto dog, mouse \mapsto dog\}$

4. $\{cat \mapsto cat, dog \mapsto dog, mouse \mapsto mouse\}$

$\square$

Exercise 8.52 Give the transitive closure of the following relation.

$$\{a \mapsto b, b \mapsto c, c \mapsto d, d \mapsto e\}$$

$\square$

8.10.3 Reflexive transitive closures

To form the reflexive transitive closure of a relation R—which we shall denote R^*—we need to find the smallest relation containing R as a subset which is both transitive and reflexive.

Assuming that R is of type $X \leftrightarrow X$, we can define R^* as follows.

$$R^* = \bigcup\{n : \mathbb{N} \bullet R^n\}$$

Recall that the relation $Id[X]$ is defined for some set X by

$$Id[X] = \{x : X \bullet x \mapsto x\}$$

As such, we assume that $R^0 = Id[X]$.

Example 8.47 Consider the set $People = \{john, richard, duncan, carsten\}$, together with the following relation, $likes$.

$$likes = \{john \mapsto richard, richard \mapsto duncan, duncan \mapsto carsten\}$$

Here, $Id[People]$ is given by

$$Id[People] = \{john \mapsto john, richard \mapsto richard,$$
$$duncan \mapsto duncan, carsten \mapsto carsten\}$$

Furthermore, $likes^+$ is given by

$$likes^+ = \{john \mapsto richard, richard \mapsto duncan, duncan \mapsto carsten,$$
$$john \mapsto duncan, richard \mapsto carsten, john \mapsto carsten\}$$

As such, *likes** is given by

$$likes^* = \{john \mapsto richard, richard \mapsto duncan, duncan \mapsto carsten,$$
$$john \mapsto duncan, richard \mapsto carsten, john \mapsto carsten,$$
$$john \mapsto john, richard \mapsto richard, duncan \mapsto duncan,$$
$$carsten \mapsto carsten\}$$

□

Exercise 8.53 Assume the set $Pet = \{cat, dog, mouse\}$. Give the reflexive transitive closure of each of the following relations of type $Pet \leftrightarrow Pet$.

1. $\emptyset$
2. $\{cat \mapsto dog, dog \mapsto mouse\}$
3. $\{cat \mapsto dog, mouse \mapsto dog\}$
4. $\{cat \mapsto cat, dog \mapsto dog, mouse \mapsto mouse\}$

□

8.10.4   Symmetric closures

To form the symmetric closure of a relation R—which we shall denote R^S—we need to find the smallest relation which contains R as a subset and is symmetric.

Recall that a relation $R \in X \leftrightarrow X$ is symmetric if, and only if, for any elements $x, y \in X$, if $(x, y) \in R$ then $(y, x) \in R$.

Recall also that, given any relation R of type $X \leftrightarrow X$, we may define the inverse of R, denoted $R^\sim$, by

$$R^\sim = \{x, y : X \mid (x, y) \in R \bullet (y, x)\}$$

As such, the smallest relation containing R which is symmetric is given by

$$R^S = R \cup R^\sim$$

Example 8.48 Consider again the set $People = \{john, richard, duncan, carsten\}$, together with the relation *likes*, which is defined by

$$likes = \{john \mapsto richard, richard \mapsto duncan, duncan \mapsto carsten\}$$

The relational inverse of *likes* is given by

$$likes^\sim = \{richard \mapsto john, duncan \mapsto richard, carsten \mapsto duncan\}$$

As such, $likes^S$ is given by

$$likes^S = \{john \mapsto richard, richard \mapsto duncan, duncan \mapsto carsten$$
$$richard \mapsto john, duncan \mapsto richard, carsten \mapsto duncan\}$$

□

Exercise 8.54 Assume the set $Pet = \{cat, dog, mouse\}$. Give the symmetric closure of each of the following relations of type $Pet \leftrightarrow Pet$.

1. $\emptyset$

2. $\{cat \mapsto dog, dog \mapsto mouse\}$

3. $\{cat \mapsto dog, mouse \mapsto dog\}$

4. $\{cat \mapsto cat, dog \mapsto dog, mouse \mapsto mouse\}$

□

8.11   n-ary relations

Recall that in Chapter 6 it was argued that sometimes pairs may not be a natural representation of the data with which we are concerned; sometimes larger tuples, such as triples or quadruples may be more appropriate. For example, the Cartesian product $Name \times Age \times Email$ allowed us to represent the set of all tuples of the form (n, a, e), such that $n \in Name$, $a \in Age$, and $e \in Email$.

Then, at the start of this chapter, it was argued that a relation is a subset of a Cartesian product. As our relational operators rely on relations being binary, we are forced to place a restriction on our definition of relations: relations are strictly sets of *pairs*, rather than sets of tuples of any size.

Returning to our example of names, ages, and email addresses, we find that it is illegal for us to define a relation of type $Name \leftrightarrow Age \leftrightarrow Email$. We could define a relation of type $Name \leftrightarrow (Age \leftrightarrow Email)$, but a typical element of this relation would be

$$\{(jack, \{(65, jack@the\_hill.com), (60, jill@the\_hill.com)\})\}$$

which, of course, is not what we require.

On the other hand, relations of type $Name \leftrightarrow (Age \times Email)$ and $(Name \times Age) \leftrightarrow Email$ are perfectly legitimate; the former consists of elements of the form $(n, (a, e))$, while the latter consists of elements of the form $((n, a), e)$. Both of these relations fulfil the requirements of binary relations, and allow us to represent the information with which we are concerned.

Exercise 8.55 Consider the following sets.

$$Yes\_No = \{yes, no\}$$
$$Name = \{dave, jim, jon\}$$

Give the type of each of the following relations.

1. $\{(dave, yes), (jim, no)\}$

2. $\{((dave, yes), yes), ((jim, no), no)\}$

3. $\{((yes, yes), (dave, yes)), ((no, yes), (dave, no))\}$

4. $\{(yes, (dave, jim)), (no, (dave, jon))\}$

□

Exercise 8.56 Assuming the relation R, which is defined by

$$R = \{((1,1),(0,1)),((1,0),(0,0)),((0,1),(1,0)),((0,0),(1,1))\}$$

calculate the following.

1. dom R

2. ran R

3. $\{(1,1)\} \lhd R$

4. $R^{\sim}$

5. $R \,\substack{\circ \\ \circ}\, R$

6. $\{r : R \bullet (r.1.1, r.2)\}$

$\square$

8.12   Further exercises

Exercise 8.57   Assume the following relations.

$$parent = \{p, q : Person \mid p \text{ is a parent of } q\}$$
$$sibling = \{p, q : Person \mid p \text{ is a brother or sister of } q\}$$
$$married = \{p, q : Person \mid p \text{ is married to } q\}$$

Assume also the predicates *male* and *female*, such that *male* (p) is equivalent to *true* if, and only if, p is male, and *female* (p) is equivalent to *true* if, and only if, p is female.

Given the above, define the following relations.

1. *husband*, such that $(p, q) \in husband$ denotes the fact that p is q's husband.

2. *wife*, such that $(p, q) \in wife$ denotes the fact that p is q's wife.

3. *mother*, such that $(p, q) \in mother$ denotes the fact that p is q's mother.

4. *father*, such that $(p, q) \in father$ denotes the fact that p is q's father.

5. *son*, such that $(p, q) \in son$ denotes the fact that p is q's son.

6. *daughter*, such that $(p, q) \in daughter$ denotes the fact that p is q's daughter.

7. *brother*, such that $(p, q) \in brother$ denotes the fact that p is q's brother.

8. *sister*, such that $(p, q) \in sister$ denotes the fact that p is q's sister.

$\square$

Exercise 8.58 Given the sets and predicates of the previous exercise, provide definitions for the following relations.

1. *aunt*, such that $(p, q) \in aunt$ denotes the fact that p is q's aunt.

2. *uncle*, such that $(p, q) \in uncle$ denotes the fact that p is q's uncle.

3. *nephew*, such that $(p, q) \in nephew$ denotes the fact that p is q's nephew.

4. *niece*, such that $(p, q) \in niece$ denotes the fact that p is q's niece.

5. *cousin*, such that $(p, q) \in cousin$ denotes the fact that p is q's cousin.

6. *father_in_law*, such that $(p, q) \in father\_in\_law$ denotes the fact that p is q's father-in-law. This relationship holds if p is the father of q's spouse.

7. *mother_in_law*, such that $(p, q) \in mother\_in\_law$ denotes the fact that p is q's mother-in-law. This relationship holds if p is the mother of q's spouse.

8. *son_in_law*, such that $(p, q) \in son\_in\_law$ denotes the fact that p is q's son-in-law. This relationship holds if p is the husband of q's daughter.

9. *daughter_in_law*, such that $(p, q) \in daughter\_in\_law$ denotes the fact that p is q's daughter-in-law. This relationship holds if p is the wife of q's son.

□

Exercise 8.59 Given the sets and predicates of Exercise 8.57, define the following relations.

1. *stepfather*, such that $(p, q) \in stepfather$ denotes the fact that p is q's stepfather.

2. *stepmother*, such that $(p, q) \in stepmother$ denotes the fact that p is q's stepmother.

3. *stepdaughter*, such that $(p, q) \in stepdaughter$ denotes the fact that p is q's stepdaughter.

4. *stepson*, such that $(p, q) \in stepson$ denotes the fact that p is q's stepson.

5. *stepbrother*, such that $(p, q) \in stepbrother$ denotes the fact that p is q's stepbrother.

6. *stepsister*, such that $(p, q) \in stepsister$ denotes the fact that p is q's stepsister.

7. *grandmother*, such that $(p, q) \in grandmother$ denotes the fact that p is q's grandmother.

8. *grandfather*, such that $(p, q) \in grandfather$ denotes the fact that p is q's grandfather.

□

Exercise 8.60 Given the relations of the previous few questions, define the following in terms of the above relations and our relational operators.

1. The relation *child*, such that $(p, q) \in child$ if, and only if, p is a child of q.

2. The set of grandchildren of a given person, p.

3. The set of great grandparents of a given person, p.

4. The set of ancestors of a given person, p.

5. The set of all mothers.

6. The set of all fathers who have daughters.

7. The set of all males who have brothers.

8. The set of all grandmothers whose grandchildren are all girls.

9. The set of all grandfathers whose children are all girls.

□

Exercise 8.61 Assume the relations R and S, which are defined as follows.

$$R = \{(1,2), (2,4), (3,6), (4,8)\}$$
$$S = \{(1,1), (2,2), (3,3), (4,4)\}$$

You may assume that both of these relations are of type $\mathbb{N} \leftrightarrow \mathbb{N}$.

1. Define the elements R and S using set comprehension.

2. What is dom R?

3. What is ran S?

4. What is $R^\sim$?

5. What is $R \, _9^\circ \, S$?

6. What is $S \, _9^\circ \, R$?

□

Exercise 8.62 Assume the following relations.

$$id = \{1 \mapsto 1, 2 \mapsto 2, 3 \mapsto 3, 4 \mapsto 4\}$$
$$plus\_one = \{1 \mapsto 2, 2 \mapsto 3, 3 \mapsto 4, 4 \mapsto 5\}$$
$$square = \{1 \mapsto 1, 2 \mapsto 4, 3 \mapsto 9, 4 \mapsto 16\}$$
$$double = \{1 \mapsto 2, 2 \mapsto 4, 3 \mapsto 6, 4 \mapsto 8\}$$

Assuming that each of the above relations is of type $\mathbb{N} \leftrightarrow \mathbb{N}$, calculate each of the following.

1. dom id

2. ran $square$

3. $double^\sim$

4. $double \, _9^\circ \, square$

5. $square \, _9^\circ \, double$

6. $\{1,2\} \lhd plus\_one$

7. $\{1,2\} \vartriangleleft plus\_one$

8. $plus\_one \rhd \{1,2\}$

9. $plus\_one \vartriangleright \{1,2\}$

10. $square (\!| \, \{1,2\} \, |\!)$

11. $double (\!| \, \{1,2\} \, |\!)$

12. *double* $^\sim$ ({1, 2})

□

Exercise 8.63 Assume the relation *loves* ∈ *Housemates* ↔ *Housemates*, which is defined by

$$loves = \{jackie \mapsto jackie, brian \mapsto jackie\}$$

Assume further that *Housemates* = {*brian, jackie*}.

1. Is *loves* reflexive?
2. Is *loves* symmetric?
3. Is *loves* transitive?

□

Exercise 8.64 Assume the relation *loves* of the previous exercise.

1. Give the reflexive closure of *loves*.
2. Give the symmetric closure of *loves*.

□

Exercise 8.65 Consider the following relation.

$$noises = \{fido \mapsto bark, fido \mapsto howl,$$
$$lassie \mapsto bark, bramble \mapsto miaow, daisy \mapsto moo\}$$

Given the above definition of *noises*, calculate the following.

1. {*fido*} ◁ *noises*
2. {*fido, lassie*} ◁ *noises*
3. {*daisy*} ◁ *noises*
4. {*daisy, bramble*} ◁ *noises*
5. *noises* ▷ {*bark*}
6. *noises* ▷ {*howl, moo*}
7. *noises* ▷ {*miaow*}
8. *noises* ▷ {*bark, miaow*}
9. *noises* ({*fido*})
10. *noises* (∅)

□

Exercise 8.66 Which of the following relations are transitive or symmetric?

1. $\emptyset$

2. $\{(jim, joe), (joe, bill), (jim, bill), (steve, rick)\}$

3. $\{(jim, joe), (joe, bill), (jim, bill)\}$

4. $\{(rick, steve), (steve, rick), (jim, bill), (bill, jim)\}$

☐

Exercise 8.67 Give the symmetric closures of the relations of Exercise 8.66. ☐

Exercise 8.68 Give the transitive closures of the relations of Exercise 8.66. ☐

Exercise 8.69 Which of the following statements are true?

1. If R and S are reflexive, then so is $R \mathbin{\overset{\circ}{,}} S$.

2. If R and S are transitive, then so is $R \mathbin{\overset{\circ}{,}} S$.

3. If R and S are symmetric, then so is $R \mathbin{\overset{\circ}{,}} S$.

☐

Exercise 8.70 Explain why $\Leftrightarrow$ is an equivalence relation on the set of all propositions. ☐

Exercise 8.71 Can $(R^+)^+$ be simplified? ☐

8.13 Solutions

Solution 8.1

$$\{\emptyset, \{0 \mapsto x\}, \{0 \mapsto y\}, \{1 \mapsto x\}, \{1 \mapsto y\}, \{0 \mapsto x, 0 \mapsto y\},$$
$$\{0 \mapsto x, 1 \mapsto x\}, \{0 \mapsto x, 1 \mapsto y\}, \{1 \mapsto x, 0 \mapsto y\}, \{1 \mapsto x, 1 \mapsto y\},$$
$$\{0 \mapsto y, 1 \mapsto y\}, \{0 \mapsto x, 0 \mapsto y, 1 \mapsto x\}, \{0 \mapsto x, 0 \mapsto y, 1 \mapsto y\},$$
$$\{0 \mapsto x, 1 \mapsto x, 1 \mapsto y\}, \{0 \mapsto y, 1 \mapsto x, 1 \mapsto y\},$$
$$\{0 \mapsto x, 0 \mapsto y, 1 \mapsto x, 1 \mapsto y\}\}$$

☐

Solution 8.2

1.

$$\begin{array}{|l}
\hline
equals : \mathbb{N} \leftrightarrow \mathbb{N} \\
\hline
equals = \{m : \mathbb{N} \bullet (m, m)\} \\
\end{array}$$

2.

> $looks\_like : People \leftrightarrow People$
> ___
> $looks\_like = \{p, q : People \mid p \text{ looks like } q \bullet (p, q)\}$

3.

> $likes : People \leftrightarrow Food$
> ___
> $likes = \{p : People;\ f : Food \mid p \text{ likes } f \bullet (p, f)\}$

4.

> $double\_of\_prime : \mathbb{N} \leftrightarrow \mathbb{N}$
> ___
> $double\_of\_prime = \{n : \mathbb{N} \mid prime(n) \bullet (2 \times n, n)\}$

□

Solution 8.3

$$(x, y) \in \{m : \mathbb{N} \bullet (m, m)\} \Leftrightarrow x = y$$
$$\Leftrightarrow (x, y) \in \{m, n : \mathbb{N} \mid m = n\}$$

□

Solution 8.4

1. domain = $\{1, 2, 3\}$
    range = $\{2, 3, 4\}$

2. domain = $\{0, 1, 2, 3, 4, 5, 6, 7, 8, 9, 10\}$
    range = $\{0, 2, 4, 6, 8, 10, 12, 14, 16, 18, 20\}$

3. domain = $\mathbb{N}$
    range = $\mathbb{N}$

4. domain = $\mathbb{N}$
    range = $\mathbb{N} \setminus \{0\}$

5. domain = $\emptyset$
    range = $\emptyset$

□

Solution 8.5

$$x \in \mathrm{dom}\,(R \cup S)$$
$$\Leftrightarrow \exists\, y : Y \bullet x \mapsto y \in R \cup S$$
$$\Leftrightarrow \exists\, y : Y \bullet x \mapsto y \in R \vee x \mapsto y \in S$$
$$\Leftrightarrow \exists\, y : Y \bullet x \mapsto y \in R \vee \exists\, y : Y \bullet x \mapsto y \in S$$
$$\Leftrightarrow x \in \mathrm{dom}\,R \vee x \in \mathrm{dom}\,S$$
$$\Leftrightarrow x \in (\mathrm{dom}\,R) \cup (\mathrm{dom}\,S)$$

□

Solution 8.6

$$y \in \mathrm{ran}\,(R \cup S)$$
$$\Leftrightarrow \exists\, x : X \bullet x \mapsto y \in R \cup S$$
$$\Leftrightarrow \exists\, x : X \bullet x \mapsto y \in R \vee x \mapsto y \in S$$
$$\Leftrightarrow \exists\, x : X \bullet x \mapsto y \in R \vee \exists\, x : X \bullet x \mapsto y \in S$$
$$\Leftrightarrow y \in \mathrm{ran}\,R \vee y \in \mathrm{ran}\,S$$
$$\Leftrightarrow y \in (\mathrm{ran}\,R) \cup (\mathrm{ran}\,S)$$

□

Solution 8.7

$$y \in \mathrm{ran}\,(R \cap \emptyset)$$
$$\Leftrightarrow \exists\, x : X \bullet x \mapsto y \in R \cap \emptyset$$
$$\Leftrightarrow \exists\, x : X \bullet x \mapsto y \in R \wedge x \mapsto y \in \emptyset$$
$$\Leftrightarrow \exists\, x : X \bullet x \mapsto y \in R \wedge \mathit{false}$$
$$\Leftrightarrow \mathit{false}$$
$$\Leftrightarrow \exists\, x : X \bullet x \mapsto y \in S \wedge \mathit{false}$$
$$\Leftrightarrow \exists\, x : X \bullet x \mapsto y \in S \wedge x \mapsto y \in \emptyset$$
$$\Leftrightarrow \exists\, x : X \bullet x \mapsto y \in S \cap \emptyset$$
$$\Leftrightarrow y \in \mathrm{ran}\,(S \cap \emptyset)$$

□

Solution 8.8

1. $\emptyset$

2. $\{(1,1),(2,2)\}$

3. $\{(2,1),(3,2),(4,3)\}$

□

Solution 8.9

1.

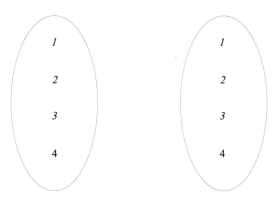

2.

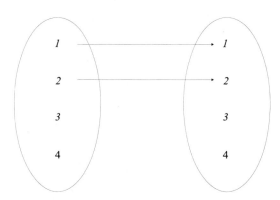

3.

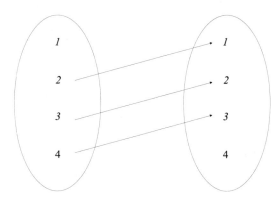

□

Solution 8.10

1. $\{(dog, andy), (parrot, rick), (goldfish, jim), (cat, jon)\}$

2. $\{(walks, andy), (cycles, jim), (drives, jim),$
   $(cycles, jon), (drives, jon), (cycles, dave)\}$

3. $\{dog, parrot, goldfish, cat\}$

4. $\{andy, rick, jim, jon\}$

5. $\{walks, cycles, drives\}$

6. $\{andy, dave, jim, jon\}$

7. $\{andy, jim, jon\}$

8. $\{andy, dave, jim, jon, rick\}$

9. $\{dave\}$

$\square$

Solution 8.11

$$x \in \operatorname{dom} R^{\sim}$$
$$\Leftrightarrow \exists\, y : Y \bullet x \mapsto y \in R^{\sim}$$
$$\Leftrightarrow \exists\, y : Y \bullet y \mapsto x \in R$$
$$\Leftrightarrow x \in \operatorname{ran} R$$

$\square$

Solution 8.12

$$(x, y) \in R^{\sim\sim}$$
$$\Leftrightarrow (y, x) \in R^{\sim}$$
$$\Leftrightarrow (x, y) \in R$$

$\square$

Solution 8.13

1. $\emptyset$

2. $\{(andy, walks), (jim, cycles), (jim, drives),$
   $(jon, cycles), (jon, drives), (dave, cycles)\}$

3. $\{(andy, walks)\}$

4. $\{(andy, walks), (dave, cycles)\}$

5. $\{(andy, walks), (dave, cycles)\}$

$\square$

Solution 8.14

$$x \in \mathrm{dom}\,(A \lhd R)$$
$$\Leftrightarrow \exists\, y : Y \bullet x \mapsto y \in A \lhd R$$
$$\Leftrightarrow x \in A \wedge \exists\, y : Y \bullet x \mapsto y \in R$$
$$\Leftrightarrow x \in A \wedge x \in \mathrm{dom}\,R$$
$$\Leftrightarrow x \in A \cap \mathrm{dom}\,R$$

$\square$

Solution 8.15

1. $\{(andy, walks), (jim, cycles), (jim, drives),$
   $(jon, cycles), (jon, drives), (dave, cycles)\}$

2. $\emptyset$

3. $\{(jim, cycles), (jim, drives), (jon, cycles), (jon, drives), (dave, cycles)\}$

4. $\{(jim, cycles), (jim, drives), (jon, cycles), (jon, drives)\}$

5. $\{(jim, cycles), (jim, drives), (jon, cycles), (jon, drives)\}$

$\square$

Solution 8.16

$$x \in \mathrm{dom}\,(A \ntriangleleft R)$$
$$\Leftrightarrow \exists\, y : Y \bullet x \mapsto y \in A \ntriangleleft R$$
$$\Leftrightarrow \exists\, y : Y \bullet x \notin A \wedge x \mapsto y \in R$$
$$\Leftrightarrow \exists\, y : Y \bullet x \mapsto y \in R \wedge x \notin A$$
$$\Leftrightarrow (\exists\, y : Y \bullet x \mapsto y \in R) \wedge x \notin A$$
$$\Leftrightarrow x \in \mathrm{dom}\,R \wedge x \notin A$$
$$\Leftrightarrow x \in \mathrm{dom}\,R \setminus A$$

$\square$

Solution 8.17

1. $\emptyset$

2. $\{(andy, walks), (jim, cycles), (jim, drives),$
   $(jon, cycles), (jon, drives), (dave, cycles)\}$

3. $\{(andy, walks)\}$

4. $\{(andy, walks), (jim, cycles), (jon, cycles), (dave, cycles)\}$

$\square$

Solution 8.18

$$(x, y) \in (A \lhd R)^{\sim}$$
$$\Leftrightarrow (y, x) \in A \lhd R$$
$$\Leftrightarrow y \in A \land (y, x) \in R$$
$$\Leftrightarrow (y, x) \in R \land y \in A$$
$$\Leftrightarrow (x, y) \in R^{\sim} \land y \in A$$
$$\Leftrightarrow (x, y) \in (R^{\sim} \rhd A)$$

□

Solution 8.19

1. $\{(andy, walks), (jim, cycles), (jim, drives),$
   $(jon, cycles), (jon, drives), (dave, cycles)\}$

2. $\emptyset$

3. $\{(jim, cycles), (jim, drives), (jon, cycles), (jon, drives), (dave, cycles)\}$

4. $\{(jim, drives), (jon, drives)\}$

□

Solution 8.20

$$(x, y) \in (R \rhd A)^{\sim}$$
$$\Leftrightarrow (y, x) \in R \rhd A$$
$$\Leftrightarrow (y, x) \in R \land x \notin A$$
$$\Leftrightarrow (x, y) \in R^{\sim} \land x \notin A$$
$$\Leftrightarrow (x, y) \in (A \lhd R^{\sim})$$

□

Solution 8.21

1. $\emptyset$

2. $\emptyset$

3. $\{drives, cycles\}$

4. $\{drives, cycles\}$

5. $\{drives, cycles, walks\}$

□

Solution 8.22

$$y \in R (\!| A |\!)$$
$$\Leftrightarrow \exists x : X \bullet x \mapsto y \in R \land x \in A$$
$$\Leftrightarrow \exists x : X \bullet x \mapsto y \in (A \lhd R)$$
$$\Leftrightarrow y \in \mathrm{ran}\,(A \lhd R)$$

$\square$

Solution 8.23 1, 4 and 5 are legal, and 2 and 3 are illegal. $\square$

Solution 8.24

1. $\{1 \mapsto 1, 2 \mapsto 2, 3 \mapsto 3\}$

2. $\{1 \mapsto 2, 2 \mapsto 3, 3 \mapsto 4\}$

3. $\{1 \mapsto 0, 2 \mapsto 0, 3 \mapsto 0\}$

4. $\{1 \mapsto 2, 2 \mapsto 3\}$

5. $\{1 \mapsto 3, 2 \mapsto 4\}$

6. $\{1 \mapsto 0, 2 \mapsto 0\}$

$\square$

Solution 8.25

1. $\{dog \mapsto bark, dog \mapsto howl, cat \mapsto miaow, cow \mapsto moo\}$

2. $\{bark, howl\}$

3. $\{dog \mapsto bark, dog \mapsto howl, cat \mapsto miaow\}$

4. $\{dog \mapsto bark, dog \mapsto howl, cat \mapsto miaow\}$

$\square$

Solution 8.26

$$x \in \mathrm{dom}\, R \,\mathring{\,}\, S$$
$$\Leftrightarrow \exists z : Z \bullet x \mapsto z \in R \,\mathring{\,}\, S$$
$$\Leftrightarrow \exists z : Z \bullet \exists y : Y \bullet x \mapsto y \in R \land y \mapsto z \in S$$
$$\Rightarrow \exists z : Z \bullet \exists y : Y \bullet x \mapsto y \in R$$
$$\Leftrightarrow \exists y : Y \bullet x \mapsto y \in R$$
$$\Leftrightarrow x \in \mathrm{dom}\, R$$

$\square$

Solution 8.27

$$x \mapsto z \in X \lhd (R \,\S\, S)$$
$$\Leftrightarrow x \in X \wedge x \mapsto z \in R \,\S\, S$$
$$\Leftrightarrow x \in X \wedge \exists\, y : Y \bullet x \mapsto y \in R \wedge y \mapsto z \in S$$
$$\Leftrightarrow \exists\, y : Y \bullet x \in X \wedge x \mapsto y \in R \wedge y \mapsto z \in S$$
$$\Leftrightarrow \exists\, y : Y \bullet x \mapsto y \in X \lhd R \wedge y \mapsto z \in S$$
$$\Leftrightarrow x \mapsto z \in (X \lhd R) \,\S\, S$$

□

Solution 8.28 A and D are homogeneous, whereas B and C are heterogeneous. □

Solution 8.29 2 and 4 are reflexive; the other two relations are not as they do not include (n, n) for every natural number n. □

Solution 8.30 2 is reflexive, while 1 is not. □

Solution 8.31 By definition, if R is reflexive then $\forall\, x : X \bullet (x, x) \in R$ holds. As such, it must be the case that

$$\{x : X \bullet (x, x)\} \subseteq R$$

holds. □

Solution 8.32 All of these relations are transitive. □

Solution 8.33 1 is transitive, while 2 is not. □

Solution 8.34 The set $R \,\S\, R$ is defined by

$$R \,\S\, R = \{x, y : X \mid \exists\, z : X \bullet x \mapsto z \in R \wedge z \mapsto y \in R\}$$

If R is transitive, then we may conclude from our definition of transitivity that the following property holds of R.

$$\forall\, x, y, z : X \bullet x \mapsto z \in R \wedge z \mapsto y \in R \Rightarrow x \mapsto y \in R$$

As this is the defining property of $R \,\S\, R$, we may conclude that $R \,\S\, R \subseteq R$. □

Solution 8.35 1, 3, and 4 are symmetric. □

Solution 8.36 1 is symmetric, while 2 is not. □

Solution 8.37 The set $R^{\sim}$ is defined by

$$R^{\sim} = \{x, y : X \mid x \mapsto y \in R \bullet y \mapsto x\}$$

If R is symmetric, then we may conclude from our definition of symmetry that the following property holds of R.

$$\forall\, x, y : X \bullet x \mapsto y \in R \Rightarrow y \mapsto x \in R$$

As this is the defining property of $R^\sim$, we may conclude that $R^\sim = R$. $\square$

Solution 8.38 1 is asymmetric. $\square$

Solution 8.39 2 is asymmetric, while 1 is not. $\square$

Solution 8.40 1, 2, and 3 are antisymmetric. $\square$

Solution 8.41 2 is antisymmetric, while 1 is not. $\square$

Solution 8.42 2 and 4 are total. $\square$

Solution 8.43 2 is total, while 1 is not. $\square$

Solution 8.44 $<$ is not a partial ordering as it is not reflexive—it is not the case that for every natural number $n \in \mathbb{N}$, $n < n$. $\square$

Solution 8.45 The relation $\subseteq$ on the subsets of X is reflexive, as $S \subseteq S$ for every $S \in \mathbb{P}\, X$. In addition, $\subseteq$ on the subsets of X is transitive, as if $R \subseteq S$ and $S \subseteq T$ then it follows that $R \subseteq T$. Finally, $\subseteq$ on the subsets of X is antisymmetric, as if $R \subseteq S$ and $S \subseteq R$ then it follows that $R = S$. As such, this relation forms a partial ordering on the subsets of X. $\square$

Solution 8.46 $=$ is not a total ordering as it is not total—it is not the case that for every pair of natural numbers, m and n, $m = n$. $\square$

Solution 8.47 For the relation $\subseteq$ to form a total ordering on the subsets of X, we would require this relation to be antisymmetric, transitive and total. Although it is both antisymmetric and transitive, it is not total. Consider two subsets of X, S and T. It is not the case that for all such subsets, $S \subseteq T$ or $T \subseteq S$ or $S = T$. $\square$

Solution 8.48 $\leq$ is not an equivalence as it is not symmetric—it is not the case that for every pair of natural numbers m and n, if $m \leq n$ then $n \leq m$. $\square$

Solution 8.49 For the relation $\subseteq$ to form an equivalence on the subsets of X, we would require this relation to be reflexive, transitive, and symmetric. Although it is both reflexive and transitive, it is not symmetric. Consider two subsets of X, S and T. It is not the case that for all such subsets, if $S \subseteq T$ then $T \subseteq S$. $\square$

Solution 8.50

1. $\{cat \mapsto cat, dog \mapsto dog, mouse \mapsto mouse\}$

2. $\{cat \mapsto dog, dog \mapsto mouse, cat \mapsto cat, dog \mapsto dog, mouse \mapsto mouse\}$

3. $\{cat \mapsto dog, mouse \mapsto dog, cat \mapsto cat, dog \mapsto dog, mouse \mapsto mouse\}$

4. $\{cat \mapsto cat, dog \mapsto dog, mouse \mapsto mouse\}$

□

Solution 8.51

1. $\emptyset$

2. $\{cat \mapsto dog, dog \mapsto mouse, cat \mapsto mouse\}$

3. $\{cat \mapsto dog, mouse \mapsto dog\}$

4. $\{cat \mapsto cat, dog \mapsto dog, mouse \mapsto mouse\}$

□

Solution 8.52

$$\{a \mapsto b, b \mapsto c, c \mapsto d, d \mapsto e\}$$
$$\cup$$
$$\{a \mapsto c, b \mapsto d, c \mapsto e\}$$
$$\cup$$
$$\{a \mapsto d, b \mapsto e\}$$
$$\cup$$
$$\{a \mapsto e\}$$
$$= \{a \mapsto b, b \mapsto c, c \mapsto d, d \mapsto e, a \mapsto c, b \mapsto d, c \mapsto e, a \mapsto d, b \mapsto e, a \mapsto e\}$$

□

Solution 8.53

1. $\{cat \mapsto cat, mouse \mapsto mouse, dog \mapsto dog\}$

2. $\{cat \mapsto dog, dog \mapsto mouse, cat \mapsto mouse,$
   $\quad cat \mapsto cat, mouse \mapsto mouse, dog \mapsto dog\}$

3. $\{cat \mapsto dog, mouse \mapsto dog,$
   $\quad cat \mapsto cat, mouse \mapsto mouse, dog \mapsto dog\}$

4. $\{cat \mapsto cat, dog \mapsto dog, mouse \mapsto mouse\}$

□

Solution 8.54

1. $\emptyset$

2. $\{cat \mapsto dog, dog \mapsto mouse, dog \mapsto cat, mouse \mapsto dog\}$

3. $\{cat \mapsto dog, mouse \mapsto dog, dog \mapsto cat, dog \mapsto mouse\}$

4. $\{cat \mapsto cat, dog \mapsto dog, mouse \mapsto mouse\}$

□

Solution 8.55

1. $Name \leftrightarrow Yes\_No$

2. $(Name \times Yes\_No) \leftrightarrow Yes\_No$

3. $(Yes\_no \times Yes\_No) \leftrightarrow (Name \times Yes\_No)$

4. $Yes\_No \leftrightarrow (Name \times Name)$

□

Solution 8.56

1. $\{(1,1),(1,0),(0,1),(0,0)\}$

2. $\{(1,1),(1,0),(0,1),(0,0)\}$

3. $\{((1,1),(0,1))\}$

4. $\{((0,1),(1,1)),((0,0),(1,0)),((1,0),(0,1)),((1,1),(0,0))\}$

5. $\{((1,1),(1,0)),((1,0),(1,1)),((0,1),(0,0)),((0,0),(0,1))\}$

6. $\{(1,(0,1)),(1,(0,0)),(0,(1,0)),(0,(1,1))\}$

□

Solution 8.57

1. $husband = \{p, q : Person \mid (p, q) \in married \wedge male(p)\}$

2. $wife = \{p, q : Person \mid (p, q) \in married \wedge female(p)\}$

3. $mother = \{p, q : Person \mid (p, q) \in parent \wedge female(p)\}$

4. $father = \{p, q : Person \mid (p, q) \in parent \wedge male(p)\}$

5. $son = \{p, q : Person \mid (q, p) \in parent \wedge male(p)\}$

6. $daughter = \{p, q : Person \mid (q, p) \in parent \wedge female(p)\}$

7. $brother = \{p, q : Person \mid (p, q) \in sibling \wedge male(p)\}$

8. $sister = \{p, q : Person \mid (p, q) \in sibling \wedge female(p)\}$

□

Solution 8.58

1. $aunt = \{p, q : Person \mid \exists r : Person \bullet (p, r) \in sister \wedge (r, q) \in parent\}$

2. $uncle = \{p, q : Person \mid \exists r : Person \bullet (p, r) \in brother \wedge (r, q) \in parent\}$

3. $nephew = \{p, q : Person \mid \exists r : Person \bullet (p, r) \in son \wedge (r, q) \in sibling\}$

4. $niece = \{p, q : Person \mid \exists r : Person \bullet (p, r) \in daughter \wedge (r, q) \in sibling\}$

5. $cousin = \{p, q : Person \mid \exists r, s : Person \bullet (r, p) \in parent \wedge (s, q) \in parent \wedge$
$(r, s) \in sibling\}$

6. $father\_in\_law = \{p, q : Person \mid \exists\, r : Person \bullet (p, r) \in father\, \wedge$
$$(r, q) \in married\}$$

7. $mother\_in\_law = \{p, q : Person \mid \exists\, r : Person \bullet (p, r) \in mother\, \wedge$
$$(r, q) \in married\}$$

8. $son\_in\_law = \{p, q : Person \mid \exists\, r : Person \bullet (p, r) \in husband\, \wedge$
$$(r, q) \in daughter\}$$

9. $daughter\_in\_law = \{p, q : Person \mid \exists\, r : Person \bullet (p, r) \in wife\, \wedge$
$$(r, q) \in son\}$$

$\square$

Solution 8.59

1. $stepfather = \{p, q : Person \mid$
$$\exists\, r : Person \bullet (p, r) \in husband\, \wedge$$
$$(r, q) \in mother\, \wedge (p, q) \notin father\}$$

2. $stepmother = \{p, q : Person \mid$
$$\exists\, r : Person \bullet (p, r) \in wife\, \wedge$$
$$(r, q) \in father\, \wedge (p, q) \notin mother\}$$

3. $stepdaughter = \{p, q : Person \mid$
$$\exists\, r : Person \bullet (p, r) \in daughter\, \wedge (r, q) \in married\, \wedge$$
$$(p, q) \notin daughter\}$$

4. $stepson = \{p, q : Person \mid$
$$\exists\, r : Person \bullet (p, r) \in son\, \wedge (r, q) \in married\, \wedge$$
$$(p, q) \notin son\}$$

5. $stepbrother = \{p, q : Person \mid$
$$\exists\, r, s : Person \bullet (r, s) \in married\, \wedge (p, r) \in son\, \wedge$$
$$(s, q) \in parent\, \wedge (p, q) \notin brother\, \wedge p \neq q\}$$

6. $stepsister = \{p, q : Person \mid$
$$\exists\, r, s : Person \bullet (r, s) \in married\, \wedge (p, r) \in daughter\, \wedge$$
$$(s, q) \in parent\, \wedge (p, q) \notin sister\, \wedge p \neq q\}$$

7. $grandmother = \{p, q : Parent \mid \exists\, r : Person \bullet (p, r) \in mother\, \wedge (r, q) \in parent\}$

8. $grandfather = \{p, q : Parent \mid \exists\, r : Person \bullet (p, r) \in father\, \wedge (r, q) \in parent\}$

$\square$

Solution 8.60

1. $child = parent^{\sim}$

2. $parent \,_9^\circ\, parent \,(\!|\, \{p\} \,|\!)$

3. $child \, _9^\circ \, child \, _9^\circ \, child \, (\!| \, \{p\} \, |\!)$

4. $child^+ \, (\!| \, \{p\} \, |\!)$

5. dom *mother*

6. dom *father* $\rhd \{p : People \mid female(p)\}$

7. dom *brother* $\cap$ ran *brother*

8. dom *grandmother* $\rhd \{p : People \mid male(p)\}$

9. dom *grandfather* $\cap$ (dom *father* $\rhd \{p : People \mid male(p)\}$)

$\square$

Solution 8.61

1. $R = \{n : \mathbb{N} \mid n > 0 \land n < 5 \bullet (n, 2 \times n)\}$
   $S = \{n : \mathbb{N} \mid n > 0 \land n < 5 \bullet (n, n)\}$

2. dom $R = \{1, 2, 3, 4\}$

3. ran $S = \{1, 2, 3, 4\}$

4. $R^\sim = \{(2,1), (4,2), (6,3), (8,4)\}$

5. $R \, _9^\circ \, S = \{(1,2), (2,4)\}$

6. $S \, _9^\circ \, R = R$

$\square$

Solution 8.62

1. $\{1, 2, 3, 4\}$

2. $\{1, 4, 9, 16\}$

3. $\{2 \mapsto 1, 4 \mapsto 2, 6 \mapsto 3, 8 \mapsto 4\}$

4. $\{1 \mapsto 4, 2 \mapsto 16\}$

5. $\{1 \mapsto 2, 2 \mapsto 8\}$

6. $\{1 \mapsto 2, 2 \mapsto 3\}$

7. $\{3 \mapsto 4, 4 \mapsto 5\}$

8. $\{1 \mapsto 2\}$

9. $\{2 \mapsto 3, 3 \mapsto 4, 4 \mapsto 5\}$

10. $\{1, 4\}$

11. $\{2, 4\}$

12. $\{1\}$

$\square$

Solution 8.63

1. *loves* is not reflexive, as *brian* $\mapsto$ *brian* $\notin$ *loves*.

2. *loves* is not symmetric, as *brian* $\mapsto$ *jackie* $\in$ *loves* but *jackie* $\mapsto$ *brian* $\notin$ *loves*.

3. *loves* is transitive, as $R^+ = R$.

$\square$

Solution 8.64

1. $\{brian \mapsto brian, brian \mapsto jackie, jackie \mapsto jackie\}$

2. $\{jackie \mapsto brian, brian \mapsto jackie, jackie \mapsto jackie\}$

$\square$

Solution 8.65

1. $\{fido \mapsto bark, fido \mapsto howl\}$

2. $\{fido \mapsto bark, fido \mapsto howl, lassie \mapsto bark\}$

3. $\{fido \mapsto bark, fido \mapsto howl, lassie \mapsto bark, bramble \mapsto miaow\}$

4. $\{fido \mapsto bark, fido \mapsto howl, lassie \mapsto bark\}$

5. $\{fido \mapsto bark, lassie \mapsto bark\}$

6. $\{fido \mapsto howl, daisy \mapsto moo\}$

7. $\{fido \mapsto bark, fido \mapsto howl, lassie \mapsto bark, daisy \mapsto moo\}$

8. $\{fido \mapsto howl, daisy \mapsto moo\}$

9. $\{bark, howl\}$

10. $\emptyset$

$\square$

Solution 8.66 1 is both transitive and symmetric, 2 and 3 are transitive but not symmetric, and 4 is symmetric but not transitive. $\square$

Solution 8.67

1. $\emptyset$

2. $\{(jim, joe), (joe, bill), (jim, bill), (steve, rick),$
   $(joe, jim), (bill, joe), (bill, jim), (rick, steve)\}$

3. $\{(jim, joe), (joe, bill), (jim, bill),$
   $(joe, jim), (bill, joe), (bill, jim)\}$

4. $\{(rick, steve), (steve, rick), (jim, bill), (bill, jim)\}$

$\square$

Solution 8.68

1. ∅

2. $\{(jim, joe), (joe, bill), (jim, bill), (steve, rick)\}$

3. $\{(jim, joe), (joe, bill), (jim, bill)\}$

4. $\{(rick, steve), (steve, rick), (jim, bill), (bill, jim),$
   $(rick, rick), (steve, steve), (jim, jim), (bill, bill)\}$

□

Solution 8.69

1. True.

2. False; a counter-example is:

$$\{a \mapsto b, c \mapsto e\} \, \mathbin{\substack{\circ \\ \circ}} \, \{b \mapsto c, e \mapsto d\} = \{a \mapsto c, c \mapsto d\}$$

3. False; a counter-example is:

$$\{a \mapsto b, b \mapsto a\} \, \mathbin{\substack{\circ \\ \circ}} \, \{b \mapsto c, c \mapsto b\} = \{a \mapsto c\}$$

□

Solution 8.70 For $\Leftrightarrow$ to be an equivalence on the set of all propositions, we require $\Leftrightarrow$ to be reflexive, symmetric, and transitive. Given any proposition p, $p \Leftrightarrow p$ is true. As such, we may conclude that $\Leftrightarrow$ is reflexive. Given any propositions p and q, if $p \Leftrightarrow q$ is true then $q \Leftrightarrow p$ is true. As such, we may conclude that $\Leftrightarrow$ is symmetric. Given any propositions p, q and r, if $p \Leftrightarrow q$ is true and $q \Leftrightarrow r$ is true then $p \Leftrightarrow r$ is true. As such, we may conclude that $\Leftrightarrow$ is transitive. Therefore, $\Leftrightarrow$ is an equivalence relation on the set of all propositions. □

Solution 8.71 Yes. R^+ is transitive, so taking its transitive closure leaves it unchanged. Therefore, $(R^+)^+$ is the same as R^+. □

Chapter 9

Functions

In the previous chapter we considered the notion of a relation. There, we saw that in a relation any element of a source set can map to any number of elements of a target set. In this chapter we consider a special category of relations, called *functions*.

9.1   A special kind of relation

Recall that the relation *less_than*, which is of type $\mathbb{N} \leftrightarrow \mathbb{N}$, can be defined by

$$less\_than = \{m, n : \mathbb{N} \mid m < n\}$$

Recalling the notation for relational image, we might write $2 \in less\_than (\!| \{1\} |\!)$; equivalently, we might write $(1, 2) \in less\_than$ or $1 \mapsto 2 \in less\_than$.

The relation *one_less_than*, which is also of type $\mathbb{N} \leftrightarrow \mathbb{N}$, might be defined by

$$one\_less\_than = \{m, n : \mathbb{N} \mid m + 1 = n\}$$

Again, we might write $2 \in one\_less\_than (\!| \{1\} |\!)$; indeed, it is the case that

$$one\_less\_than (\!| \{1\} |\!) = \{2\}$$

Furthermore, it is clear that $one\_less\_than \subseteq less\_than$.

In this second relation, every element of the source set maps to *exactly one* element of the target set: 0 maps to 1, 1 maps to 2, 2 maps to 3, and so on. We shall call any relation which has the special property that every element of the source maps to at most one element of the target (or, equivalently, that every element of the domain maps to exactly one element of the range) a *function*.

Example 9.1 Consider the relation *holidays*, which is defined in the following way.

$$holidays = \{jim \mapsto kenya, jon \mapsto egypt, jeff \mapsto spain\}$$

The fact that we can reply to the question "Where did Jeff take his holiday?" with a *unique* answer (if one exists), makes this relation a function.

Now consider the relation *souvenir*, which is defined by

$$souvenir = \{jon \mapsto pyramid, jeff \mapsto donkey, jeff \mapsto hat\}$$

Here, if we ask the question "What souvenir did Jeff return with?", we get two possible answers: a donkey, and a hat. As such, this relation does not possess the special property required of a function—each element of the domain does not map to a unique element of the range. □

Example 9.2 The identity relation $Id\,[X] : X \leftrightarrow X$ is a function.

$$Id\,[X] = \{x : X \bullet x \mapsto x\}$$

Here, every element of the source maps to exactly one element of the target. □

We denote the set of all functions from X to Y by $X \nrightarrow Y$. It should be noted that $X \nrightarrow Y$ is a subset of $X \leftrightarrow Y$. Indeed, we may define the set of all functions from X to Y in terms of the set of all relations from X to Y in the following way.

$$X \nrightarrow Y = \{r : X \leftrightarrow Y \mid \forall x : \mathrm{dom}\ r \bullet \exists_1\ y : \mathrm{ran}\ r \bullet x \mapsto y \in r\}$$

In our visual representation of relations, determining whether a given relation is a function is a relatively straightforward task. As an example, consider the relation *likes*.

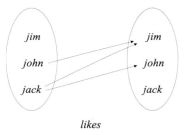

likes

This relation *is not* a function, as *jack* maps to both *jim* and *john*: the relation has *diverging arrows*.

On the other hand, consider the relation *hates*.

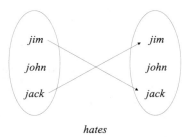

hates

This relation *is* a function, as every element of the domain maps to exactly one element of the range; it has no diverging arrows.

Thus, we may determine visually if a given relation is a function by checking whether or not it has diverging arrows.

Exercise 9.1 Which of the following are functions?

1. $\emptyset$

2. $\{(1,2),(2,3),(3,1)\}$

3. $\{(1,2),(2,1)\}$

4. $\{(1,2),(1,1)\}$

□

Exercise 9.2 List all the functions from $\{1,2\}$ to $\{a,b\}$. □

Exercise 9.3 Is the inverse of a function necessarily a function? □

Exercise 9.4 Is a subset of a function necessarily a function? □

Exercise 9.5 Given functions f and g, both of which are of the same type, are the following necessarily functions?

1. $f \cap g$

2. $f \cup g$

3. $f \setminus g$

□

9.2 Total functions

Recall our function *holidays*, which was defined as

$$holidays = \{jim \mapsto kenya, jon \mapsto egypt, jeff \mapsto spain\}$$

Assume further that the source and target of this function are given by *People* and *Places* respectively, such that

$$People = \{jim, jon, jeff, jack\}$$
$$Places = \{kenya, egypt, spain, greece\}$$

The key to this relation being a function was that, if we were asked a question of the form "Where did p take his holiday?" (where p is one of *jim*, *jon*, or *jeff*), then we could respond with a *unique* answer. If, however, we were asked the question "Where did Jack take his holiday?" then we wouldn't be in a position to respond—*jack* appears in the *source* of the function, but not in the *domain*.

If, in a function f, *every* element of the source maps to exactly one element of the target, then we may say that f is a *total function*. If $f \in X \nrightarrow Y$, this holds exactly when dom $f = X$.

We denote the set of all total functions from X to Y by $X \rightarrow Y$. Note that

$$X \rightarrow Y \subseteq X \nrightarrow Y$$

We can define the set of total functions from X to Y by

$$X \rightarrow Y = \{r : X \leftrightarrow Y \mid r \in X \nrightarrow Y \land \text{dom } r = X\}$$

Again, in our visual representation of relations, it is easy to determine whether or not a function is total—a function is total if, and only if, every element of the source has an arrow emerging from it.

Example 9.3 Consider the relation *hates*.

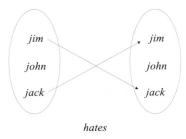

hates

This relation *is not* a total function, as *john* does not map to an element of the target. On the other hand, the relation *looks_like*, which is given by

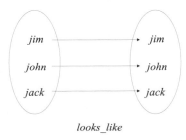

looks_like

is a total function—every element of the source maps to exactly one element of the target. □

Example 9.4 Consider the following homogeneous relation on the set $\{1, 2, 3\}$.

$$\{1 \mapsto 1, 1 \mapsto 2, 2 \mapsto 3\}$$

This relation *is not* a total function, as 3 does not map to an element of the target. On the other hand, the following homogeneous relation on $\{1, 2, 3\}$ *is* a total function, as every element of the source maps to exactly one element of the target.

$$\{1 \mapsto 1, 3 \mapsto 2, 2 \mapsto 3\}$$

□

Exercise 9.6 Assuming the sets *People* and *Mode*, defined by

$$People = \{barry, nigel, david\}$$
$$Mode = \{car, bike, train\}$$

which of the following relations of type *People* $\leftrightarrow$ *Mode* are total functions?

1. $\emptyset$

2. $\{barry \mapsto car, nigel \mapsto car\}$

3. $\{barry \mapsto car, nigel \mapsto car, david \mapsto train\}$

4. $\{barry \mapsto car, barry \mapsto bike, nigel \mapsto car, david \mapsto train\}$

□

Exercise 9.7 List all the total functions from $\{1, 2\}$ to $\{a, b\}$. □

Exercise 9.8 Is the inverse of a total function necessarily a function? □

Exercise 9.9 Given total functions f and g which are of the same type, is it necessarily the case that $f \cap g$ is a total function? □

9.3 Function application

As we have already seen, functions have the virtue of mapping each element of their source to at most one element of the target. As such, the value of $f (\!| \{x\} |\!)$ for some function f and some element x is always either the empty set or a singleton set containing the sole value of the range that x maps to. Because of this, we can consider the *application* of a function to a particular element; this is simply a specialisation of the relational image operation.

Given a function $f \in X \nrightarrow Y$, together with some value $x \in X$, we shall denote the function application of f to x by $f\, x$. Here, $f\, x = y$ if, and only if, $(x, y) \in f$.[1]

Example 9.5 Consider the function *pets*, which is given by

$$pets = \{emily \mapsto goldfish, richard \mapsto gerbil, michael \mapsto cat\}$$

Here,

$$pets\ emily = goldfish$$

and

$$pets\ richard = gerbil$$

□

Of course, we may still refer to the relational images associated with this function, but when we are dealing with functions it makes more sense to consider function applications.

We may define function application formally as follows. Given a function, $f \in X \nrightarrow Y$, together with some element $x \in X$, the function application of f to x is given by

$$f\, x = (\mu\, y : Y \mid x \mapsto y \in f)$$

Of course, if x does not appear in the domain of f then $f\, x$ is undefined. Note also that we may only perform a function application on a function; function applications on relations are undefined.

---

[1] In many texts, the function application of a function f to the term x is written $f(x)$ as opposed to $f\, x$. Both forms are valid, but we choose to use the latter in this book.

Exercise 9.10 Consider the following functions.

$$likes = \{ken \mapsto duncan, duncan \mapsto john,$$
$$richard \mapsto becky, becky \mapsto richard\}$$
$$dislikes = \{ken \mapsto becky, becky \mapsto ken, duncan \mapsto ken\}$$

Which of the following are defined?

1. *likes ken*

2. $(likes ^\sim) \, ken$

3. *dislikes ken*

4. $(dislikes ^\sim) \, ken$

5. $(likes \, {}^\circ_\circ \, likes) \, ken$

6. $(dislikes \, {}^\circ_\circ \, dislikes) \, ken$

☐

Exercise 9.11 Consider the function *pets*.

$$pets = \{emily \mapsto goldfish, richard \mapsto gerbil, michael \mapsto cat\}$$

1. What is the value of *pets richard*?

2. What is the value of *pets michael*?

3. What is the value of $pets ^\sim goldfish$?

4. What is the value of $(\{emily\} \lhd pets) \, emily$?

☐

Exercise 9.12 Assuming the relation *pets* of the previous exercise, calculate the results of the following set comprehensions.

1. $\{p : pets\}$

2. $\{p : pets \bullet p.1\}$

3. $\{p : pets \mid p.2 = goldfish\}$

4. $\{p : pets \mid p.2 = goldfish \bullet p.1\}$

5. $\{p : pets \mid p.1 = emily \bullet pets \, p.1\}$

6. $\{p : pets \mid p.1 = emily \bullet p.2\}$

☐

9.4 Overriding

If we are to use functions to model real-world entities, then it is desirable to have a means of updating such a view of the world. After all, although the values associated with some functions may remain the same throughout time (e.g., addition on the natural numbers), we may expect the values associated with other functions—especially those concerned with the modelling of some real-life system—to change as and when necessary.

Take, as an example, the following function.

$$wage = \{thomas \mapsto 23000, evans \mapsto 18750,$$
$$jones \mapsto 18750, brown \mapsto 28000\}$$

If *evans* was to get a pay increase of 50 pounds per year, then we would want our new view of *wage* to be as follows.

$$new\_wage = \{thomas \mapsto 23000, evans \mapsto 18800,$$
$$jones \mapsto 18750, brown \mapsto 28000\}$$

Rather than have to define the function *new_wage* from nothing, it would be preferable to define *new_wage* in terms of *wage*; the *function override* operator, $\oplus$, allows us to do exactly this.

Given two functions f and g of the same type, $f \oplus g$ denotes the overriding of f by g. We may say that f is 'overridden' by g. This operation effectively updates f according to the values of g. As such,

$$new\_wage = wage \oplus \{evans \mapsto 18800\}$$

reflects the change that we desire. Here, *wage* is updated by the function $\{evans \mapsto 18800\}$: we expect the wage associated with *evans* to change, but all others to remain the same.

We may define overriding formally as follows.

$$f \oplus g = (\text{dom } g \lhd f) \cup g$$

This definition becomes clear if we consider the fact that there are three classes of elements from the source type to consider: those which appear in the domain of f and do not appear in the domain of g; those which appear in the domain of f and in the domain of g; and those which do not appear in the domain of f but do appear in the domain of g. We consider each case in turn.

If a maplet $a_1 \mapsto b_1$ appears in f, and a_1 does not appear in g, then the mapping from a_1 to b_1 is not affected by the overriding of f by g. As such, $a_1 \mapsto b_1$ appears in $f \oplus g$.

On the other hand if a maplet $a_2 \mapsto b_2$ appears in f and a_2 *does* appear in g, then the mapping from a_2 to b_2 *is* affected by the overriding of f by g. As such, $a_2 \mapsto g\, a_2$ appears in $f \oplus g$.

Finally, if a maplet, $a_3 \mapsto b_3$ appears in g and a_3 does not appear in f, then the mapping from a_3 to b_3 is added to the resulting function. As such, $a_3 \mapsto b_3$ appears in $f \oplus g$.

Example 9.6 Consider the functions f and g:

$$f = \{0 \mapsto 0, 1 \mapsto 1\}$$
$$g = \{0 \mapsto 1, 2 \mapsto 3\}$$

If we consider each of the elements of the source in turn, we have the following.

- 0 maps to 0 in f and 0 maps to 1 in g, so $(f \oplus g)\,0 = 1$.
- 1 maps to 1 in f and 1 does not appear in the domain of g, so $(f \oplus g)\,1 = 1$.
- 2 does not appear in the domain of f and 2 maps to 3 in g, so $(f \oplus g)\,2 = 3$.

As such,

$$f \oplus g = \{0 \mapsto 1, 1 \mapsto 1, 2 \mapsto 3\}$$

□

Example 9.7 Consider the function f, given below.

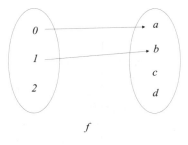

Now consider the function g.

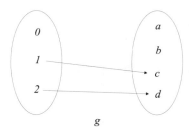

The function $f \oplus g$ overrides f with g to give the following result.

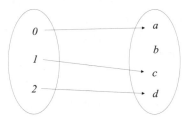

□

Exercise 9.13 Assume the following functions:

$$A = \{jon \mapsto coffee, dave \mapsto orange\}$$
$$B = \{steve \mapsto tea, dave \mapsto coffee\}$$
$$C = \{jon \mapsto orange, steve \mapsto coffee\}$$

Calculate the following.

1. $A \oplus B$
2. $B \oplus A$
3. $B \oplus C$
4. $(A \oplus B) \oplus C$
5. $A \oplus (B \oplus C)$

□

Exercise 9.14 Prove each of the following.

1. $f \oplus f = f$
2. $f \oplus \emptyset = f$
3. $\emptyset \oplus f = f$

□

9.5   Properties of functions

Just as functions are special types of relations, so we can define special types of functions. In this section, we consider three special types of functions: injections, surjections, and bijections.

9.5.1   Injections

We say that a function $f \in X \nrightarrow Y$ is an *injective function*, or *injection*, if, and only if, every element of Y is mapped to by at most one element of X. We denote the set of all injective functions of type $X \leftrightarrow Y$ by $X \rightarrowtail\!\!\!\!\rightarrow Y$. Of course,

$$X \rightarrowtail\!\!\!\!\rightarrow Y \subseteq X \nrightarrow Y$$

Furthermore, we denote the set of all total injections of type $X \leftrightarrow Y$ by $X \rightarrowtail Y$. We define these sets formally as follows.

$$X \rightarrowtail\!\!\!\!\rightarrow Y = \{r : X \leftrightarrow Y \mid$$
$$r \in X \nrightarrow Y \wedge \forall y : \operatorname{ran} r \bullet (\exists_1 x : \operatorname{dom} r \bullet (x, y) \in r)\}$$

$$X \rightarrowtail Y = X \rightarrowtail\!\!\!\!\rightarrow Y \cap X \rightarrow Y$$

Example 9.8 Consider the sets *Person* and *Colour*, given by

$$Person = \{robin, matt, simba\}$$
$$Colour = \{green, blue, yellow\}$$

and the function *favourite*, which is given by

$$favourite = \{robin \mapsto green, matt \mapsto blue, simba \mapsto yellow\}$$

Here, *favourite* is clearly injective—every element of the target set is mapped to by at most one element of the source. Furthermore, this function is a total injection, as dom *favourite* = *Person*.

On the other hand, the function *car_colour* ∈ *Person* ↛ *Colour* is not injective, as both *robin* and *matt* map to *blue*.

$$car\_colour = \{robin \mapsto blue, matt \mapsto blue\}$$

□

We may use potato diagrams as a means of determining whether or not a given relation is injective. As an example, consider the potato diagram for *favourite* ∈ *Person* ↛ *Colour*, which is given below.

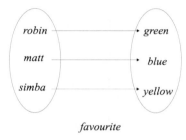

favourite

Here, as there are no *converging arrows* we may conclude that this is an injective function. The potato diagram for *car_colour*, on the other hand, does have converging arrows—two arrows point to *blue*—and, as such, we can conclude that this function is not injective.

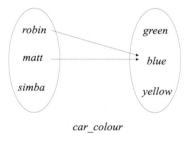

car_colour

Thus, the presence or absence of converging arrows may be used as a means of determining whether or not a given function is injective.

Exercise 9.15 Which of the following are injective functions?

1. ∅
2. $\{1 \mapsto a, 1 \mapsto b\}$
3. $\{1 \mapsto a, 2 \mapsto a\}$
4. $\{1 \mapsto a, 2 \mapsto b\}$

□

Exercise 9.16 Give the set of all injections from $\{0, 1\}$ to $\{a, b\}$. □

Exercise 9.17 Give the set of all total injections from $\{0, 1\}$ to $\{a, b\}$. □

Exercise 9.18 Is the inverse of an injection necessarily an injection? □

Exercise 9.19 Is the inverse of a total injection necessarily a total injection? □

9.5.2 Surjections

We say that a function $f \in X \nrightarrow Y$ is a *surjective function*, or *surjection*, if, and only if, every element of Y is mapped to by some element of X. We denote the set of all surjective functions of type $X \leftrightarrow Y$ by $X \twoheadrightarrow Y$. Of course,

$$X \twoheadrightarrow Y \subseteq X \nrightarrow Y$$

Furthermore, we denote the set of all total surjections of type $X \leftrightarrow Y$ by $X \twoheadrightarrow Y$. We define these sets formally in the following way.

$$X \twoheadrightarrow Y = \{r : X \leftrightarrow Y \mid r \in X \nrightarrow Y \wedge \operatorname{ran} r = Y\}$$

$$X \twoheadrightarrow Y = X \twoheadrightarrow Y \cap X \rightarrow Y$$

Example 9.9 Consider again the sets *Person* and *Colour*, given by

$$Person = \{robin, matt, simba\}$$
$$Colour = \{green, blue, yellow\}$$

and the function *favourite* $\in$ *Person* $\twoheadrightarrow$ *Colour*.

$$favourite = \{robin \mapsto green, matt \mapsto blue, simba \mapsto yellow\}$$

Here, *favourite* is surjective—every element of the target set is mapped to by some element of the source. Furthermore, this function is a total surjection, as dom *favourite* = *Person*.

Recall also the function *car_colour* $\in$ *Person* $\nrightarrow$ *Colour*.

$$car\_colour = \{robin \mapsto blue, matt \mapsto blue\}$$

This is not surjective, as neither *green* nor *yellow* are mapped to by some element of the domain.
□

Again, we may use potato diagrams as a means of determining whether or not a given relation is surjective—this property only holds if every element of the target set is mapped to. This property certainly holds of *favourite*, as we can see from the following potato diagram.

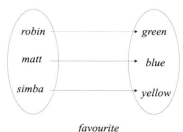

favourite

However, this property does not hold of *car_colour*.

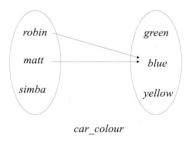

car_colour

Thus, we may establish whether or not a given function is surjective by determining whether or not every element of the target is mapped to.

Exercise 9.20 Assuming that all of the following are of type $\{a, b, c\} \leftrightarrow \{0, 1\}$, which are surjective functions?

1. $\emptyset$

2. $\{a \mapsto 0, b \mapsto 1\}$

3. $\{a \mapsto 0, b \mapsto 1, c \mapsto 1\}$

4. $\{a \mapsto 0, b \mapsto 0, b \mapsto 1\}$

$\square$

Exercise 9.21 Give the set of all surjections from $\{0, 1\}$ to $\{a, b\}$. $\square$

Exercise 9.22 Give the set of all total surjections from $\{0, 1\}$ to $\{a, b\}$. $\square$

Exercise 9.23 If X has m elements and Y has n elements, what must be true of the relationship between m and n for $f \in X \twoheadrightarrow Y$ to be surjective. $\square$

Exercise 9.24 Assume X and Y both have n elements. Assume further that $f \in X \twoheadrightarrow Y$ is a total surjection. Is it necessarily the case that f is injective? $\square$

Exercise 9.25 Is the inverse of a surjection necessarily a surjection? $\square$

Exercise 9.26 Is the inverse of a total injection necessarily a surjection? $\square$

9.5.3 Bijections

We say that a function $f \in X \twoheadrightarrow Y$ is a *bijective function*, or *bijection*, if, and only if, f is both injective and surjective. We denote the set of all bijective functions of type $X \leftrightarrow Y$ by $X \rightarrowtail\!\!\!\!\rightarrow Y$. Of course,

$$X \rightarrowtail\!\!\!\!\rightarrow Y \subseteq X \twoheadrightarrow Y$$

Furthermore, we denote the set of all total bijections of type $X \leftrightarrow Y$ by $X \rightarrowtail\kern-1.8ex\rightarrow Y$. We define these sets formally as follows.

$$X \rightarrowtail\kern-1.8ex\rightarrow Y = X \rightarrowtail Y \cap X \twoheadrightarrow Y$$

$$X \rightarrowtail Y = X \rightarrowtail Y \cap X \rightarrow Y$$

Example 9.10 Recall our function, *favourite* $\in$ *Person* $\twoheadrightarrow$ *Colour*, which was defined as

$$favourite = \{robin \mapsto green, matt \mapsto blue, simba \mapsto yellow\}$$

This is a bijection as it is both injective and surjective. Furthermore, it is a total bijection as dom *favourite* = *Person*. On the other hand, *car_colour* $\in$ *Person* $\twoheadrightarrow$ *Colour*, which was defined as

$$car\_colour = \{robin \mapsto blue, matt \mapsto blue\}$$

is not bijective, as it is neither injective nor surjective. $\square$

Example 9.11 Recall our discussion of isomorphic Boolean algebras from Chapter 5. A function that transforms one Boolean algebra to another must be a total bijection for the two Boolean algebras to be isomorphic. $\square$

Exercise 9.27 Given that the following are all of type $\{a, b, c, d\} \leftrightarrow \{0, 1, 2\}$, which are bijective functions?

1. $\emptyset$
2. $\{a \mapsto 0, b \mapsto 1\}$
3. $\{a \mapsto 0, b \mapsto 0, c \mapsto 1, d \mapsto 2\}$
4. $\{a \mapsto 0, b \mapsto 1, c \mapsto 2\}$

$\square$

Exercise 9.28 Give the set of all bijections from $\{0, 1\}$ to $\{a, b\}$. $\square$

Exercise 9.29 If X has m elements and Y has n elements, what must be true of the relationship between m and n if $f \in X \rightarrowtail\kern-1.8ex\rightarrow Y$ is to be a total bijection? $\square$

Exercise 9.30 Is the inverse of a bijection necessarily a bijection? $\square$

Exercise 9.31 Is the inverse of a total bijection necessarily a total bijection? $\square$

9.6 Recursively defined functions

Ironically, one of the more advanced topics covered in this book—that of recursively defined functions—was first introduced in Chapter 2. There, we saw a function, *mult*, which was defined for Peano arithmetic as follows.

$$mult\,(0, n) = 0$$
$$mult\,(0 + 1, n) = n$$
$$mult\,(m + 1, n) = n + mult(m, n)$$

Here, if the number by which n is to be multiplied is 0 or $0+1$, then the result is 0 or n respectively. On the other hand, if the number by which n is to be multiplied is greater than $0 + 1$ then the result provided by *mult* is defined in terms of *mult*.

Any function which is defined in terms of itself (with *mult* being a prime example), may be referred to as a *recursively defined function*. For example, the function *digits*, which calculates the number of digits contained in a given natural number, is defined as follows.

$$digits\,(n) = 1 \qquad\qquad \text{if } n < 10$$
$$digits\,(n) = 1 + digits\,(n \text{ div } 10) \quad \text{otherwise}$$

Here, $digits\,(397)$ can be calculated as follows.

$$
\begin{aligned}
digits\,(397) &= 1 + digits\,(39) \\
&= 1 + (1 + digits\,(3)) \\
&= 1 + (1 + 1) \\
&= 3
\end{aligned}
$$

Thus, $digits\,(397)$ is defined in terms of $digits\,(39)$, which, in turn, is defined in terms of $digits\,(3)$, which—finally—is given by 1. We can see that each invocation of *digits* contributes to the overall solution of $digits\,(397)$.

As a further example, we may construct a new definition of *mult* in terms of the function *add*, which is itself recursively defined.

$$mult\,(0, n) = 0$$
$$mult\,(0 + 1, n) = n$$
$$mult\,(m + 1, n) = add\,(n, mult\,(m, n))$$

$$add\,(0, n) = n$$
$$add\,((m + 1), n) = add\,(m, n) + 1$$

Thus, the multiplication of $0 + 1 + 1$ by $0 + 1 + 1 + 1$ via Peano arithmetic can be evaluated recursively in the following way.

$$
\begin{aligned}
mult\,(0 + 1 + 1, 0 + 1 + 1 + 1) &= add\,(0 + 1 + 1 + 1, mult\,(0 + 1, 0 + 1 + 1 + 1)) \\
&= add\,(0 + 1 + 1 + 1, 0 + 1 + 1 + 1) \\
&= add\,(0 + 1 + 1, 0 + 1 + 1 + 1) + 1 \\
&= add\,(0 + 1, 0 + 1 + 1 + 1) + 1 + 1 \\
&= add\,(0, 0 + 1 + 1 + 1) + 1 + 1 + 1 \\
&= 0 + 1 + 1 + 1 + 1 + 1 + 1
\end{aligned}
$$

The general form of a recursive function is given by

$$f(n) = x \qquad \text{if } n = m$$
$$= y + f(n') \quad \text{otherwise}$$

Typically, there are one or more simple, or 'base' cases, and one or more complex cases, which are defined in terms of the function itself. Here, $n = m$ acts as the 'base' case—when we get to this stage, the recursion terminates. In the more complex case, $f(n)$ is defined in terms of $y + f(n')$. The general idea is that n' is somehow 'smaller' than n and, in some sense, 'nearer' to m.

It is only by careful selection of the base case (or cases) and by ensuring that the recursive definition moves every value closer to the base case (or cases) at each stage that we can guarantee that such recursion will not continue forever. As such, we shall say that a recursively defined function f is *well-defined* if it possesses the following properties.

1. There are certain values, called *base values*, for which the function does not refer to itself.

2. Every time the function does refer to itself, the argument of the function is closer to a base value.

In our *digit* example,

$$digits(n) = 1 \text{ if } n < 10$$

took the role of the base case, and

$$digits(n) = 1 + digits(n \text{ div } 10)$$

defined $digits(n)$ in terms of $digits(n \text{ div } 10)$, with n div 10 being closer to the base case than n.

Example 9.12 The factorial function, !, can be defined recursively as follows.

$$0! = 1$$
$$1! = 1$$
$$n! = n \times (n-1)! \text{ if } n > 1$$

What is more, this recursive function is well-defined. □

Exercise 9.32 Why is the following not a well-defined recursive function?

$$f(0) = 1$$
$$f(n) = f(n+1) + 1 \text{ if } n > 0$$

□

Exercise 9.33 Why is the following not a well-defined recursive function?

$$f(1) = 1$$
$$f(n) = f(n-2) + 1 \text{ if } n > 1$$

□

Exercise 9.34 Define a recursive function which counts the number of 0s that appear in a given natural number. ☐

Exercise 9.35 Define a recursive function that adds together all of the digits which appear in a natural number. As an example, applying the function to 347 would give the result $3 + 4 + 7 = 14$. ☐

Exercise 9.36 Define a recursive function that multiplies together all of the digits which appear in a natural number. As an example, applying the function to 347 would give the result $3 \times 4 \times 7 = 84$. ☐

Exercise 9.37 Define a recursive function that returns the largest digit appearing in a given natural number. As an example, applying the function to 347 would return the answer 7. You may assume the existence of a function *greater*, such that *greater* (m, n) returns the larger of m and n as a result. ☐

9.7 Further exercises

Exercise 9.38 A school database is concerned with keeping track of subjects chosen by students, and the students for which each teacher is responsible. Give the type of two functions that may represent this information. ☐

Exercise 9.39 The function f associates every phrase in the English language to the number of distinct letters used to spell that phrase. What are the values of the following?

1. f (simple)

2. f (slightly harder)

3. f (c'est impossible)

☐

Exercise 9.40 Which of the following are functions?

1. $\emptyset$

2. $\{fred \mapsto cat, bob \mapsto dog, fred \mapsto goldfish\}$

3. $\{fred \mapsto cat, bob \mapsto dog, joe \mapsto dog\}$

☐

Exercise 9.41 What is the difference between a partial and a total function? Illustrate this difference using an example. ☐

Exercise 9.42 Assume the following functions.

$$holiday = \{\, jim \mapsto iceland, emily \mapsto tenerife, dave \mapsto singapore,$$
$$steve \mapsto iceland, ali \mapsto greece, becky \mapsto greece\,\}$$
$$likes = \{\, jim \mapsto \{cold\}, emily \mapsto \emptyset, dave \mapsto \{cold, hot, humid\},$$
$$steve \mapsto \{cold, hot, humid\}, ali \mapsto \{hot\}, becky \mapsto \{cold\}\,\}$$
$$climate = \{\, iceland \mapsto cold, tenerife \mapsto hot,$$
$$singapore \mapsto humid, greece \mapsto hot\,\}$$

Calculate the following.

1. $likes \,(\!|\; \{jim, emily\}\; |\!)$

2. $(holiday \,\fatsemi\, climate) \rhd \{cold\}$

3. $holiday \oplus \{emily \mapsto greece\}$

☐

Exercise 9.43 How might one represent the set of people who went on holiday to a destination which has the climate that he or she likes, using the functions from the previous exercise? ☐

Exercise 9.44 Consider the following sets.

$$A = \{(1, a), (2, b), (3, c), (4, d)\}$$
$$B = \{(2, b), (4, b)\}$$
$$C = \{x : A \mid x.1 = 1 \lor x.1 = 3 \bullet (x.1, c)\}$$

Calculate the following.

1. $A \oplus B$

2. $A \oplus C$

3. $A \oplus (C \oplus B)$

4. $(A \oplus C) \oplus B$

☐

Exercise 9.45 Assume $A = \{1, 2\}$ and $B = \{x, y\}$.

1. Give the set of all total functions from A to B.

2. Give the set of all injections from A to B.

3. Give the set of all total injections from A to B.

4. Give the set of all surjections from A to B.

5. Give the set of all bijections from A to B.

☐

Exercise 9.46 Consider again the *parent* relation of Exercise 8.57.

$$parent = \{p, q : Person \mid p \text{ is a parent of } q\}$$

Define the following functions in terms of *parent*.

1. *children* : *Person* $\twoheadrightarrow$ $\mathbb{P}$ *Person*, such that *children* (p) denotes the set containing p's children.

2. *number_of_children* : *Person* $\twoheadrightarrow$ $\mathbb{N}$, such that *number_of_children* (p) represents the number of children that p has.

☐

Exercise 9.47 Consider the following sets.

$$People = \{rachel, dominic, emily\}$$
$$Fruit = \{apple, banana\}$$

Which of the following functions are injective, which are surjective, and which are bijective?

1. $\emptyset$

2. $\{rachel \mapsto apple, dominic \mapsto apple\}$

3. $\{emily \mapsto apple, dominic \mapsto banana, emily \mapsto banana\}$

4. $\{emily \mapsto apple, dominic \mapsto banana\}$

☐

Exercise 9.48 The *Fibonacci sequence* of numbers is given as follows.

$$0, 1, 1, 2, 3, 5, 8, 13, 21, 34, 55, 89, 144, \ldots$$

Give a recursively defined function that generates this sequence of numbers. ☐

Exercise 9.49 The *Ackermann function* is given as follows.

$$A(0, n) = n + 1$$
$$A(m, 0) = A(m - 1, 1) \qquad \text{if } m > 0$$
$$A(m, n) = A(m - 1, A(m, n - 1)) \quad \text{if } m > 0 \text{ and } n > 0$$

Calculate the following.

1. $A(0, 2)$

2. $A(2, 0)$

☐

9.8   Solutions

Solution 9.1 1, 2 and 3 are functions, while 4 is not. □

Solution 9.2 The functions from $\{1, 2\}$ to $\{a, b\}$ are

$$\emptyset, \{1 \mapsto a\}, \{1 \mapsto b\}, \{2 \mapsto a\}, \{2 \mapsto b\}, \{1 \mapsto a, 2 \mapsto a\},$$
$$\{1 \mapsto a, 2 \mapsto b\}, \{1 \mapsto b, 2 \mapsto a\}, \text{ and } \{1 \mapsto b, 2 \mapsto b\}$$

□

Solution 9.3 No. Consider the function

$$f = \{0 \mapsto 1, 1 \mapsto 1\}$$

Here,

$$f^{\sim} = \{1 \mapsto 0, 1 \mapsto 1\}$$

which is clearly not a function. □

Solution 9.4 Yes. If f is a function, then every element of the source maps to at most one element of the target. If g is a subset of f, then the properties of $\subseteq$ ensure that g will also satisfy the properties necessary of a function. □

Solution 9.5

1. Yes. f and g are both functions. As $f \cap g$ is a subset of both f and g, and any subset of a function must also be a function, then $f \cap g$ must be a function.

2. No. Consider $f = \{0 \mapsto 0\}$ and $g = \{0 \mapsto 1\}$. Here, f and g are both functions, but $f \cup g$ is not.

3. Yes. f is a function. As $f \setminus g$ is a subset of f and any subset of a function is also a function, then $f \setminus g$ must be a function.

□

Solution 9.6 3 is a total function. 1 and 2 are partial functions, while 4 is not a function. □

Solution 9.7 The total functions from $\{1, 2\}$ to $\{a, b\}$ are

$$\{1 \mapsto a, 2 \mapsto a\}, \{1 \mapsto a, 2 \mapsto b\}, \{1 \mapsto b, 2 \mapsto a\}, \text{ and } \{1 \mapsto b, 2 \mapsto b\}$$

□

Solution 9.8 No. Consider the total function $f : X \to X$ such that $X = \{0, 1\}$ and

$$f = \{0 \mapsto 1, 1 \mapsto 1\}$$

Here,

$$f^{\sim} = \{1 \mapsto 0, 1 \mapsto 1\}$$

which is not a function. $\square$

Solution 9.9 No. Consider the total functions f and g, such that f and g are defined as follows.

$$f = \{0 \mapsto 0, 1 \mapsto 1\}$$
$$g = \{0 \mapsto 1, 1 \mapsto 0\}$$

Here, $f \cap g = \emptyset$, which is clearly not a total function. $\square$

Solution 9.10 1, 3, 5, and 6 are defined. $\square$

Solution 9.11

1. *gerbil*

2. *cat*

3. *emily*

4. undefined

$\square$

Solution 9.12

1. $\{emily \mapsto goldfish, richard \mapsto gerbil, michael \mapsto cat\}$

2. $\{emily, richard, michael\}$

3. $\{emily \mapsto goldfish\}$

4. $\{emily\}$

5. $\{goldfish\}$

6. $\{goldfish\}$

$\square$

Solution 9.13

1. $\{jon \mapsto coffee, dave \mapsto coffee, steve \mapsto tea\}$

2. $\{jon \mapsto coffee, dave \mapsto orange, steve \mapsto tea\}$

3. $\{jon \mapsto orange, dave \mapsto coffee, steve \mapsto coffee\}$

4. $\{jon \mapsto orange, dave \mapsto coffee, steve \mapsto coffee\}$

5. $\{jon \mapsto orange, dave \mapsto coffee, steve \mapsto coffee\}$

$\square$

Solution 9.14

1. $f \oplus f = (\operatorname{dom} f \vartriangleleft f) \cup f$
$= \emptyset \cup f$
$= f$

2. $f \oplus \emptyset = (\operatorname{dom} \emptyset \vartriangleleft f) \cup \emptyset$
$= (\emptyset \vartriangleleft f) \cup \emptyset$
$= f \cup \emptyset$
$= f$

3. $\emptyset \oplus f = (\operatorname{dom} f \vartriangleleft \emptyset) \cup f$
$= \emptyset \cup f$
$= f$

$\square$

Solution 9.15 Only 1 and 4 are injective functions. 2 is not a function, whereas 3 is a function, but not injective. $\square$

Solution 9.16 The set of injections from $\{0, 1\}$ to $\{a, b\}$ is

$$\{\emptyset, \{0 \mapsto a\}, \{0 \mapsto b\}, \{1 \mapsto a\}, \{1 \mapsto b\}, \{0 \mapsto a, 1 \mapsto b\}, \{0 \mapsto b, 1 \mapsto a\}\}$$

$\square$

Solution 9.17 The set of total injections from $\{0, 1\}$ to $\{a, b\}$ is

$$\{\{0 \mapsto a, 1 \mapsto b\}, \{0 \mapsto b, 1 \mapsto a\}\}$$

$\square$

Solution 9.18 Yes. The function has no converging or diverging arrows, and, as such, its inverse will have no such arrows either. $\square$

Solution 9.19 No. Consider the function $f \in \{0, 1\} \rightarrowtail \{a, b, c\}$, such that

$$f = \{0 \mapsto a, 1 \mapsto b\}$$

The inverse of f, $\{a \mapsto 0, b \mapsto 1\}$, is an injection, but it is not a total injection. $\square$

Solution 9.20 Only 2 and 3 are surjective functions. 4 is not a function, whereas 1 is a function, but not surjective. $\square$

Solution 9.21 The set of surjections is

$$\{\{0 \mapsto a, 1 \mapsto b\}, \{0 \mapsto b, 1 \mapsto a\}\}$$

$\square$

Solution 9.22 The set of total surjections is

$$\{\{0 \mapsto a, 1 \mapsto b\}, \{0 \mapsto b, 1 \mapsto a\}\}$$

□

Solution 9.23 $m \geq n$ must hold for f to be surjective. □

Solution 9.24 Yes. As X and Y are of the same cardinality, all elements of X are mapped from and all elements of Y are mapped to, it follows that every element of X must map to a unique element of Y. □

Solution 9.25 No: this certainly will not be the case if the cardinality of the source type is strictly greater than the cardinality of the target type. □

Solution 9.26 Yes. If f is a total injection, then every element of the source type is mapped from and there are no converging or diverging arrows. As such, it follows that every element of the target type of $f^\sim$ is mapped to and $f^\sim$ has no diverging arrows. As such, $f^\sim$ is a surjective function. □

Solution 9.27 1 and 2 are injective but not surjective. 3 is surjective but not injective. 4 is bijective. □

Solution 9.28 The set of bijections is

$$\{\{0 \mapsto a, 1 \mapsto b\}, \{0 \mapsto b, 1 \mapsto a\}\}$$

□

Solution 9.29 $m = n$ must hold if f is to be a total bijection. □

Solution 9.30 No. Consider $f \in \{a, b, c\} \twoheadrightarrow \{0, 1\}$, such that

$$f = \{a \mapsto 0, b \mapsto 1\}$$

This is both injective and surjective and, therefore, bijective. However, $f^\sim$ is not surjective, and, therefore, is not bijective. □

Solution 9.31 Yes. Consider a total bijection, f. By definition, this has no converging or diverging arrows, and all elements of the source are mapped from and all elements of the target are mapped to. It follows that the inverse of f will also have these properties. As such, $f^\sim$ will be a total bijection. □

Solution 9.32 This is not a well-defined recursive function as the recursive case moves the argument away from the base case, rather than towards it. Consider $f(4)$. This gives

$$f(4) = f(5) + 1$$
$$= f(6) + 1 + 1$$
$$\vdots$$

□

Solution 9.33 This is not a well-defined recursive function as it is possible that the base case will not be arrived at. Consider $f(4)$. This gives

$$
\begin{aligned}
f(4) &= f(2) + 1 \\
&= f(0) + 1 + 1 \\
&= f(-2) + 1 + 1 + 1 \\
&\quad\vdots
\end{aligned}
$$

□

Solution 9.34

$$
\begin{array}{ll}
zero\_count\,(n) = 1 & \text{if } n = 0 \\
zero\_count\,(n) = 0 & \text{if } n > 0 \wedge n \leq 9 \\
zero\_count\,(n) = 1 + zero\_count\,(n \text{ div } 10) & \text{if } n \geq 10 \wedge n \bmod 10 = 0 \\
zero\_count\,(n) = zero\_count\,(n \text{ div } 10) & \text{if } n \geq 10 \wedge n \bmod 10 \neq 0
\end{array}
$$

□

Solution 9.35

$$
\begin{array}{ll}
digit\_add\,(n) = n & \text{if } n < 10 \\
digit\_add\,(n) = n \bmod 10 + digit\_add\,(n \text{ div } 10) & \text{otherwise}
\end{array}
$$

□

Solution 9.36

$$
\begin{array}{ll}
digit\_mult\,(n) = n & \text{if } n < 10 \\
digit\_mult\,(n) = n \bmod 10 \times digit\_mult\,(n \text{ div } 10) & \text{otherwise}
\end{array}
$$

□

Solution 9.37

$$
\begin{array}{ll}
largest\,(n) = n & \text{if } n < 10 \\
largest\,(n) = greater\,(n \bmod 10, largest\,(n \text{ div } 10)) & \text{otherwise}
\end{array}
$$

□

Solution 9.38

$$
\begin{aligned}
&subjects : Student \nrightarrow \mathbb{P}\ Subject \\
&tutees : Teacher \nrightarrow \mathbb{P}\ Student
\end{aligned}
$$

□

Solution 9.39

1. 6

2. 11

3. undefined, as the phrase is not in English

□

Solution 9.40 1 and 3 are functions; 2 is not a function as it has diverging arrows. □

Solution 9.41 A total function maps every element of the source set to exactly one element of the target set, while this property does not necessarily hold of a partial function. For example, the square function on the natural numbers is a total function, while division on the natural numbers is partial. □

Solution 9.42

1. $\{\{cold\}, \emptyset\}$

2. $\{emily \mapsto hot, dave \mapsto humid, ali \mapsto hot, becky \mapsto hot\}$

3. $\{jim \mapsto iceland, emily \mapsto greece, dave \mapsto singapore,$
   $steve \mapsto iceland, ali \mapsto greece, becky \mapsto greece\}$

□

Solution 9.43

$$\{n : \mathrm{dom}\ holiday \mid holiday \mathbin{\S} weather (\!| \{n\} |\!) \subseteq likes\ n\}$$

□

Solution 9.44

1. $\{(1, a), (2, b), (3, c), (4, b)\}$

2. $\{(1, c), (2, b), (3, c), (4, d)\}$

3. $\{(1, c), (2, b), (3, c), (4, b)\}$

4. $\{(1, c), (2, b), (3, c), (4, b)\}$

□

Solution 9.45

1. $\{\{(1, x), (2, x)\}, \{(1, x), (2, y)\}, \{(1, y), (2, x)\}, \{(1, y), (2, y)\}\}$

2. $\{\emptyset, \{(1, x)\}, \{(1, y)\}, \{(2, x)\}, \{(2, y)\}, \{(1, x), (2, y)\}, \{(1, y), (2, x)\}\}$

3. $\{\{(1, x), (2, y)\}, \{(1, y), (2, x)\}\}$

4. $\{\{(1, x), (2, y)\}, \{(1, y), (2, x)\}\}$

5. $\{\{(1, x), (2, y)\}, \{(1, y), (2, x)\}\}$

□

Solution 9.46

1. $children = \{p : Person; \; q : \mathbb{P} \; Person \mid (\forall x : q \bullet (p, x) \in parent) \bullet (p, q)\}$

2. $number\_of\_children = \{p : Person \bullet (p, \# \; children \; p)\}$

□

Solution 9.47 1 is injective and 4 is bijective. 3 is not a function, while 2 is a function, but neither injective nor surjective. □

Solution 9.48

$$F(0) = 0$$
$$F(1) = 1$$
$$F(n) = F(n-1) + F(n-2) \; \text{if} \; n > 1$$

□

Solution 9.49

1. $A(0, 2) = 2 + 1$
   $\qquad\quad = 3$

2. $A(2, 0) = A(1, 1)$
   $\qquad\quad = A(0, A(1, 0))$
   $\qquad\quad = A(1, 0) + 1$
   $\qquad\quad = A(0, 1) + 1$
   $\qquad\quad = 1 + 1 + 1$
   $\qquad\quad = 3$

□

Chapter 10

Sequences

In Chapter 4 we introduced the subject of set theory. There, we defined sets to be unordered collections of objects—there was no notion of order or repetition in the context of set theory.

In this chapter we introduce a more structured means of representing information—a means of representing information in such a way that not only can elements appear more than once, but the order in which they appear is also important. Thus, both repetition and order have a role to play.

Before introducing this structure—which we shall term a *sequence*—we investigate a structure in which repetition, but not order, is important.

10.1 Bags

Sets are structures in which neither order nor repetition are important. For example, we have seen that all of the following sets are equal.

$$\{1, 1, 3\} = \{1, 3, 1\} = \{3, 1, 1\} = \{1, 3\} = \{3, 1\}$$

Sometimes, however, especially when modelling real-world entities, we are concerned with the repetition of elements.

Consider, as an example, the following set.

$$pets = \{dog, cat, hamster\}$$

We may conclude from the elements of this set that the person in question has three pets: a dog, a cat, and a hamster. Suppose, however, that the person adopted a second cat. We might update our view of the set of pets in the following way.

$$pets\_now = \{dog, cat, hamster\} \cup \{cat\}$$

Unfortunately, however, as there is no notion of repetition for sets, we have

$$pets\_now = \{dog, cat, hamster\}$$

There is no way of determining how many cats the person now owns if we represent this information in terms of a set—all we know is that he owns at least one cat, at least one dog, and at least one hamster.

If the number of occurrences of elements is important to us, then we need to consider representing our information using a different structure; in this case, a suitable structure is a *bag*. If, however, repetition is not important, then sets are perfectly satisfactory.

First, we consider the notation for denoting bags.

We enclose the elements of a bag between square brackets: $[\![$ and $]\!]$. As such, we may represent the set of pets in terms of a bag by

$$[\![dog, cat, hamster, cat]\!]$$

Just as we could perform operations on pairs of sets using the $\cup$, $\cap$ and $\setminus$ operators, so we have similar operators for bags.

Assuming two bags $B = [\![a, a, b, c]\!]$ and $C = [\![b, b, c, d]\!]$, the bag operators work on B and C in the following way.

The bag union operator—just like the set union operator—merges the elements of the two contributing bags. However, unlike set union, bag union preserves repeated elements. As such,

$$B \cup C = [\![a, a, b, b, b, c, c, d]\!]$$

The intersection of bags B and C gives the following result.

$$B \cap C = [\![b, c]\!]$$

Here, the first occurrence of b appears in both bags, but the second occurrence of b appears only in C. As such, one copy of b appears in the intersection. In addition, c appears once in both B and C. As such, c appears in the intersection.

Finally, the results of applying bag difference to B and C are given below.

$$B \setminus C = [\![a, a]\!]$$
$$C \setminus B = [\![b, d]\!]$$

Here, a appears twice in B and not at all in C. In addition, d appears in C but not in B, and b appears twice in C and only once in B.

Exercise 10.1 Which of the following bags are equal?

1. $[\![a, a, b, b]\!]$
2. $[\![a, b, a, b]\!]$
3. $[\![a, b, b, a]\!]$

$\square$

Exercise 10.2 Suppose the following bags were to be represented in terms of sets. What would be the resulting set in each case?

1. $[\![ ]\!]$
2. $[\![1, 2]\!]$

3. $[\![1, 1, 2, 2]\!]$

4. $[\![\{1, 2\}, \{2\}]\!]$

$\square$

Exercise 10.3 Assume the following bags.

$$B = [\![2, 3, 4]\!]$$
$$C = [\![2, 2]\!]$$
$$D = [\![3, 3, 4]\!]$$

Calculate the following.

1. $C \cup D$

2. $B \cap D$

3. $C \setminus B$

4. $C \cap (B \cup D)$

5. $(C \cap B) \cup D$

$\square$

10.2   A need for order

As we have seen, it is sometimes the case that we are concerned with the repetition of elements; in such cases, bags may be adequate for our modelling needs. There are, however, some circumstances under which we need to consider the ordering of elements as well as their repetition. Consider, as an example, a real-time scheduler that keeps track of processes that require resources. A scheduling algorithm will determine which job is to be executed next and, when the relevant resources become available, allocate those resources to that job. Modelling such an entity in terms of a set or a bag containing those jobs which are to be processed would be problematic, as all the hard work performed by the scheduling algorithm in determining the order in which jobs should be processed would be lost due to our representation having no notion of order. Thus, we need a slightly more sophisticated means of representing this information: we need *sequences*.

The elements of a sequence are enclosed within angular brackets: $\langle$ and $\rangle$.

Example 10.1 The sequence $\langle a, b \rangle$ has two elements: a and b. What is more, as the elements of a sequence are ordered, we may say that a is at position 1 of the sequence and b is at position 2. As order is important, this sequence is *not* the same as $\langle b, a \rangle$. Furthermore, as repetition is also important, the sequence $\langle a, b, b \rangle$ is also a very different entity to $\langle a, b \rangle$ (and, for that matter, $\langle b, a \rangle$). $\square$

Exercise 10.4 Which of the following sequences are equal?

1. $\langle a, a, b, b \rangle$

2. $\langle a, b, a, b \rangle$

3. $\langle a, b, b, a \rangle$

□

Exercise 10.5 Suppose the following sequences were to be represented in terms of sets. What would be the resulting set in each case?

1. $\langle \rangle$

2. $\langle 1, 2 \rangle$

3. $\langle 1, 1, 2, 2 \rangle$

4. $\langle \{1, 2\}, \{2\} \rangle$

□

We denote the set of all sequences of type X by seq X.

Example 10.2 The set of all sequences of natural numbers is denoted seq $\mathbb{N}$, with $\langle 1, 2, 3 \rangle$ being an element of seq $\mathbb{N}$. As such, we may write $\langle 1, 2, 3 \rangle \in$ seq $\mathbb{N}$. □

Exercise 10.6 To which sets do the following sequences belong?

1. $\langle 7, 11, 13, 17, 19, 23 \rangle$

2. $\langle (7, 11), (13, 17), (19, 23) \rangle$

3. $\langle \{7, 11\}, \{13, 17\}, \{19, 23\} \rangle$

4. $\langle \{(7, 11), (13, 17)\}, \{(19, 23)\} \rangle$

□

10.3   Modelling sequences

Just as a function is a special kind of relation and a relation is a special kind of set, so a sequence may be viewed as being a special kind of function.

Consider a Formula One Grand Prix race, for which we wish to keep track of the drivers occupying the first three positions. If Schumacher is in first place then we would like to associate him with the value 1; if Irvine is in second place, then we would like to associate him with the value 2; and if Hakkinen is in third place then we would like to associate him with the value 3. With regards to our sequence notation, this gives us

$\langle schumacher, irvine, hakkinen \rangle$

A different solution might be to define a function *position* which maps natural numbers (representing positions) to drivers. So, assuming the existence of the type *Driver*, we might have

$position : \mathbb{N} \nrightarrow Driver$

with, for the above scenario,

$$position = \{1 \mapsto schumacher, 2 \mapsto irvine, 3 \mapsto hakkinen\}$$

This is effectively the same information as is provided by our sequence: in both cases, position 1 is associated with Schumacher, position 2 is associated with Irvine, and position 3 is associated with Hakkinen.

As such, we may think of a sequence of type X as being a partial function from a contiguous subsequence of the natural numbers, starting from 1. That is,

$$\text{seq } X \subseteq \mathbb{N} \setminus \{0\} \nrightarrow X$$

Of course, as sequences are functions, all of the operators which could legitimately be applied to functions may also be applied to sequences.

Exercise 10.7 Define seq X formally, in terms of set comprehension. □

Exercise 10.8 Which of the following functions are also sequences?

1. $\emptyset$

2. $\{a \mapsto 1, b \mapsto 2, c \mapsto 3\}$

3. $\{0 \mapsto a, 1 \mapsto b, 2 \mapsto c\}$

4. $\{1 \mapsto a, 3 \mapsto b, 4 \mapsto c\}$

5. $\{1 \mapsto a, 2 \mapsto b, 3 \mapsto c\}$

□

Exercise 10.9 Calculate the following.

1. dom $\langle london, edinburgh, cardiff, belfast \rangle$

2. ran $\langle london, edinburgh, cardiff, belfast \rangle$

3. $\langle london, edinburgh, cardiff, belfast \rangle \rhd \{cardiff, belfast\}$

4. $\{3, 4\} \lhd \langle london, edinburgh, cardiff, belfast \rangle$

5. $\langle london, edinburgh, cardiff, belfast \rangle \oplus \langle manchester, glasgow \rangle$

6. $\langle london, edinburgh, cardiff, belfast \rangle \, (\!| \, \{1, 3\} \, |\!)$

7. $\langle london, edinburgh, cardiff, belfast \rangle \,^{\sim}$

8. $\langle london, edinburgh, cardiff, belfast \rangle \, 3$

□

10.4 The empty sequence

Just as the empty set was a special set, so the empty sequence is a special sequence; we denote the empty sequence by $\langle\rangle$. This is the sequence that contains no elements.

Example 10.3 Consider the following function.

$$id\_seq(n) = \{m : 1 \mathinner{\ldotp\ldotp} n \bullet m \mapsto m\}$$

If $n = 3$, then

$$id\_seq(n) = \langle 1, 2, 3 \rangle$$

Furthermore, if $n = 0$, then

$$id\_seq(n) = \langle\rangle$$

□

10.5 Length

The cardinality operator, #, was defined in terms of sets. Given some set S, then $\# S$ denotes the number of elements which appear in S. As sequences are defined in terms of functions, which are themselves sets of pairs, it follows that the cardinality operator is also applicable to sequences. In this context, however, we refer to # as the *length* operator. Given some sequence s, then $\# s$ denotes the length of s.

Example 10.4 The length of the empty sequence is 0. □

Law 10.1

$$\# \langle\rangle = 0$$

□

Example 10.5

$$\# \langle a, b, c \rangle = 3$$

□

Exercise 10.10 What are the lengths of the following sequences?

1. $\langle a, b, b, a \rangle$

2. $id\_seq(10)$

3. $\langle london, edinburgh, cardiff, belfast \rangle \oplus \langle manchester, glasgow \rangle$

□

Exercise 10.11 Prove that, for any sequence s,

$$\# \, id\_seq \, (\# \, s) = \# \, s$$

☐

Exercise 10.12 Which of the following are true in general?

1. $\# \, s = \# \, (\text{dom } s)$
2. $\# \, s = \# \, (\text{ran } s)$

☐

10.6 Concatenation

Thus far, we have considered how to represent sequences and denote their length. We have not, however, seen any means of combining two sequences; we remedy this now.

The *concatenation* operator, $\frown$, allows us to add one sequence to the end of another one to produce a sequence that contains the elements of both sequences in their original order. Given two sequences s and t, the concatenation of s and t is written $s \frown t$.

Example 10.6

$$\langle schumacher, irvine, hakkinen \rangle \frown \langle button, coulthard \rangle =$$
$$\langle schumacher, irvine, hakkinen, button, coulthard \rangle$$

☐

Example 10.7

$$\langle london, edinburgh \rangle \frown \langle cardiff, belfast \rangle =$$
$$\langle london, edinburgh, cardiff, belfast \rangle$$

☐

In function notation, the concatenation of two sequences s and t is defined formally as follows.

$$s \frown t = s \cup \{x : (\# \, s + 1) \, . . \, (\# \, s + \# \, t) \bullet x \mapsto t \, (x - \# \, s)\}$$

Example 10.8 If $s = \langle a, b \rangle$ and $t = \langle c, d \rangle$, then the above definition gives

$$s \frown t = s \cup \{x : 3 \, . . \, 4 \bullet x \mapsto \langle c, d \rangle \, (x - 2)\}$$

which, of course, is equivalent to $\langle a, b, c, d \rangle$. ☐

There are a number of laws associated with concatenation. First, the length of $s \frown t$ is defined in terms of the lengths of its contributing parts.

Law 10.2 For any sequences s and t,

$$\#\,(s \frown t) = (\#\,s) + (\#\,t)$$

□

Second, the concatenation of the empty sequence with any other sequence s, is equal to s.

Law 10.3 For any sequence s,

$$\langle\rangle \frown s = s$$

and

$$s \frown \langle\rangle = s$$

□

Finally, the concatenation operator is associative.

Law 10.4 For any sequences s, t and u,

$$s \frown (t \frown u) = (s \frown t) \frown u$$

□

Exercise 10.13 Which of the following are legal applications of $\frown$?

1. $\{1 \mapsto a, 2 \mapsto b\} \frown \{\}$
2. $\{1 \mapsto a, 2 \mapsto b\} \frown \langle\rangle$
3. $\{1 \mapsto a, 2 \mapsto b\} \frown \{1 \mapsto c\}$
4. $\{1 \mapsto a, 2 \mapsto b\} \frown \{1 \mapsto 1\}$
5. $\{1 \mapsto a, 2 \mapsto b\} \frown \{3 \mapsto c\}$

□

Exercise 10.14 Which of the following statements are true in general?

1. $s \frown t = t \frown s$
2. $s \frown s = s$
3. $s \frown (t \frown u) = (s \frown t) \frown (s \frown u)$

□

Exercise 10.15 Calculate the following.

1. $\langle a, b \rangle \frown \langle c, d \rangle$
2. $\langle a, b \rangle \frown \langle\rangle$

3. $\langle a, b \rangle \frown (\langle c, d \rangle \frown \langle e, f \rangle)$

4. $\# (id\_seq\,(300) \frown id\_seq\,(200))$

5. $\# (\langle\rangle \frown id\_seq\,(200))$

6. $\# ((id\_seq\,(300) \frown id\_seq\,(200)) \setminus id\_seq\,(100))$

$\square$

Exercise 10.16 Calculate the following.

1. $\langle london, edinburgh, cardiff, belfast \rangle \frown \langle manchester, glasgow \rangle$

2. $\langle london, edinburgh, cardiff, belfast \rangle \oplus \langle manchester, glasgow \rangle$

3. $\langle london, edinburgh, cardiff, belfast \rangle \cup \langle manchester, glasgow \rangle$

$\square$

10.7   Head and tail

If we have gone to the trouble of storing information in a sequence as opposed to a set, it is only natural to try and make use of the fact that this information is ordered. In many applications in which information is stored in a sequence, it is often useful to reason about the first element of the sequence. In the context of real-time scheduling, if we store the jobs to be executed in the order in which they are to be processed, the element at the head of the sequence is the job that is to be executed next: this is clearly an important and unique element given the context. We refer to this first element of a sequence as the *head* of the sequence. Given any non-empty sequence s, the head of s is denoted *head s*.

Example 10.9 Given the sequence $\langle a, b, c \rangle$,

$$head \langle a, b, c \rangle = a$$

$\square$

There is a further operator which complements the head operator—the *tail* operator. This denotes the part of the sequence which follows the head. Given any non-empty sequence s, the tail of s is denoted *tail s*.

Example 10.10 Given the sequence $\langle a, b, c \rangle$,

$$tail \langle a, b, c \rangle = \langle b, c \rangle$$

$\square$

As further examples,

$$tail\ tail \langle a, b, c \rangle = \langle c \rangle$$

and

$$head\ tail \langle a, b, c \rangle = b$$

Note that our head and tail operators are only defined on non-empty sequences. The result of applying either operator to $\langle\rangle$ is undefined in each case.

Of course, given that we consider sequences to be defined in terms of functions, we may define *head s* for any non-empty sequence *s* as follows.

$$head\ s = s\,1$$

Furthermore, we may define the tail operator as follows.

$$tail\ s = \{n : 1 \mathbin{..} \#(s - 1) \bullet n \mapsto s\,(n + 1)\}$$

Example 10.11

$$tail\ \langle a, b, c \rangle = \{n : 1 \mathbin{..} 2 \bullet n \mapsto \langle a, b, c \rangle\,(n + 1)\}$$

$\square$

Note that the type of *head* for sequences of type X is seq $X \nrightarrow X$ and the type of *tail* for sequences of type X is seq $X \nrightarrow$ seq X—the former returns an element as a result, whereas the latter returns a sequence.

Our first law for *head* and *tail* states that any sequence is equivalent to the concatenation of its head and its tail.

Law 10.5 For any non-empty sequence s,

$$s = \langle head\ s \rangle \frown (tail\ s)$$

$\square$

Example 10.12

$$\langle a, b, b, a \rangle = \langle head\ \langle a, b, b, a \rangle \rangle \frown (tail\ \langle a, b, b, a \rangle)$$

$\square$

Secondly, the size of any sequence is one greater than the size of its tail.

Law 10.6 For any non-empty sequence s,

$$\#\,s = 1 + (\#\,tail\ s)$$

$\square$

Example 10.13

$$\#\,\langle a, b, b, a \rangle = 1 + (\#\,tail\ \langle a, b, b, a \rangle)$$

$\square$

Exercise 10.17 Which of the following are defined?

1. *head* $\langle\rangle$

2. *tail* $\langle a \rangle$

3. *head tail* $\langle a \rangle$

4. *head id_seq* (0)

$\square$

Exercise 10.18 Calculate the following.

1. *head tail id_seq* (3)

2. *tail tail tail id_seq* (3)

3. *head* $\langle id\_seq\,(0) \rangle$

4. *head* $\langle id\_seq\,(1), id\_seq\,(2), id\_seq\,(3) \rangle$

5. *tail* $\langle head\ id\_seq\,(1), head\ id\_seq\,(2), head\ id\_seq\,(3) \rangle$

6. $\#\ tail\ id\_seq\,(100)$

7. $\#\,(tail\ id\_seq\,(100) \frown tail\ id\_seq\,(100))$

$\square$

10.8   Restrictions

Given a sequence s, we may wish to consider only some particular subset of its elements. The *restriction* operator allows us to do exactly this. Given a sequence s, together with a set of elements X, $s \upharpoonright X$ denotes the *restriction* of s to those elements which appear in X. The result of this operation is a sequence featuring only those elements which appear in both s and X, with the ordering of these elements determined by their order in s.

Example 10.14 Consider the following sequence.

$$\langle rachel, duncan, john, richard, emily, andy \rangle$$

Here, six elements are maintained in a sequence, with *rachel* appearing at the head of the list. The restriction of this sequence to the set $\{emily, rachel\}$ is denoted

$$\langle rachel, duncan, john, richard, emily, andy \rangle \upharpoonright \{emily, rachel\}$$

This gives the result $\langle rachel, emily \rangle$. $\square$

We may define this operator recursively as follows.

$$\langle\rangle \upharpoonright X = \langle\rangle$$
$$(\langle x \rangle \frown s) \upharpoonright X = \langle x \rangle \frown (s \upharpoonright X) \text{ if } x \in X$$
$$(\langle x \rangle \frown s) \upharpoonright X = s \upharpoonright X \qquad\quad \text{ if } x \notin X$$

We may apply this definition to the above example in the following fashion.

$$\langle rachel, duncan, john, richard, emily, andy\rangle \upharpoonright \{emily, rachel\}$$
$$= \langle rachel\rangle \frown (\langle duncan, john, richard, emily, andy\rangle \upharpoonright \{emily, rachel\})$$
$$= \langle rachel\rangle \frown (\langle john, richard, emily, andy\rangle \upharpoonright \{emily, rachel\})$$
$$= \langle rachel\rangle \frown (\langle richard, emily, andy\rangle \upharpoonright \{emily, rachel\})$$
$$= \langle rachel\rangle \frown (\langle emily, andy\rangle \upharpoonright \{emily, rachel\})$$
$$= \langle rachel\rangle \frown \langle emily\rangle \frown (\langle andy\rangle \upharpoonright \{emily, rachel\})$$
$$= \langle rachel\rangle \frown \langle emily\rangle \frown (\langle\rangle \upharpoonright \{emily, rachel\})$$
$$= \langle rachel\rangle \frown \langle emily\rangle \frown \langle\rangle$$
$$= \langle rachel, emily\rangle$$

The first law associated with this operator states that the empty sequence restricted to any set is the empty sequence.

Law 10.7 For any set X,

$$\langle\rangle \upharpoonright X = \langle\rangle$$

□

Example 10.15

$$\langle\rangle \upharpoonright \{emily, rachel\} = \langle\rangle$$

□

Next, any sequence restricted to the empty set gives the empty sequence.

Law 10.8 For any sequence s,

$$s \upharpoonright \emptyset = \langle\rangle$$

□

Example 10.16

$$\langle rachel, duncan, john, richard, emily, andy\rangle \upharpoonright \emptyset = \langle\rangle$$

□

Finally, $\upharpoonright$ distributes through $\frown$.

Law 10.9 For any sequences $s, t \in \text{seq } X$ and any set $S \in \mathbb{P}\, X$,

$$(s \frown t) \upharpoonright S = (s \upharpoonright S) \frown (t \upharpoonright S)$$

□

Example 10.17 We can show that

$$(\langle a, b, c \rangle \frown \langle b, c, d \rangle) \restriction \{b\} = (\langle a, b, c \rangle \restriction \{b\}) \frown (\langle b, c, d \rangle \restriction \{b\})$$

holds.

Here, the left-hand side can be simplified as follows.

$$(\langle a, b, c \rangle \frown \langle b, c, d \rangle) \restriction \{b\} = \langle a, b, c, b, c, d \rangle \restriction \{b\}$$
$$= \langle b, b \rangle$$

In addition, the right-hand side can be simplified thus.

$$(\langle a, b, c \rangle \restriction \{b\}) \frown (\langle b, c, d \rangle \restriction \{b\}) = \langle b \rangle \frown \langle b \rangle$$
$$= \langle b, b \rangle$$

As such, the two sides are equal. $\square$

Exercise 10.19 Calculate the following.

1. $\langle tea, coffee, tea, tea, coffee \rangle \restriction \{tea\}$
2. $\langle tea, coffee, tea, tea, coffee \rangle \restriction \{coffee\}$
3. $\langle tea, coffee, tea, tea, coffee \rangle \restriction \{tea, coffee\}$
4. $\langle tea, coffee, tea, tea, coffee \rangle \restriction \emptyset$

$\square$

Exercise 10.20 Calculate the following.

1. $head\,(\langle a, b, c \rangle \restriction \{b, c\})$
2. $tail\,(\langle a, b, c \rangle \restriction \{b, c\})$
3. $tail\,(tail\,(\langle a, b, c \rangle \restriction \{b, c\}))$

$\square$

Exercise 10.21 Assuming that $A \in \text{seq}\,(People \times People)$, and that

$$A = \langle (jim, jon), (dave, duncan), (emily, elizabeth) \rangle$$

calculate the following.

1. $A \restriction \{p : A \mid p.1 = jim\}$
2. $A \restriction \{p : tail\,A \mid p.1 = jim\}$
3. $A \restriction \{p, q : Person \mid (p, q) \in A \land p = emily\}$
4. $A \restriction \{p : \{jim, dave\};\ q : Person \mid (p, q) \in A\}$

$\square$

Exercise 10.22 Consider the set A of the previous exercise. Show how the following can be generated from A via set comprehension.

1. $\langle jim, dave, emily \rangle$

2. $\langle jim, duncan \rangle$

3. $\langle dave, emily \rangle$

□

10.9 Reversing

Given a sequence s, the *reverse* of s is also a sequence, but one in which the order in which the elements appear is reversed, i.e., the first element of s is the last element of the reverse of s, the second element of s is the penultimate element of the reverse of s, and so on. Given a sequence s, we denote the reverse of s by *reverse s*.

Example 10.18 The reverse of the sequence $\langle a, b, c \rangle$ is given by $\langle c, b, a \rangle$. □

Example 10.19

$$reverse \langle 1, 2, 3, 4, 5 \rangle = \langle 5, 4, 3, 2, 1 \rangle$$

□

We may define *reverse* formally in terms of a recursively defined function.

$$reverse \langle \rangle = \langle \rangle$$
$$reverse \langle x \rangle = \langle x \rangle$$
$$reverse \langle x \rangle \frown s = (reverse\ s) \frown \langle x \rangle$$

Example 10.20 *reverse* $\langle a, b, c \rangle$ may be calculated as follows.

$$
\begin{aligned}
reverse \langle a, b, c \rangle &= (reverse \langle b, c \rangle) \frown \langle a \rangle \\
&= ((reverse \langle c \rangle) \frown \langle b \rangle) \frown \langle a \rangle \\
&= (((\langle c \rangle) \frown \langle b \rangle) \frown \langle a \rangle \\
&= \langle c, b, a \rangle
\end{aligned}
$$

□

The first law associated with *reverse* is that the reverse of the reverse of a sequence s is s.

Law 10.10 Given any sequence s,

$$reverse\ reverse\ s = s$$

□

Example 10.21

$$reverse\ reverse\ \langle 1, 2, 3 \rangle = \langle 1, 2, 3 \rangle$$

□

Next, if we wish to reverse and restrict a sequence, it makes no difference in which order these operations are performed.

Law 10.11 Given any sequence $s \in$ seq X and any set $S \in \mathbb{P}\ X$,

$$(reverse\ s) \restriction S = reverse\ (s \restriction S)$$

□

Example 10.22 We can show that

$$(reverse\ \langle a, b, c \rangle) \restriction \{a, b\} = reverse\ (\langle a, b, c \rangle \restriction \{a, b\})$$

holds.

The left-hand side can be evaluated as follows.

$$\begin{aligned}
(reverse\ \langle a, b, c \rangle) \restriction \{a, b\} &= \langle c, b, a \rangle \restriction \{a, b\} \\
&= \langle b, a \rangle
\end{aligned}$$

Furthermore, the right-hand side can be evaluated thus.

$$\begin{aligned}
reverse\ (\langle a, b, c \rangle \restriction \{a, b\}) &= reverse\ \langle a, b \rangle \\
&= \langle b, a \rangle
\end{aligned}$$

As such, the two sides are equal. □

Finally, the size of the reverse of s is the same as the size of s.

Law 10.12 Given any sequence s,

$$\#\ reverse\ s = \#\ s$$

□

Example 10.23

$$\#\ \langle a, b, c \rangle = \#\ \langle c, b, a \rangle$$

□

Exercise 10.23 Prove that *reverse* does not distribute through $^\frown$. □

Exercise 10.24 Calculate the following.

1. *reverse id_seq* (3)

2. *reverse tail id_seq* (3)

3. *head reverse id_seq* (3)

4. *reverse* (*id_seq* (10) ↾ {*n* : ℕ | *n* is odd})

5. (*reverse id_seq* (3)) ⌢ (*reverse id_seq* (3))

□

10.10   Injective sequences

Recall from Chapter 9 that an injective function was defined to be a function in which each element of the target is mapped to by at most one element of the source. As such, each element of the target appears at most once in the range. For example, i, which is given below is an injective function.

$$i = \{a \mapsto 1, b \mapsto 2, c \mapsto 3, d \mapsto 4\}$$

Here, no element of the target appears more than once in the range.

On the other hand, the function j is not an injective function.

$$j = \{a \mapsto 1, b \mapsto 2, c \mapsto 1, d \mapsto 2\}$$

Here, both 1 and 2 appear in the range more than once.

As a sequence is a special kind of function, we may also consider the notion of an *injective sequence*. Given a sequence s, we may say that s is injective if, and only if, no element appears more than once in s. We denote the set of all injective sequences of type X by iseq X. We define this set formally as follows.

$$\text{iseq } X = \{s : \text{seq } X \mid (\forall x, y : \text{dom } s \mid x \neq y \bullet s\,x \neq s\,y)\}$$

Equivalently,

$$\text{iseq } X = (\text{seq } X) \cap (\mathbb{N} \setminus \{0\} \rightarrowtail X)$$

Example 10.24 $\langle a, b, c, d \rangle$ is an injective sequence: it contains four elements which occur exactly once. On the other hand, $\langle a, b, b, a \rangle$ is not an injective sequence: it contains two elements, both of which appear twice. □

Example 10.25 *id_seq* (*n*), for any $n \in \mathbb{N}$, is an injective sequence. □

Exercise 10.25 Which of the following sequences are injective?

1. $\langle \rangle$

2. *id_seq* (3)

3. $id\_seq\,(3) \frown id\_seq\,(5)$

4. $id\_seq\,(3) \frown (tail\ tail\ tail\ id\_seq\,(5))$

□

Exercise 10.26 Prove that a sequence s is injective exactly when

$$\# s = \# \text{ ran } s$$

□

10.11 Recursively defined functions revisited

In the previous chapter we introduced the notion of recursively defined functions. There, we showed that the general form of a recursively defined function is given by

$$f(n) = x \qquad \text{if } n = m$$
$$\quad\; = y + f(n') \quad \text{otherwise}$$

Furthermore, we said that a recursively defined function f is *well-defined* if it possesses the following properties.

1. There are certain values, called *base values*, for which the function does not refer to itself.

2. Every time the function does refer to itself, the argument of the function is closer to a base value.

The structure of our sequences means that many functions that we may wish to define upon sequences may be defined recursively. Consider, for example, the following definition of the *reverse* operator.

$$reverse\ \langle\rangle = \langle\rangle$$
$$reverse\ \langle x \rangle = \langle x \rangle$$
$$reverse\ \langle x \rangle \frown s = (reverse\ s) \frown \langle x \rangle$$

Here, not only do we find that *reverse* can be defined in terms of itself, but there are also two base cases—the empty sequence and the singleton sequence—at which point the recursion terminates. Furthermore, the recursive definition moves the argument closer to the base cases. As such, not only is *reverse* a recursively defined function, but it is also well-defined.

As a further example, we may define the length operator, $\#$, recursively as follows.

$$\#\ \langle\rangle = 0$$
$$\#\ \langle x \rangle \frown s = 1 + \#\ s$$

Again, this function is well-defined.

Exercise 10.27 Define a recursive function of type seq $\mathbb{N} \nrightarrow \mathbb{N}$ which returns the largest natural number appearing in a sequence of natural numbers. You may assume that the result of applying this function to $\langle\rangle$ is 0. □

Exercise 10.28 Show how the function of the previous exercise might be defined non-recursively in terms of the μ operator. $\square$

Exercise 10.29 Define a recursive function of type seq $\mathbb{N} \nrightarrow$ seq $\mathbb{N}$ which, when given a sequence of natural numbers, s, returns a sequence in which every element is doubled. So, for example, applying this function to $\langle 1, 2, 3 \rangle$ would give $\langle 2, 4, 6 \rangle$. $\square$

Exercise 10.30 Show how the function of the previous exercise might be defined non-recursively in terms of set comprehension. $\square$

10.12   Further exercises

Exercise 10.31 Assume some sequence $s \in$ seq X, such that $\# s = n$. How many elements do the following sequences have?

1. $\langle \rangle$
2. $s \mathbin{\frown} s$
3. *tail s*
4. *reverse s*
5. $s \upharpoonright \emptyset$
6. $s \upharpoonright X$
7. *head* $\langle s, s \rangle$
8. *tail* $\langle s, s \rangle$

$\square$

Exercise 10.32 What is the value of each of the following?

1. $\# \langle \langle \rangle, \langle m, a, d \rangle, \langle a, m \rangle \rangle$
2. ran $\langle m, a, d, a, m \rangle$
3. dom $\langle m, a, d, a, m \rangle$
4. $\langle m, a, d, a, m \rangle \oplus \langle a, d, a, m \rangle$

$\square$

Exercise 10.33 Given that *square* and *double* are defined by

$$square = \langle 1, 4, 9, 16 \rangle$$
$$double = \langle 2, 4, 6, 8 \rangle$$

calculate the following.

1. *square* $\mathbin{\frown}$ *double*
2. (*square* $\mathbin{\frown}$ *double*) $\upharpoonright \{1, 3, 5, 7, 9\}$

3. $head\ ((square \frown double) \upharpoonright \{1, 3, 5, 7, 9\})$

4. $tail\ ((square \frown double) \upharpoonright \{1, 3, 5, 7, 9\})$

5. $\#\ ((square \frown double) \upharpoonright \{1, 3, 5, 7, 9\})$

6. $reverse\ ((square \frown double) \upharpoonright \{1, 3, 5, 7, 9\})$

□

Exercise 10.34 Consider the following sequences.

$$r = \langle athens, barcelona, cologne, dieppe \rangle$$
$$s = \langle athens, atlanta, barcelona \rangle$$
$$t = \langle atlanta, boston, chicago, denver \rangle$$

Given the above sequences, calculate the following.

1. $r \cap s$

2. $r \cup s$

3. $r \setminus s$

4. dom t

5. ran s

6. $r^\sim$

7. $\{1, 2\} \lhd r$

8. $s \rhd \{atlanta\}$

9. $(t \oplus s) (\!| \{1, 2\} |\!)$

10. $r\ 3$

□

Exercise 10.35 Given the sequences of the previous exercise, calculate the following.

1. $r \frown s$

2. $\#\ r$

3. $reverse\ s$

4. $r \upharpoonright ran\ s$

5. $(r \frown s) \upharpoonright \{atlanta, athens\}$

6. $head\ t$

7. $tail\ t$

8. $reverse\ tail\ t$

   9. *head tail t*

  10. $\langle head\ t \rangle \frown \langle head\ r \rangle$

$\square$

Exercise 10.36 Define a predicate $P(s)$ that is equivalent to *true* if, and only if, the sequence s has no duplicates. $\square$

Exercise 10.37 Define a predicate $P(s, x)$ that is equivalent to *true* if, and only if, the element x occurs in the sequence s. $\square$

Exercise 10.38 Define a predicate $P(s, x, n)$ which is equivalent to *true* if, and only if, the element x occurs at position n in sequence s. $\square$

Exercise 10.39 Define a function of type $X \times \text{seq } X \nrightarrow \mathbb{P}\mathbb{N}$, which, when given a sequence $s \in \text{seq } X$ and an element $x \in X$, returns all the indices at which x occurs in s. $\square$

Exercise 10.40 Define a function of type $X \times \text{seq } X \nrightarrow \mathbb{N}$, which, when given a sequence $s \in \text{seq } X$ and an element $x \in X$, returns the first index at which x occurs in s. $\square$

Exercise 10.41 Define a function of type $X \times \text{seq } X \nrightarrow \mathbb{N}$, which, when given a sequence $s \in \text{seq } X$ and an element $x \in X$, returns the last index at which x occurs in s. $\square$

Exercise 10.42 Define a recursive function of type $X \times \text{seq } X \rightarrow \mathbb{N}$, which, when given a sequence $s \in \text{seq } X$ and an element $x \in X$, returns the number of times x occurs in s. $\square$

Exercise 10.43 Give a non-recursive definition of the function of the previous exercise. $\square$

Exercise 10.44 Define a function of type $X \times \text{seq } X \times \mathbb{N} \nrightarrow \mathbb{N}$, which, when given a sequence $s \in \text{seq } X$, an element $x \in X$ and a natural number n, returns the position at which the nth occurrence of x appears in s. $\square$

Exercise 10.45 Define a recursive function *repeated*, which, for any sequence s, returns as a result the number of elements which occur more than once in s. $\square$

Exercise 10.46 Consider the function *flatten*, such that

$$\textit{flatten }\langle\langle a \rangle, \langle b \rangle, \langle c \rangle\rangle = \langle a, b, c \rangle$$

and

$$\textit{flatten }\langle\langle a, b, c \rangle, \langle \rangle, \langle d, e \rangle\rangle = \langle a, b, c, d, e \rangle$$

Give a recursive definition of *flatten*. $\square$

Exercise 10.47 Little Hampton train station has four platforms. Each of the platforms has a display which informs passengers of the next few departures from that platform. This information is represented in terms of the function *display*, the type of which is

$$\textit{display} : \textit{Platform} \nrightarrow \text{seq}\,(\textit{Time} \times \textit{Destination})$$

For example, if platform 2's display is currently indicating that the 12.20 train to Norwich and the 13.30 train to Bristol are the next departures from that platform, then

> *display p2* = ⟨(12.20, *Norwich*), (13.30, *Bristol*)⟩

1. Write a predicate which states that all sequences which appear on the displays are ordered by time, with the earliest timed train appearing at the head of the sequence.

2. Write a predicate which states that all displays contain unique information; that is, no specific (*time*, *destination*) pair can appear on more than one display.

3. Write a set comprehension to define a set containing all the destinations which are currently displayed.

4. Write a set comprehension to define a set containing all the times of departures, together with the relevant platform, to Swansea.

5. Define a function that adds a given (*time*, *destination*) pair to the end of a given platform's display.

6. Define a function that removes a given (*time*, *destination*) pair from a given platform's display.

□

Exercise 10.48 A departure board at an airport shows the following information: departure times, flight numbers, destinations, gate numbers, and status. An example state of the board is

Departure time	Flight number	Destination	Gate	Status
10.30	BA11	Singapore	29	Boarding
10.40	KL27	Amsterdam	27	Delayed
10.45	QA09	Melbourne	13	Boarding

We may model such a departure board formally as

> *b* : seq (*Time* × *Number* × *Destination* × *Gate* × *Status*)

So, in the above example, we have

> *b* = ⟨(10.30, BA11, Singapore, 29, Boarding),
> (10.40, KL27, Amsterdam, 27, Delayed),
> (10.45, QA09, Melbourne, 13, Boarding)⟩

Furthermore, we assume that there is an ordering, <, on *Time*, and we assume the existence of a variable, *current* : *Time*, which represents the current time. So, for example, if the current time is 10.37, we have 10.30 < *current* and *current* < 10.40.

1. Write a predicate which states that all tuples which appear in *b* are ordered according to their time of departure.

2. Write a predicate which states that only delayed flights or those which have a departure time of later than the current time may appear in *b*.

3. Write a predicate which states that if a flight with number BA11 appears in b, then it must be destined for Singapore.

4. Write a set comprehension to generate the details of all flights which are delayed.

5. Write a set comprehension to generate the destinations of all flights in b.

6. Write a set comprehension to generate (destination, time) pairs of all flights in b.

7. Write a set comprehension to generate (destination, time) pairs of all flights which are delayed.

□

10.13 Solutions

Solution 10.1 They are all equal. □

Solution 10.2

1. $\emptyset$
2. $\{1, 2\}$
3. $\{1, 2\}$
4. $\{\{1, 2\}, \{2\}\}$

□

Solution 10.3

1. $[\![2, 2, 3, 3, 4]\!]$
2. $[\![3, 4]\!]$
3. $[\![2]\!]$
4. $[\![2]\!]$
5. $[\![2, 3, 3, 4]\!]$

□

Solution 10.4 They are all different. □

Solution 10.5

1. $\emptyset$
2. $\{1, 2\}$
3. $\{1, 2\}$
4. $\{\{1, 2\}, \{2\}\}$

□

Solution 10.6

1. seq $\mathbb{N}$

2. seq $(\mathbb{N} \times \mathbb{N})$

3. seq $\mathbb{P}\,\mathbb{N}$

4. seq $\mathbb{P}\,(\mathbb{N} \times \mathbb{N})$

$\square$

Solution 10.7

$$\text{seq } X = \{s : \mathbb{N} \setminus \{0\} \nrightarrow X \mid (\exists\, n : \mathbb{N} \bullet \text{dom } s = 1 \mathrel{.\,.} n)\}$$

$\square$

Solution 10.8 Only 1 and 5 are also sequences. The others fail to satisfy the criteria for them to be sequences: 2 is not of the correct type, the domain of 3 includes 0, and 4 has a 'gap' in the domain. $\square$

Solution 10.9

1. $\{1, 2, 3, 4\}$

2. $\{london, edinburgh, cardiff, belfast\}$

3. $\langle london, edinburgh \rangle$

4. $\{3 \mapsto cardiff, 4 \mapsto belfast\}$

5. $\langle manchester, glasgow, cardiff, belfast \rangle$

6. $\{london, cardiff\}$

7. $\{london \mapsto 1, edinburgh \mapsto 2, cardiff \mapsto 3, belfast \mapsto 4\}$

8. $cardiff$

$\square$

Solution 10.10

1. 4

2. 10

3. 4

$\square$

Solution 10.11 Given some sequence s, the length of s is denoted by $\#\,s$. The domain of $id\_seq\,(\#\,s)$ is a contiguous subsequence of the natural numbers $1 \mathrel{.\,.} \#\,s$. As such, the length of this sequence is $\#\,s$. $\square$

Solution 10.12 1 is always true. 2 is not always true. Consider, for example, the sequence $\langle a, a \rangle$. Here, $\# \langle a, a \rangle = 2$ and $\# (\mathrm{ran} \langle a, a \rangle) = 1$. □

Solution 10.13 1, 2 and 3 are legal applications of $\frown$. 4 is not a legal application of $\frown$, as the sequences are of different types, and 5 is not a legal application of $\frown$ as the second argument is not a sequence. □

Solution 10.14 None of these statements are true in general. □

Solution 10.15

1. $\langle a, b, c, d \rangle$
2. $\langle a, b \rangle$
3. $\langle a, b, c, d, e, f \rangle$
4. 500
5. 200
6. 400

□

Solution 10.16

1. $\langle london, edinburgh, cardiff, belfast, manchester, glasgow \rangle$
2. $\langle manchester, glasgow, cardiff, belfast \rangle$
3. $\{1 \mapsto london, 2 \mapsto edinburgh, 3 \mapsto cardiff,$
   $4 \mapsto belfast, 1 \mapsto manchester, 2 \mapsto glasgow\}$

□

Solution 10.17 Only 2 is defined. □

Solution 10.18

1. 2
2. $\langle \rangle$
3. $\langle \rangle$
4. $\langle 1 \rangle$
5. $\langle 1, 1 \rangle$
6. 99
7. 198

□

Solution 10.19

1. $\langle tea, tea, tea \rangle$

2. $\langle coffee, coffee \rangle$

3. $\langle tea, coffee, tea, tea, coffee \rangle$

4. $\langle \rangle$

□

Solution 10.20

1. b

2. $\langle c \rangle$

3. $\langle \rangle$

□

Solution 10.21

1. $\langle (jim, jon) \rangle$

2. $\langle \rangle$

3. $\langle (emily, elizabeth) \rangle$

4. $\langle (jim, jon), (dave, duncan) \rangle$

□

Solution 10.22

1. $\{n : \text{dom } A \bullet n \mapsto (A\,n).1\}$

2. $\{n : \text{dom } A \mid n < 3 \bullet n \mapsto (A\,n).n\}$

3. $\{n : \text{dom } tail\,A \bullet n \mapsto (A\,n).1\}$

□

Solution 10.23 For *reverse* to distribute through $\frown$, the following would have to hold.

$$reverse\,(s \frown t) = (reverse\,s) \frown (reverse\,t)$$

Consider $s = \langle a, b \rangle$ and $t = \langle c, d \rangle$. Here,

$$reverse\,(\langle a, b \rangle \frown \langle c, d \rangle) = \langle d, c, b, a \rangle$$

and

$$(reverse\,\langle a, b \rangle) \frown (reverse\,\langle c, d \rangle) = \langle b, a, d, c \rangle$$

As such, *reverse* does not distribute through $\frown$. □

Solution 10.24

1. $\langle 3, 2, 1 \rangle$

2. $\langle 3, 2 \rangle$

3. 3

4. $\langle 9, 7, 5, 3, 1 \rangle$

5. $\langle 3, 2, 1, 3, 2, 1 \rangle$

□

Solution 10.25 1, 2 and 4 are injective, whereas 3 is not. □

Solution 10.26 By definition, s is injective if, and only if, every element of s appears exactly once. As such, $\# \operatorname{dom} s = \# \operatorname{ran} s$. We have previously seen that, for all sequences s, $\# s = \# \operatorname{dom} s$. By transitivity of equals, it follows that $\# s = \# \operatorname{ran} s$ must hold if s is an injective sequence. □

Solution 10.27

$$f\left(\langle \rangle\right) = 0$$
$$f\left(\langle x \rangle \frown s\right) = max\left(x, f\left(s\right)\right)$$

such that $max\left(x, y\right)$ is defined for natural numbers x and y by

$$max\left(x, y\right) = x \quad \text{if } x > y$$
$$= y \quad \text{otherwise}$$

□

Solution 10.28 This function may be defined non-recursively as follows.

$$f\left(s\right) = (\mu\, n : \operatorname{dom} s \mid (\forall\, m : \operatorname{dom} s \bullet (s\, n > s\, m) \vee (s\, n = s\, m \wedge n < m)) \bullet s\, n$$

□

Solution 10.29

$$f\left(\langle \rangle\right) = \langle \rangle$$
$$f\left(\langle x \rangle \frown s\right) = \langle x \times 2 \rangle \frown f\left(s\right)$$

□

Solution 10.30 This function may also be defined in the following way.

$$f\left(s\right) = \{n : \operatorname{dom} s \bullet n \mapsto 2 \times (s\, n)\}$$

□

Solution 10.31

1. 0

2. $2 \times n$

3. $n - 1$

4. n

5. 0

6. n

7. n

8. 1

□

Solution 10.32

1. 3

2. $\{m, a, d\}$

3. $\{1, 2, 3, 4, 5\}$

4. $\langle a, d, a, m, m \rangle$

□

Solution 10.33

1. $\langle 1, 4, 9, 16, 2, 4, 6, 8 \rangle$

2. $\langle 1, 9 \rangle$

3. 1

4. $\langle 9 \rangle$

5. 2

6. $\langle 9, 1 \rangle$

□

Solution 10.34

1. $\langle athens \rangle$

2. $\{1 \mapsto athens, 2 \mapsto barcelona, 2 \mapsto atlanta,$
   $\qquad 3 \mapsto cologne, 3 \mapsto barcelona, 4 \mapsto dieppe\}$

3. $\{2 \mapsto barcelona, 3 \mapsto cologne, 4 \mapsto dieppe\}$

4. $\{1, 2, 3, 4\}$

5. $\{athens, atlanta, barcelona\}$

6. $\{athens \mapsto 1, barcelona \mapsto 2, cologne \mapsto 3, dieppe \mapsto 4\}$

7. $\langle athens, barcelona \rangle$

8. $\{1 \mapsto athens, 3 \mapsto barcelona\}$

9. $\{athens, atlanta\}$

10. $cologne$

$\square$

Solution 10.35

1. $\langle athens, barcelona, cologne, dieppe, athens, atlanta, barcelona \rangle$

2. 4

3. $\langle barcelona, atlanta, athens \rangle$

4. $\langle athens, barcelona \rangle$

5. $\langle athens, athens, atlanta \rangle$

6. $atlanta$

7. $\langle boston, chicago, denver \rangle$

8. $\langle denver, chicago, boston \rangle$

9. $boston$

10. $\langle atlanta, athens \rangle$

$\square$

Solution 10.36

$$P(s) \Leftrightarrow \forall m, n : \text{dom } s \bullet m \neq n \Rightarrow s\,m \neq s\,n$$

$\square$

Solution 10.37

$$\forall s : \text{seq } X; \; x : X \bullet P(s, x) \Leftrightarrow x \in \text{ran } s$$

$\square$

Solution 10.38

$$\forall s : \text{seq } X; \; x : X; \; n : \mathbb{N} \bullet P(s, x, n) \Leftrightarrow s\,n = x$$

$\square$

Solution 10.39

$$\forall x : X; \; s : \text{seq } X \bullet f(x, s) = \{i : \text{dom } s \mid s\,i = x\}$$

$\square$

Solution 10.40

$$\forall x : X; \; s : \text{seq } X \bullet$$
$$f(x, s) = (\mu \, n : \mathbb{N} \mid s \, n = x \wedge (\forall m : \mathbb{N} \mid s \, m = x \bullet n \leq m))$$

□

Solution 10.41

$$\forall x : X; \; s : \text{seq } X \bullet$$
$$f(x, s) = (\mu \, n : \mathbb{N} \mid s \, n = x \wedge (\forall m : \mathbb{N} \mid s \, m = x \bullet n \geq m))$$

□

Solution 10.42

$$f(x, \langle \rangle) = 0$$
$$f(x, \langle y \rangle \frown s) = 1 + f(x, s) \quad \text{if } x = y$$
$$f(x, \langle y \rangle \frown s) = f(x, s) \qquad \text{otherwise}$$

□

Solution 10.43

$$\forall x : X; \; s : \text{seq } X \bullet f(x, s) = \# \{n : \mathbb{N} \mid s \, n = x\}$$

□

Solution 10.44

$$\forall x : X; \; s : \text{seq } X; \; n : \mathbb{N} \bullet$$
$$f(x, s, n) = (\mu \, m : \mathbb{N} \mid s \, m = x \wedge \# \{l : \mathbb{N} \mid s \, l = x \wedge l \leq m\} = n)$$

□

Solution 10.45

$$\textit{repeated } \langle \rangle = 0$$
$$\textit{repeated } \langle x \rangle \frown s = 1 + \textit{repeated } (s \restriction \text{ran } s \setminus \{x\}) \quad \text{if } x \in \text{ran } s$$
$$\textit{repeated } \langle x \rangle \frown s = \textit{repeated } s \qquad \qquad \qquad \text{otherwise}$$

□

Solution 10.46

$$\textit{flatten } \langle \rangle = \langle \rangle$$
$$\textit{flatten } \langle x \rangle \frown s = x \frown \textit{flatten } s$$

□

Solution 10.47

1. $\forall s : \operatorname{ran} display \bullet \forall p, q : \operatorname{dom} s \mid p < q \bullet p.1 \le q.1$

2. $\forall s, t : \operatorname{ran} display \bullet \operatorname{ran} s \cap \operatorname{ran} t = \emptyset$

3. $\bigcup \{ s : \operatorname{ran} display \bullet \{ p : \operatorname{ran} s \bullet p.2 \} \}$

4. $\{ p : Platform; \ s : \operatorname{seq} (Time \times Destination) \mid$
   $\qquad (p, s) \in display \wedge (\exists x : \operatorname{ran} s \bullet x.2 = swansea) \bullet x.1 \mapsto p \}$

5. $\forall d : Platform \nrightarrow \operatorname{seq} (Time \times Destination); \ p : Platform; \ q : Time \times Destination \bullet$
   $\qquad add (d, p, q) = d \oplus \{ p \mapsto (d\, p) \frown \langle q \rangle \}$

6. $\forall d : Platform \nrightarrow \operatorname{seq} (Time \times Destination); \ p : Platform; \ q : Time \times Destination \bullet$
   $\qquad remove (d, p, q) = d \oplus \{ p \mapsto (d\, p) \upharpoonright \{ q \} \}$

$\square$

Solution 10.48

1. $\forall s, t : \operatorname{dom} b \bullet s < t \Leftrightarrow (b\, s).1 \le (b\, t).1$

2. $\forall s : \operatorname{ran} b \bullet s.5 = delayed \vee s.1 > current$

3. $\forall s : \operatorname{ran} b \bullet s.2 = BA11 \Rightarrow s.3 = Singapore$

4. $\{ s : \operatorname{ran} b \mid s.5 = delayed \}$

5. $\{ s : \operatorname{ran} b \bullet s.3 \}$

6. $\{ s : \operatorname{ran} b \bullet (s.3, s.1) \}$

7. $\{ s : \operatorname{ran} b \mid s.5 = delayed \bullet (s.3, s.1) \}$

$\square$

Chapter 11

Induction

Thus far in this book, we have considered a number of ways of proving theorems. In Chapter 3, we saw that theorems of propositional logic could be established via truth tables, substitution, equational reasoning, or natural deduction. When we considered predicate logic, we saw that the techniques of equational reasoning and natural deduction could be extended from propositional logic to predicate logic. Furthermore, we saw that we could establish whether two sets (or relations, or functions, or sequences) are equivalent by using our system of equational reasoning. Together, the techniques of equational reasoning and natural deduction allow us to reason about, and prove properties of, most formal descriptions that we may develop.

If, on the other hand, we wished to prove some property of *all* sequences of a given type, or *all* natural numbers, then we need a different technique—that technique is called *induction*. Induction is a powerful mathematical technique, which is central to many computing applications. Furthermore, it is closely related to recursion, which we met in Chapters 9 and 10. Indeed, induction is often the only means of establishing that certain classes of recursive algorithms are correct.

In this chapter we consider two types of induction: mathematical induction, which allows us to establish properties of all natural numbers, and structural induction, which allows us to establish certain properties of all elements of a given inductively-defined type (such as, for example, sequences). We consider mathematical induction first.

Two texts that illustrate the relevance of mathematical induction are [You64] and [Wan80]. The former is more concerned with mathematics, whereas the latter is more concerned with the application of mathematical induction to programming.

11.1 Mathematical induction

In Chapter 2 we considered the topic of Peano arithmetic—a treatment of the natural numbers originally proposed by Giuseppe Peano. There, we saw that the natural numbers can be defined in terms of five axioms, the first four of which we considered in Chapter 2. These were stated as follows.

1. 0 is a natural number.

2. If x is a natural number, then $x + 1$ is also a natural number.

3. There is no natural number, z, such that $z + 1 = 0$.

4. Given two natural numbers x and y, if $x + 1 = y + 1$ then $x = y$.

Peano's fifth axiom, which we consider now, can be stated in the following way.

> If one can prove that if a property holds for some natural number x then it holds for $x + 1$, and if you can prove that the same property holds for 0, then the property will hold for all natural numbers.

That is, given some property p, if we can establish that $p(0)$ holds, and if we can also establish that $p(n) \Rightarrow p(n + 1)$ holds for every natural number n, then it follows that $p(n)$ holds for all natural numbers. We can write this axiom in the style of our natural deduction rules as follows.

$$\frac{p(0) \quad \forall n : \mathbb{N} \bullet p(n) \Rightarrow p(n + 1)}{\forall n : \mathbb{N} \bullet p(n)} \text{ [mathematical induction]}$$

Thus the proof that such a property holds for all natural numbers is divided into the proof of two cases: the *base case*—in which we establish that the property holds for 0—and the *inductive step*—in which we establish that if the property holds for n, then it also holds for $n + 1$. This is achieved by assuming the property holds of n—this is termed the *inductive hypothesis*—and using this assumption to prove that the property holds of $n + 1$. We may state the *principle of mathematical induction* formally as follows.

> Assume some property p, which is either true or false for each natural number. If
>
> 1. $p(0)$ is true and
> 2. $p(n) \Rightarrow p(n + 1)$ is true,
>
> then p is true for all natural numbers.

We start with a relatively simple example.

Example 11.1 We wish to prove that the theorem

$$\forall n : \mathbb{N} \bullet n < n + 1$$

holds.

To prove every natural number n is less than $n + 1$, we have to prove the base case and the inductive step. We consider the base case first. Here, we have

$$0 < 0 + 1$$

which is obviously true, by definition of $<$ and $+$.

Next, we consider the inductive step. To prove this, we first assume that the theorem is true for some value x, i.e., that

$$x < x + 1$$

This is the inductive hypothesis. From this, we have to prove

$$(x + 1) < (x + 1) + 1$$

By arithmetic, if $m < n$ then it follows that $m + 1 < n + 1$. Given that the inductive hypothesis states that $x < x + 1$, we may use this to conclude

$$(x + 1) < (x + 1) + 1$$

As we have established both the base case and the inductive step, we can conclude that

$$\forall\, n : \mathbb{N} \bullet n < n + 1$$

holds. $\square$

Many discrete mathematics texts motivate mathematical induction by means of falling dominoes. This analogy is given by considering a row of dominoes lined up one behind another. If the dominoes have been correctly spaced, then knocking over the first one will result in the second one being knocked over, which will result in the third one being knocked over, and so on. This establishes that all of the dominoes will be knocked over. The way that mathematical induction works is that if the theorem holds for 0, then it holds for 1; if it holds for 1, then it holds for 2; if it holds for 2, then it holds for 3; and so on. This analogy is certainly appropriate for the above example.

We now consider a second example.

Example 11.2 We can establish that the formula

$$\sum_{i=1}^{n} i = \frac{n \times (n + 1)}{2}$$

holds for all natural numbers n, by means of an inductive proof. To convince oneself of the validity of this formula consider the case of $n = 3$, which is given below.

$$\sum_{i=1}^{3} i = \frac{3 \times 4}{2}$$
$$= 6$$

As a further example, consider the case of $n = 5$, which is given below.

$$\sum_{i=1}^{5} i = \frac{5 \times 6}{2}$$
$$= 15$$

To establish the base case of $n = 0$, we arrive at

$$\sum_{i=1}^{0} i = \frac{0 \times 1}{2}$$
$$= 0$$

which, of course, is true. (Recall that $\sum_{i=n}^{m} = 0$ if $m < n$.)

We attempt to establish the inductive step by first assuming that the theorem is true for some value x, i.e.,

$$\sum_{i=1}^{x} i = \frac{x \times (x + 1)}{2}$$

This is the inductive hypothesis. The inductive step is proved if we can establish that

$$\sum_{i=1}^{x+1} i = \frac{(x + 1) \times ((x + 1) + 1)}{2}$$

follows from

$$\sum_{i=1}^{x} i = \frac{x \times (x + 1)}{2}$$

When $n = x + 1$, the left-hand side of the equation is given by

$$1 + 2 + 3 + .. + x + (x + 1)$$

The inductive hypothesis allows us to rewrite this as

$$\frac{x \times (x + 1)}{2} + (x + 1)$$

We can continue with our proof as follows.

$$\frac{x \times (x + 1)}{2} + (x + 1) = (x + 1) \times (\frac{x}{2} + 1)$$
$$= \frac{(x + 1) \times (x + 2)}{2}$$
$$= \frac{(x + 1) \times ((x + 1) + 1)}{2}$$

As such, we have proved the inductive step.

Given that the base case and the inductive step both hold, we can conclude that

$$\sum_{i=1}^{n} i = \frac{n \times (n + 1)}{2}$$

holds for all natural numbers, n. $\square$

Exercise 11.1 Prove, by mathematical induction, that

$$\forall\, n : \mathbb{N} \bullet (n^2 + n) \text{ is even}$$

is true. $\square$

11.2   Structural induction

Mathematical induction is a technique for establishing that certain properties hold of all natural numbers. Structural induction is a technique for establishing that certain properties hold of all elements of an inductively-defined type. The only type of object which may be defined in terms of induction that we have considered thus far is a sequence. In order to establish that we can indeed define sequences inductively, consider the following definition of seq $\mathbb{N}$, the set of all sequences of natural numbers.

1. $\langle\rangle$ is an element of seq $\mathbb{N}$.

2. For any $s \in$ seq $\mathbb{N}$ and any $x \in \mathbb{N}$, $\langle x \rangle \frown s$ is an element of seq $\mathbb{N}$.

The above is sufficient to define seq $\mathbb{N}$.

Example 11.3 $\langle\rangle$ is a sequence of natural numbers. As $\langle\rangle \in$ seq $\mathbb{N}$, we may conclude that $\langle 0 \rangle \in$ seq $\mathbb{N}$ and $\langle 1 \rangle \in$ seq $\mathbb{N}$ and $\langle 2 \rangle \in$ seq $\mathbb{N}$, and so on. Furthermore, as $\langle 0 \rangle \in$ seq $\mathbb{N}$, we may conclude that $\langle 0, 0 \rangle \in$ seq $\mathbb{N}$ and $\langle 0, 1 \rangle \in$ seq $\mathbb{N}$ and $\langle 0, 2 \rangle \in$ seq $\mathbb{N}$, and so on. Thus, *any* sequence of natural numbers may be defined using this process. $\square$

Exercise 11.2 Assuming the set *Car*, give an inductive definition of the set of all sequences of cars. $\square$

As was the case with mathematical induction, a proof via structural induction consists of two sub-proofs. To prove that a property p holds for sequences of type X, we have to establish that two cases hold.

1. The base case: $p(\langle\rangle)$

2. The inductive step: $\forall x : X; \ s : \text{seq } X \bullet p(s) \Rightarrow p(\langle x \rangle \frown s)$

That is, we need to prove that p holds of the empty sequence, and we also prove that $p(\langle x \rangle \frown s)$ follows from $p(s)$. This second case generally involves using $p(s)$—the inductive hypothesis—as an assumption in the proof of $p(\langle x \rangle \frown s)$.

We may represent this in the style of our laws of natural deduction as follows.

$$\frac{p(\langle\rangle) \quad \forall s : \text{seq } X; \ x : X \bullet p(s) \Rightarrow p(\langle x \rangle \frown s)}{\forall s : \text{seq } X \bullet p(s)} \ \text{[structural induction]}$$

Here, establishing $p(\langle\rangle)$ is the base case, and establishing

$$\forall s : \text{seq } X; \ x : X \bullet p(s) \Rightarrow p(\langle x \rangle \frown s)$$

is the inductive step.

We may state the *principle of structural induction for sequences* formally as follows.

Assume some property p, which is either true or false for each sequence of some type, X. If

1. $p(\langle\rangle)$ is true and
2. $\forall\, s : \operatorname{seq} X;\ x : X \bullet p(s) \Rightarrow p(\langle x\rangle \frown s)$ is true,

then p is true for all sequences of type X.

Exercise 11.3 Suppose we wished to prove some property p of all sequences of cars. State the base case and the inductive step for such a proof. □

As we saw in Chapter 10, there are some properties of sequences which hold of *all* sequences, irrespective of their type. For example,

$$\#(s \frown t) = (\#\, s) + (\#\, t)$$

holds for all sequences s and t, no matter if they are of type $\operatorname{seq} \mathbb{N}$ or $\operatorname{seq} \mathit{Car}$. The only condition is that s and t must both be sequences of the *same* type.

Such properties can also be established via structural induction.

Example 11.4 Assume some type X, such that $s \in \operatorname{seq} X$ and $t \in \operatorname{seq} X$. To establish that

$$\forall\, s, t : \operatorname{seq} X \bullet \#(s \frown t) = (\#\, s) + (\#\, t)$$

holds, we need to establish that the two sub-cases hold. First, that this property holds for the empty sequence, i.e., that

$$\forall\, t : \operatorname{seq} X \bullet \#(\langle\rangle \frown t) = (\#\langle\rangle) + (\#\, t)$$

holds.

Second, we need to establish that if this property holds for some sequence s, then it also holds for $\langle x\rangle \frown s$, where x is some arbitrary element of X, i.e., that

$$\begin{aligned}
\forall\, s, t &: \operatorname{seq} X;\ x : X \bullet \\
&\#(s \frown t) = (\#\, s) + (\#\, t) \Rightarrow \\
&\quad \#(((\langle x\rangle \frown s) \frown t) = (\#((\langle x\rangle \frown s)) + (\#\, t)
\end{aligned}$$

holds.

We consider the base case first. Here, we have the following.

$$\begin{aligned}
&\#(\langle\rangle \frown t) \\
={}& \#\, t \\
={}& 0 + \#\, t \\
={}& \#\langle\rangle + \#\, t
\end{aligned}$$

Next, we consider the inductive step. Here, we wish to prove that

$$\#(((\langle x\rangle \frown s) \frown t) = \#(\langle x\rangle \frown s) + (\#\, t)$$

follows from

$$\# (s \frown t) = (\# s) + (\# t)$$

First, consider

$$\# (((\langle x \rangle \frown s) \frown t))$$

By the associativity of $\frown$, this is the same as

$$\# ((\langle x \rangle \frown (s \frown t)))$$

By our definition of $\#$, this is equal to

$$1 + \# (s \frown t)$$

By the inductive hypothesis, this is equal to

$$1 + (\# s) + (\# t)$$

Finally, by our definition of $\#$, this is equal to

$$\# ((\langle x \rangle \frown s) + (\# t)$$

$\square$

Exercise 11.4 Prove that for any sequence s,

$$reverse \, (reverse \, s) = s$$

holds. $\square$

11.3 Further exercises

Exercise 11.5 Prove, by mathematical induction, that

$$\forall \, n : \mathbb{N} \mid n \geq 1 \bullet 1^2 + 2^2 + 3^2 + .. + n^2 = \frac{n \times (n + 1) \times (2n + 1)}{6}$$

is true. Note that this is a property of all natural numbers greater than 0. As such, the base case is not 0, but 1. $\square$

Exercise 11.6 Prove, by mathematical induction, that

$$\forall \, n : \mathbb{N} \mid n \geq 1 \bullet 1 + 4 + 7 + .. + (3n - 2) = \frac{n \times (3n - 1)}{2}$$

is true. $\square$

Exercise 11.7 Prove, by mathematical induction, that

$$\forall n : \mathbb{N} \mid n \geq 4 \bullet n! \geq 2^n$$

is true.

Recall that the *factorial* of n, denoted $n!$, is defined by

$$n! = n \times (n - 1)!$$

if $n > 1$, and

$$n! = 1$$

if $n = 0$ or $n = 1$. $\square$

Exercise 11.8 Prove, by mathematical induction, that

$$\forall n : \mathbb{N} \mid n \geq 3 \bullet n^2 \geq 2n + 1$$

is true. $\square$

Exercise 11.9 Prove, by structural induction, that

$$\forall s : \text{seq } X; \ A : \mathbb{P} \ X \bullet (s \frown t) \restriction A = (s \restriction A) \frown (t \restriction A)$$

is true for some arbitrary type X. $\square$

Exercise 11.10 Prove, by structural induction, that

$$\forall s : \text{seq } X; \ A : \mathbb{P} X \bullet \textit{reverse } (s \restriction A) = (\textit{reverse } s) \restriction A$$

is true for some arbitrary type X. $\square$

11.4   Solutions

Solution 11.1 We prove the base case first. Here,

$$0^2 + 0 = 0$$

We know that 0 is even, and, as such, the property holds for 0.

Next, we prove the inductive step. To do this we assume that $x^2 + x$ is even, for some value x. Considering $x + 1$, we have

$$(x + 1)^2 + (x + 1) = x^2 + 2x + 1 + (x + 1)$$
$$= x^2 + x + 2x + 2$$

By the inductive hypothesis, we know that $x^2 + x$ is even. Furthermore, $2x + 2$ is obviously even for any x. As the product of two even numbers is always even, we may conclude that $x^2 + x + 2x + 2$ must be even.

Therefore, by mathematical induction,

$$\forall\, n : \mathbb{N} \bullet n^2 + n \text{ is even}$$

is true. □

Solution 11.2

1. $\langle\rangle$ is an element of seq *Car*.

2. For any $s \in$ seq *Car* and any $c \in$ *Car*, $\langle c \rangle \frown s$ is an element of seq *Car*.

□

Solution 11.3 The base case is given by $p\,(\langle\rangle)$, while the inductive step is given by

$$\forall\, s : \text{seq } Car;\ c : Car \bullet p\,(s) \Rightarrow p\,(\langle c \rangle \frown s)$$

□

Solution 11.4 The base case can be established as follows.

$$
\begin{aligned}
&reverse\,(reverse\,\langle\rangle) \\
&= reverse\,\langle\rangle &&[\text{Definition of } reverse] \\
&= \langle\rangle &&[\text{Definition of } reverse]
\end{aligned}
$$

The inductive step can be established as follows.

$$
\begin{aligned}
&reverse\,(reverse\,\langle x \rangle \frown s) \\
&= reverse\,((reverse\ s) \frown \langle x \rangle) &&[\text{Definition of } reverse] \\
&= \langle x \rangle \frown (reverse\ reverse\ s) &&[\text{Definition of } reverse] \\
&= \langle x \rangle \frown s &&[\text{Inductive hypothesis}]
\end{aligned}
$$

As such,

$$reverse\,(reverse\ s) = s$$

holds for all sequences. □

Solution 11.5 We prove the base case first. Here,

$$1^2 = \frac{1 \times 2 \times 3}{6}$$

As such, the base case holds.

Next, we prove the inductive step. To do this we assume that

$$1^2 + 2^2 + 3^2 + .. + x^2 = \frac{x \times (x + 1) \times (2x + 1)}{6}$$

holds.

Considering $x + 1$, we have

$$1^2 + 2^2 + .. + x^2 + (x + 1)^2 = \frac{x \times (x + 1) \times (2x + 1)}{6} + (x + 1)^2$$

$$= \frac{x \times (x + 1) \times (2x + 1) + 6(x + 1)^2}{6}$$

$$= \frac{(x + 1) \times ((2x^2 + x) + (6x + 6))}{6}$$

$$= \frac{(x + 1) \times (2x^2 + 7x + 6)}{6}$$

$$= \frac{(x + 1) \times (x + 2) \times (2x + 3)}{6}$$

$$= \frac{(x + 1) \times (x + 2) \times (2(x + 1) + 1)}{6}$$

Therefore, by mathematical induction,

$$\forall n : \mathbb{N} \bullet 1^2 + 2^2 + 3^2 + .. + n^2 = \frac{n \times (n + 1) \times (2n + 1)}{6}$$

is true. □

Solution 11.6 We prove the base case first. Here, the left-hand side of the equation is equal to 1. In addition, the right-hand side is given by

$$\frac{1 \times (3 - 1)}{2} = \frac{1 \times 2}{2}$$

$$= 1$$

As such, the two sides of the equation are equal.

Next, we prove the inductive step. To do this we assume that

$$1 + 4 + 7 + .. + (3x - 2) = \frac{x \times (3x - 1)}{2}$$

holds.

Considering $x + 1$, we have

$$1 + 4 + 7 + .. + (3(x+1) - 2) = \frac{x \times (3x - 1)}{2} + (3x + 1)$$

$$= \frac{x \times (3x - 1) + 2 \times (3x + 1)}{2}$$

$$= \frac{3x^2 + 5x + 2}{2}$$

$$= \frac{(x + 1) \times (3x + 2)}{2}$$

$$= \frac{(x + 1) \times (3(x + 1) - 1)}{2}$$

Therefore, by mathematical induction,

$$\forall\, n : \mathbb{N} \bullet 1 + 4 + 7 + .. + (3n - 2) = \frac{n \times (3n - 1)}{2}$$

is true. $\square$

Solution 11.7 First, we consider the base case, where $n = 4$. Here, we have

$$4! = 24$$

and

$$2^4 = 16$$

and, as such, the base case holds.

Next we consider the inductive step. The inductive hypothesis gives us

$$x! \geq 2^x$$

As $x + 1 > 0$, it follows that

$$(x + 1) \times x! \geq (x + 1) \times 2^x$$

By the definition of !, this is equal to

$$(x + 1)! \geq (x + 1) \times 2^x$$

As $x + 1 > 2$, we can conclude

$$(x + 1)! \geq 2 \times 2^x$$

It follows that

$$(x + 1)! \geq 2^{x+1}$$

As such, the inductive step holds.

Therefore, by mathematical induction,

$$\forall\, n : \mathbb{N} \mid n \geq 4 \bullet n! \geq 2^n$$

is true. $\square$

Solution 11.8 First, we consider the base case, where $n = 3$. Here, we have

$$3^2 \geq (2 \times 3) + 1$$

and, as such, the base case holds.

Next we consider the inductive step. First, we can expand $(x + 1)^2$ as follows.

$$(x + 1)^2 = x^2 + 2x + 1$$

By the inductive hypothesis, we have

$$x^2 + 2x + 1 \geq (2x + 1) + 2x + 1$$

As $(2x + 1) + 2x + 1 = 4x + 2$, we may rewrite the above as

$$x^2 + 2x + 1 \geq 4x + 2$$

Finally, as $x \geq 0$ holds, we have

$$x^2 + 2x + 1 \geq 2(x + 1) + 1$$

As such, we may conclude that the inductive step holds.

Therefore, by mathematical induction,

$$\forall\, n : \mathbb{N} \mid n \geq 3 \bullet n^2 \geq 2n + 1$$

is true. $\square$

Solution 11.9 We consider the base case first. Here,

$$\begin{aligned}
(\langle\rangle \frown t) \restriction A &= t \restriction A \\
&= \langle\rangle \frown (t \restriction A) \\
&= (\langle\rangle \restriction A) \frown (t \restriction A)
\end{aligned}$$

Thus, the base case holds.

We now consider the inductive step. The inductive hypothesis is that

$$(s \frown t) \restriction A = (s \restriction A) \frown (t \restriction A)$$

From this, we have to prove

$$(((\langle x \rangle \frown s) \frown t) \restriction A = (((\langle x \rangle \frown s) \restriction A) \frown (t \restriction A)$$

We do this by first establishing the existence of two sub-cases.

$$((\langle x \rangle \frown s) \frown t) \restriction A = \langle x \rangle \frown ((s \frown t) \restriction A) \text{ if } x \in A$$
$$= (s \frown t) \restriction A \qquad \text{otherwise}$$

Assume that it is the case $x \in A$. Here, by the inductive hypothesis,

$$\langle x \rangle \frown ((s \frown t) \restriction A) = \langle x \rangle \frown (s \restriction A) \frown (t \restriction A)$$
$$= ((\langle x \rangle \frown s) \restriction A) \frown (t \restriction A)$$

Now assume that it is the case that $x \notin A$. Here, by the inductive hypothesis,

$$(s \frown t) \restriction A = (s \restriction A) \frown (t \restriction A)$$
$$= ((\langle x \rangle \frown s) \restriction A) \frown (t \restriction A)$$

Therefore, the inductive step holds. As such,

$$\forall s : \text{seq } X; \ A : \mathbb{P} \ X \bullet (s \frown t) \restriction A = (s \restriction A) \frown (t \restriction A)$$

is true for some arbitrary type X. $\square$

Solution 11.10 We consider the base case first. Here,

$$reverse \ (\langle \rangle \restriction A) = reverse \ \langle \rangle$$
$$= (reverse \ \langle \rangle) \restriction A \ .$$

We now consider the inductive step. The inductive hypothesis is

$$reverse \ (s \restriction A) = (reverse \ s) \restriction A$$

From this, we need to prove

$$reverse \ ((\langle x \rangle \frown s) \restriction A) = (reverse \ \langle x \rangle \frown s) \restriction A$$

We do this in the following way. First,

$$reverse \ ((\langle x \rangle \frown s) \restriction A) = reverse \ (\langle x \rangle \frown (s \restriction A)) \text{ if } x \in A$$
$$= reverse \ (s \restriction A) \qquad \text{otherwise}$$

Assume it is the case that $x \in A$. By the inductive hypothesis, we have

$$reverse \ (\langle x \rangle \frown (s \restriction A)) = reverse \ (s \restriction A) \frown \langle x \rangle$$
$$= ((reverse \ s) \restriction A) \frown \langle x \rangle$$
$$= (reverse \ \langle x \rangle \frown s) \restriction A$$

Now assume that it is the case that $x \notin A$. Again, by the inductive hypothesis, we have

$$reverse \ (s \restriction A) = (reverse \ s) \restriction A$$
$$= (reverse \ \langle x \rangle \frown s) \restriction A$$

Therefore, the inductive step holds. As such,

$$\forall\, s : \mathrm{seq}\ X;\ A : \mathbb{P}\,X \bullet \mathit{reverse}\,(s \restriction A) = (\mathit{reverse}\ s) \restriction A$$

is true for some arbitrary type X. $\square$

Chapter 12

Graph theory

In Chapter 1, it was postulated that the teaching of discrete mathematics to computing students at Higher Education institutions is performed for one of two reasons. The first is to describe the mathematical theory that underpins formal description techniques, such as Z (see, for example, [Spi92] or [WD96]) or CSP (see, for example, [Hoa85] or [Ros97]). The other is to describe the mathematical theory which underpins theoretical computer science. The subjects of this chapter, graph theory, and the next, combinatorics, belong to this second category. In this chapter, we provide a necessarily brief introduction to the main concepts of graph theory. For a more in-depth description of this topic, the interested reader is referred to [Wes95], [Wil96], or [Mer00]. For further exercises on graph theory, the interested reader might consult [Bal97].

Graphs and, consequently, graph theory, are central to many areas of computer science. For example, the means by which computers are linked or networked together form structures which may be represented as graphs. As a further example, the binary tree structure, which is a fundamental data structure in high level programming, is a special class of graph. In a more general sense, graphs can form a visual representation of relations. We start by considering exactly what it is that we mean by the term *graph*.

12.1 Graphs

Consider the figure below.

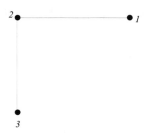

Here, we have a graph which consists of three *vertices* (or *nodes*, or *points*—all of these terms are equally valid): 1, 2, and 3. Furthermore, there are two *edges* (or *lines*—again, both terms are equally valid), the first of which connects vertex 1 and vertex 2, and the second of which connects vertex 2 and vertex 3. Note that each edge starts and ends at a vertex.

Formally, we consider a graph to consist of a non-empty set of vertices, and a (possibly empty) set of edges, with each edge starting and ending at one of the vertices.

We say that a vertex is *adjacent* to another vertex if, and only if, the two vertices are connected by an edge. So, in the above example, vertex 1 is adjacent to vertex 2 (and vice versa), and vertex 2 is adjacent to vertex 3 (and vice versa).

Exercise 12.1 Which pairs of vertices are adjacent in the following graph?

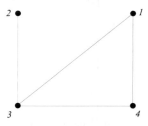

□

We say that an edge is *incident* to two vertices if, and only if, the two vertices are those appearing at either end of the edge.

Example 12.1 In the following graph—which has two edges—one edge is incident to vertices 1 and 2, while the other edge is incident to vertices 2 and 3.

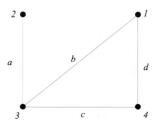

□

It is possible to label edges, as well as vertices. As an example, consider the graph of Exercise 12.1. We might choose to label the edges of this graph *a*, *b*, *c*, and *d*. This may be accomplished in the following way.

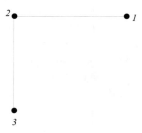

Exercise 12.2 The above graph contains four edges. For each edge, state the vertices that it is incident to. □

A graph can have more than one edge between a given pair of vertices. Consider, for example, the following figure.

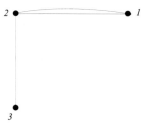

Here, there are two *parallel edges* between 1 and 2. This is a perfectly legitimate graph.

In addition, a graph may involves *loops*, that is, edges which start and end at the same vertex. As an example, the following graph involves a loop at vertex 1.

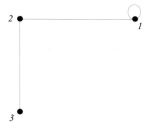

We say that a graph is *simple* if, and only if, it has no loops or parallel edges.

Exercise 12.3   Which of the following graphs are simple graphs?

1.

2.

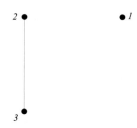

3.

☐

Exercise 12.4 Is a graph with no edges a simple graph? ☐

Exercise 12.5 Consider a graph with n vertices. What is the maximum number of edges that this graph can have if it is a simple graph? ☐

Given a simple graph G, the *complement* of G is the simple graph that has the same vertices as G and edges which are adjacent if, and only if, they are not adjacent in G. We denote the complement of G by $\overline{G}$.

Example 12.2 Consider the following simple graph.

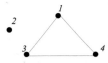

The complement of this graph is given by

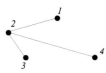

☐

Exercise 12.6 Give the complements of the following simple graphs.

1.

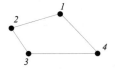

2.

□

Given any vertex v contained in a graph, we may consider the *degree* of v. The degree of v, denoted $deg\,(v)$, is the number of edges to which it is incident. Note that if a vertex v is incident to a loop, then that loop is regarded as being equivalent to *two* edges.

Example 12.3 Consider again the graphs of Exercise 12.3.

1.

2.

3.

In the first graph, we have

$$deg\,(1) = 1$$
$$deg\,(2) = 4$$
$$deg\,(3) = 1$$

In the second graph, we have

$$deg\,(1) = 0$$
$$deg\,(2) = 1$$
$$deg\,(3) = 1$$

Finally, in the third graph, we have

$$deg\,(1) = 1$$
$$deg\,(2) = 3$$
$$deg\,(3) = 2$$

□

Exercise 12.7 Calculate the degrees of the vertices of the following graph.

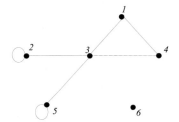

□

If, for any vertex v, $deg\,(v) = 0$, then v is said to be an *isolated* vertex.

Exercise 12.8 Which of the vertices of the graph of the previous exercise are isolated? □

Some readers will have noticed that a relationship holds between the sum of the degrees of all vertices and the number of edges in a graph—the former is exactly twice the latter. Formally, this relationship is given by

$$\left(\sum_{v \in V} deg\,(v) \right) = 2 \times \# E$$

where V represents the graph's vertices and E represents the graph's edges.

Exercise 12.9 Verify that

$$\left(\sum_{v \in V} deg\,(v) \right) = 2 \times \# E$$

holds of the graph of Exercise 12.7. □

A second theorem that is concerned with degrees of vertices, known as the 'handshaking lemma', can be stated as follows.

In any graph, the number of vertices which have a degree that is an odd number is even.

Exercise 12.10 Verify that the handshaking lemma holds of the graph of Exercise 12.7. □

A graph is said to be *connected* if it is possible to move from any vertex to any other vertex via—if necessary—one or more intermediate nodes.[1]

Example 12.4 The following is a connected graph.

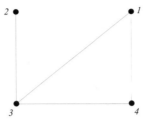

This graph is, in some sense, 'in one piece'. □

Example 12.5 The following graph, on the other hand, is not connected: there is no way of reaching, for example, vertex 2 from vertex 1 (or vice versa).

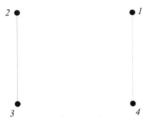

□

Exercise 12.11 Which of the following graphs are connected?

1.

---

[1] We shall be in a position to consider a more technical definition of connectedness later in the chapter. For now, however, this definition shall suffice.

2.

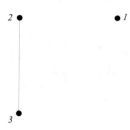

3.

□

Exercise 12.12 Consider a graph with n vertices. What is the minimum number of edges that this graph must have if it is to be connected? □

A graph is said to be *complete* if, and only if, every vertex is adjacent to every other.

Example 12.6 Consider the following graph.

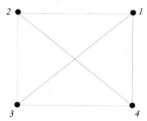

This graph is complete as every vertex is connected via an edge to every other vertex. However, if we were to remove any edge from this graph, then the resulting graph would not be complete. □

Exercise 12.13 Which of the following graphs are complete?

1.

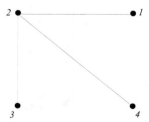

2.

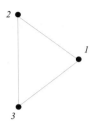

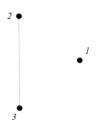

3.

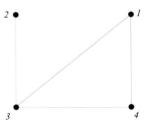

□

Exercise 12.14 Consider a graph with n vertices. What is the minimum number of edges that this graph must have if it is to be complete? □

Exercise 12.15 Prove that if a graph is complete, then it is connected. □

12.2 Representing graphs as sets and bags

We may represent graphs formally in terms of sets and bags. Recalling that a graph consists of vertices and edges, and that edges connect vertices, we may represent a graph in terms of a non-empty set of vertices and a (possibly empty) bag of edges (or pairs of vertices). We denote the set of vertices by V and the set of edges by E. Given a graph G, with vertices V and edges E, we may write $G = (V, E)$.

Consider again the following graph.

We may represent this graph formally as follows.

$V = \{1, 2, 3, 4\}$
$E = [\![(1, 3), (1, 4), (2, 3), (3, 4)]\!]$

It is important to note that there is no restriction on the size of V, except that it must be non-empty. Furthermore, there is no restriction on the pairs that may appear in E, except that they must all be of type $V \times V$. In particular, there may be elements of the form (v, v) in E; this would indicate a graph with a loop.

Example 12.7

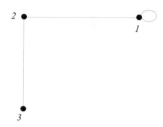

This graph is represented formally as follows.

$$V = \{1, 2, 3\}$$
$$E = [\![(1,1), (1,2), (2,3)]\!]$$

Note that the existence of the pair $(2, 3)$ in E indicates the fact that vertex 2 is adjacent to vertex 3; this may equally well be denoted by $(3, 2)$ appearing in E. This is because—for the moment—we consider only undirected graphs. Note also that the existence of an edge is recorded only once. □

It is possible for the bag E to be empty.

Example 12.8 Consider the graph given below.

This graph is represented formally as follows.

$$V = \{1, 2, 3\}$$
$$E = [\![\,]\!]$$

□

Exercise 12.16 Why is E represented by a bag, and not a set? □

Exercise 12.17 Draw a diagram for each of the following graphs.

1. $V = \{1, 2, 3, 4\}$, $E = [\![\,]\!]$
2. $V = \{1, 2, 3, 4\}$, $E = [\![(1,1), (2,2), (3,3)]\!]$

3. $V = \{1, 2, 3, 4\}$, $E = [\![(1, 1), (1, 2), (3, 4), (4, 4)]\!]$

□

Exercise 12.18 Represent each of the following graphs using sets and bags.

1.

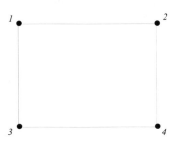

2.

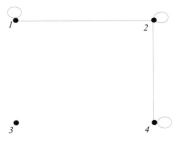

□

12.3 Representing graphs as matrices

If we wish to represent graphs in terms of a high level programming structure—that is, if we wish our graphs to be represented and manipulated by a computer—then we need ways of representing graphs other than by visual means. One way in which this can be facilitated is by means of *adjacency matrices*. This is the most common way of representing graphs in a form suitable for handling by computers.

Consider a graph $G = (V, E)$, such that $\# V = n$. Assume that the vertices in V can be indexed (and, hence, ordered) in the form $v_1, v_2, .., v_n$. In this case, the adjacency matrix for G is an $n \times n$ matrix, in which the entry in position (i, j) denotes the number of edges from vertex v_i to vertex v_j.

Consider, as an example, the following graph.

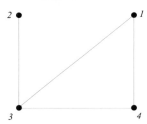

This graph has four vertices, and, as such, we can expect the adjacency matrix representing this graph to be a 4×4 matrix. Furthermore, for simplicity's sake, we assume that $v_1 = 1$, $v_2 = 2$, $v_3 = 3$, and $v_4 = 4$. In addition, this graph has four edges: $(1,3)$, $(1,4)$, $(2,3)$, and $(3,4)$. As such, the entries in the positions in the matrix corresponding to these edges will be 1, whereas the entries in the positions corresponding to other edges will be 0. Assuming that the entry in the top left-hand corner represents $(1,1)$, the one below it represents $(1,2)$, and so on, the adjacency matrix for this graph is given as follows.

$$\begin{pmatrix} 0 & 0 & 1 & 1 \\ 0 & 0 & 1 & 0 \\ 1 & 1 & 0 & 1 \\ 1 & 0 & 1 & 0 \end{pmatrix}$$

This matrix denotes the fact that vertex 1 is connected to vertex 3 (and vice versa) via one edge, vertex 1 is connected to vertex 4 (and vice versa) via one edge, vertex 2 is connected to vertex 3 (and vice versa) via one edge, and vertex 3 is connected to vertex 4 (and vice versa) via one edge.

If we were to remove the edge connecting vertices 3 and 4 from this graph, then the resulting adjacency matrix would be

$$\begin{pmatrix} 0 & 0 & 1 & 1 \\ 0 & 0 & 1 & 0 \\ 1 & 1 & 0 & 0 \\ 1 & 0 & 0 & 0 \end{pmatrix}$$

On the other hand, if we were to add a second, parallel, edge between vertices 3 and 4 to the original graph (as opposed to removing one), then the resulting adjacency matrix would be

$$\begin{pmatrix} 0 & 0 & 1 & 1 \\ 0 & 0 & 1 & 0 \\ 1 & 1 & 0 & 2 \\ 1 & 0 & 2 & 0 \end{pmatrix}$$

Recall that a graph is said to be *simple* if, and only if, it has no parallel edges or loops. As such, a matrix representation of a simple graph will have two special properties. First, all of the entries on the diagonal from the top left-hand corner to the bottom right-hand corner will be 0 (this is the 'no loops' property). Second, all of the entries in the matrix will be either 0 or 1 (this is the 'no parallel edges' property). The graph given above is simple and, as such, its adjacency matrix satisfies these properties. Adding a second edge between vertices 3 and 4 results in a graph which is not simple, and, as such, an adjacency matrix which does not satisfy this second property.

Exercise 12.19 Give the adjacency matrices of the following graphs.

1.

2.

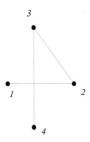

□

Exercise 12.20 Draw the graphs represented by the following adjacency matrices.

1. $\begin{pmatrix} 0 & 1 & 0 \\ 1 & 1 & 1 \\ 0 & 1 & 0 \end{pmatrix}$

2. $\begin{pmatrix} 0 & 1 & 0 & 1 \\ 1 & 0 & 1 & 0 \\ 0 & 1 & 0 & 1 \\ 1 & 0 & 1 & 0 \end{pmatrix}$

□

Exercise 12.21 Which of the following adjacency matrices denote simple graphs?

1. $\begin{pmatrix} 0 & 0 & 0 & 0 \\ 0 & 0 & 0 & 0 \\ 0 & 0 & 0 & 0 \\ 0 & 0 & 0 & 0 \end{pmatrix}$

2. $\begin{pmatrix} 1 & 1 & 0 & 0 \\ 1 & 0 & 1 & 0 \\ 0 & 1 & 0 & 0 \\ 0 & 0 & 0 & 0 \end{pmatrix}$

3. $\begin{pmatrix} 0\ 2\ 0\ 0 \\ 2\ 0\ 1\ 0 \\ 0\ 1\ 0\ 0 \\ 0\ 0\ 0\ 0 \end{pmatrix}$

□

Because we are dealing with *undirected* graphs (we consider the topic of directed graphs in Section 12.9), the entry at position (i, j) of an adjacency matrix is always the same as that at position (j, i) of the same matrix. This is due to the fact that both entries represent the number of edges between vertex v_i and vertex v_j. This, of course, is a symmetric property: if vertex v_i is adjacent to vertex v_j, then vertex v_j is adjacent to vertex v_i. Because of this, when dealing with undirected graphs, we need only consider the bottom left-hand half of the matrix: this gives us no less information than is contained in the whole of the matrix.

As an example, consider the following matrix.

$\begin{pmatrix} 0\ 0\ 1\ 1 \\ 0\ 0\ 1\ 0 \\ 1\ 1\ 0\ 1 \\ 1\ 0\ 1\ 0 \end{pmatrix}$

This may also be represented in the following way.

$\begin{pmatrix} 0 \\ 0\ 0 \\ 1\ 1\ 0 \\ 1\ 0\ 1\ 0 \end{pmatrix}$

The main advantage of this second representation is in terms of computer implementations: when storing information pertaining to a graph, it takes slightly over half as much memory to store it in the second form as it does to store it in the first form. When dealing with large graphs, this can represent a substantial saving.

Exercise 12.22 Represent the graphs of Exercise 12.19 in terms of an adjacency matrix with values contained in only the bottom left-hand half. □

Exercise 12.23 Assuming that $v_1 = 1$, $v_2 = 2$, $v_3 = 3$, and $v_4 = 4$, draw the graphs represented by the following adjacency matrices.

1. $\begin{pmatrix} 0 \\ 1\ 0 \\ 1\ 1\ 0 \\ 0\ 0\ 1\ 0 \end{pmatrix}$

2. $\begin{pmatrix} 0 & & & \\ 0 & 0 & & \\ 0 & 1 & 0 & \\ 1 & 0 & 1 & 0 \end{pmatrix}$

$\square$

Exercise 12.24 If a graph has n vertices, how many entries will an adjacency matrix with values contained in the bottom left-hand half have? $\square$

12.4 Isomorphic graphs

Recall the notion of isomorphic Boolean algebras from Chapter 5. In that chapter, it was shown that two Boolean algebras A and B could be considered to be isomorphic if they have the same structure. This involved exhibiting what we later knew to be a total bijective function, $f : A \rightarrowtail B$, which mapped terms of type A to terms of type B, while 'preserving' the operations of A. (Recall that a bijection is a function that is both injective and surjective.) As such, we could establish that, for example, the Boolean algebra of set theory and the Boolean algebra of propositional logic are isomorphic. In a similar fashion, we may also reason about two graphs being isomorphic. Furthermore, our definition of what it means for two graphs to be isomorphic is similar to our definition of what it means for two Boolean algebras to be isomorphic.

Consider two graphs $G_1 = (V_1, E_1)$ and $G_2 = (V_2, E_2)$. Formally, for G_1 and G_2 to be isomorphic, we require that there exists a total bijection, $f : V_1 \rightarrowtail V_2$, such that for any vertices, $u, v \in V_1$, u and v are adjacent in G_1 if, and only if, $f(u)$ and $f(v)$ are adjacent in G_2. Informally, two graphs G_1 and G_2 are isomorphic if we can change the names of the vertices of G_1 to those of G_2 and if we can rearrange the shape of G_1 (with these changed names) to match that of G_2.

Example 12.9 The following graphs are isomorphic.

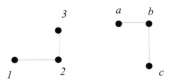

We may exhibit a total bijection, f, such that

$$f = \{1 \mapsto c, 2 \mapsto b, 3 \mapsto a\}$$

Furthermore, in the first graph 1 is adjacent to 2 and 2 is adjacent 3, while in the second graph c is adjacent to b and b is adjacent to a. $\square$

Example 12.10 The following graphs are isomorphic.

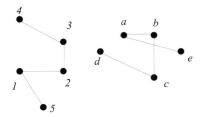

Here, we may exhibit a total bijection, f, such that

$$f = \{1 \mapsto c, 2 \mapsto b, 3 \mapsto a, 4 \mapsto e, 5 \mapsto d\}$$

In the first graph the following pairs of vertices are adjacent: 1 and 2, 1 and 5, 2 and 3, and 3 and 4. In the second graph, we find that $f(1)$ and $f(2)$ are adjacent, $f(1)$ and $f(5)$ are adjacent, $f(2)$ and $f(3)$ are adjacent, and $f(3)$ and $f(4)$ are adjacent. Furthermore, no other vertices are adjacent in the second graph. As such, the two graphs are isomorphic. □

Exercise 12.25 Consider two graphs G_1 and G_2, such that G_1 has m vertices and G_2 has n vertices. Furthermore, $m \neq n$. Is it possible for G_1 and G_2 to be isomorphic? □

Exercise 12.26 Consider two graphs G_1 and G_2, such that G_1 has m edges and G_2 has n edges. Furthermore, $m \neq n$. Is it possible for G_1 and G_2 to be isomorphic? □

Exercise 12.27 Consider sets of vertices V_1 and V_2, such that

$$V_1 = \{a, b, c, d\}$$
$$V_2 = \{1, 2, 3, 4\}$$

For each of the following values of E_1, define a bag, E_2, such that (V_1, E_1) is isomorphic to (V_2, E_2).

1. $E_1 = [\![(a, a), (a, b), (c, d)]\!]$
2. $E_1 = [\![(a, d), (c, d), (b, c), (b, d)]\!]$

□

Exercise 12.28 Which of the following graphs are isomorphic?

$$V_1 = \{1, 2, 3, 4\} \; ; \; E_1 = [\![(1, 3), (1, 4), (2, 4), (2, 2)]\!]$$
$$V_2 = \{a, b, c, d\} \; ; \; E_2 = [\![(a, a), (c, b), (d, b), (c, d)]\!]$$
$$V_3 = \{1, 2, 3, 4\} \; ; \; E_3 = [\![(4, 2), (4, 3), (1, 3), (1, 1)]\!]$$
$$V_4 = \{a, b, c, d, e\} \; ; \; E_4 = [\![(a, b), (c, d), (a, b)]\!]$$
$$V_5 = \{1, 2, 3, 4, 5\} \; ; \; E_5 = [\![(1, 2), (3, 4), (1, 3)]\!]$$

□

12.5 Paths

Consider the graph below. Here, 1 is connected to 2, and 2 is connected to 3. If we imagine the edges of this graph as, for example, commercial flights, then we can see that we are able to fly from 1 to 3 via 2, even though we cannot fly directly from 1 to 3. This is an example of a *path*.

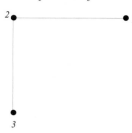

We define a path formally as follows. A path[2] of length n is a sequence of vertices $\langle v_0, v_1, \ldots, v_n \rangle$, such that each v_i for $0 \le i \le n - 1$ adjacent to v_{i+1}.

Example 12.11 Consider the following graph.

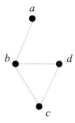

This graph contains a number of paths. For example, $\langle a, b, d \rangle$ is a path, as is $\langle a, b, c \rangle$; they are both paths of length 2. Note that there is no restriction on how many times a path may visit a given vertex. As such, $\langle a, b, d, c, b \rangle$ is a path. Furthermore, $\langle a \rangle$ is a path—it is a path of length 0. □

A *simple path* is defined to be a path in which no vertex appears twice.

Example 12.12 Considering the graph of the previous example, $\langle a, b, d \rangle$ is a simple path, whereas $\langle a, b, d, c, b \rangle$ is not a simple path. □

Exercise 12.29 List all of the paths of length 2 contained in the following graph.

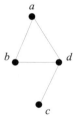

□

Exercise 12.30 List all of the simple paths of length 2 contained in the graph of the previous exercise. □

Exercise 12.31 Draw a graph with vertices a, b, c, and d that consists of the following simple paths.

$$\langle a, b, c \rangle, \langle a, b, d \rangle, \langle c, b, a \rangle, \langle c, b, d \rangle, \langle d, b, a \rangle, \text{ and } \langle d, b, c \rangle.$$

□

A *circuit* is defined to be a path $\langle v_0, v_1, \ldots, v_n \rangle$, such that $v_0 = v_n$.

---

[2]Unfortunately, the concept referred to here as a path is defined somewhat differently in some other textbooks. It should be noted that this does not mean that this or the other definitions are incorrect; it is simply that some of the terminology employed in graph theory is not standardised. This problem is also true of the following definitions of cycles and circuits.

Example 12.13 Consider again the following graph.

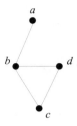

This graph contains the following circuits of length 3.

$$\langle b, c, d, b \rangle, \langle b, d, c, b \rangle, \langle c, b, d, c \rangle, \langle c, d, b, c \rangle, \langle d, b, c, d \rangle, \text{ and } \langle d, c, b, d \rangle.$$

□

Exercise 12.32 Consider the graph of Exercise 12.29. List the circuits contained in that graph that contain no repeated edges. □

We are now in a position to revisit and formalise our definition of what it means for a graph to be connected. As such, we may state that any graph $G = (V, E)$ is connected if, and only if, for all pairs of vertices, $u, v \in V$, there is a path from u to v in G.

Exercise 12.33 Which of the following graphs are connected?

1. $V = \{1, 2, 3\}$; $E = [\![(1, 2), (1, 3)]\!]$
2. $V = \{1, 2, 3, 4\}$; $E = [\![(1, 2), (2, 3), (3, 4), (4, 1)]\!]$
3. $V = \{1, 2, 3, 4, 5\}$; $E = [\![(1, 3), (2, 4), (2, 5), (3, 4)]\!]$

□

Exercise 12.34 For each of the following sets of vertices, state the maximum number of edges that a simple graph with that set of vertices can have *without* it being connected.

1. $V = \{1, 2, 3\}$
2. $V = \{1, 2, 3, 4\}$
3. $V = \{1, 2, 3, 4, 5\}$
4. $V = \{1, 2, 3, 4, 5, 6\}$

□

We now consider two special classes of paths: *Eulerian paths*[3] and *Hamiltonian paths*.[4] We consider the former first.

   An *Eulerian path* in a connected graph $G = (V, E)$ is a path in which every edge $e \in E$ is passed through exactly once. It follows that an *Eulerian circuit* is an Eulerian path that is also a circuit.

---

[3]Named after Leonhard Euler (1707 - 1783). See [Dun99] for an overview of the work of Euler.
[4]Named after Sir William Hamilton (1805 - 1865).

Example 12.14 Consider the following graph.

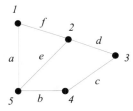

This graph has four Eulerian paths:

$$\langle 2, 5, 4, 3, 2, 1, 5 \rangle, \langle 2, 5, 1, 2, 3, 4, 5 \rangle, \langle 5, 1, 2, 3, 4, 5, 2 \rangle, \text{ and } \langle 5, 4, 3, 1, 5, 2 \rangle.$$

□

The graph of the previous exercise has four Eulerian paths, but no Eulerian circuits. A graph such as this, i.e., one that is connected, that contains an Eulerian path, but does not contain an Eulerian circuit is referred to as being *semi-Eulerian*. Those graphs which are connected and contain an Eulerian circuit are referred to as being *Eulerian*.

Example 12.15 The following graph is Eulerian, as it is connected and contains an Eulerian circuit.

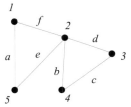

□

Exercise 12.35 List the Eulerian paths contained in the following graph.

$$V = \{1, 2, 3, 4\}$$
$$E = [\![(1, 2), (1, 3), (1, 4), (2, 3), (3, 4)]\!]$$

□

Exercise 12.36 For each of the following graphs, state whether they are Eulerian, semi-Eulerian, or neither.

1.

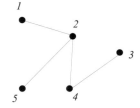

2.

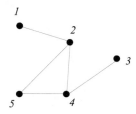

3.

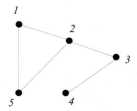

4.

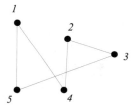

□

A *Hamiltonian path* in a connected graph $G = (V, E)$ is one in which every vertex $v \in V$ is visited exactly once (except in the case of a *Hamiltonian circuit*, in which case $v_0 = v_n$).

Example 12.16 Consider again the following graph.

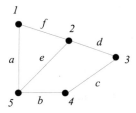

This graph has a number of Hamiltonian paths. For example, the paths $\langle 1, 2, 3, 4, 5 \rangle$, $\langle 1, 2, 5, 4, 3 \rangle$, and $\langle 1, 5, 2, 3, 4 \rangle$ all satisfy the relevant property. Furthermore, this graph also has a number of Hamiltonian circuits, with $\langle 1, 2, 3, 4, 5, 1 \rangle$ being one example. Such a graph, i.e., one which is connected and contains a Hamiltonian path is referred to as being *Hamiltonian*. □

Exercise 12.37 List all of the Hamiltonian paths contained in the graph of Exercise 12.35. □

Exercise 12.38 Consider again the graphs of Exercise 12.36. For each graph, state whether it is Hamiltonian or not. □

12.6 Cycles

In the previous section, we considered the concept of a path. In this section, we consider a special type of path called a *cycle*. A cycle has the following properties.

1. It has at least one edge.

2. It has no repeated edges.

3. There are only two repeated vertices: the first and last ones.

As such, we may consider a cycle to be a circuit in which there is only one repeated vertex.

Example 12.17 Consider the following graph.

In this graph, there are several cycles, including $\langle 2, 1, 5, 2 \rangle$ and $\langle 2, 3, 4, 2 \rangle$. □

Note that in the above example $\langle 2, 3, 4, 2, 1, 5, 2 \rangle$ is not a cycle, as vertex 2 is visited three times. However, it is a circuit. Note also that we consider the cycle $\langle 1, 5, 2, 1 \rangle$ to be equivalent to $\langle 2, 1, 5, 2 \rangle$. More generally, when we denote circuits in terms of sequences and edges, we shall consider two circuits c_1 and c_2 to be equivalent whenever ran c_1 = ran c_2.

Exercise 12.39 Is it possible to have a cycle in a graph with two edges? □

Exercise 12.40 Is it possible to have a cycle in a graph with one edge? □

Exercise 12.41 Recall that a simple graph is a graph involving no loops and no parallel edges. What is the smallest possible number of edges a simple graph with a cycle can have? □

Exercise 12.42 List all of the cycles contained in each the following graphs.

1.

2.

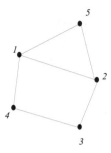

3.

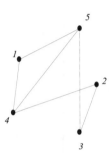

□

12.7   Trees

A tree is a particular type of graph which appears in many different situations in computing. For example, in Chapter 13, we shall meet the concept of a tree diagram; such structures allow us to illustrate the different combinations of outcomes of a sequence of events. In addition, the concept of a binary tree—which we shall meet later in this chapter—is a fundamental data structure in high-level programming.

Essentially, a graph is a tree *if it is connected and it has no cycles.*

Example 12.18 All of the following four graphs are trees, as they satisfy the required property.

1.

2.

3.

4.

□

If a graph does not contain cycles but is not connected, then it is referred to as a *forest*. The following is an example of a forest.

Essentially, a forest is a graph that is made up of a collection of trees.

Trees have the following property.

For any tree T, if T has n vertices then T has $n - 1$ edges.

This is easily verified by studying the examples of trees given above.

Exercise 12.43 Consider a graph $G = (V, E)$, such that G is a tree. If $\# V = 4$, what is $\# E$? □

Exercise 12.44 Which of the following graphs are trees?

1. $V = \{1, 2, 3, 4\}$; $E = [\![(1, 2), (2, 3), (2, 4)]\!]$
2. $V = \{1, 2, 3, 4\}$; $E = [\![(1, 2), (2, 3), (2, 4), (3, 4)]\!]$
3. $V = \{1, 2, 3, 4\}$; $E = [\![(1, 2), (1, 3), (3, 4)]\!]$
4. $V = \{1, 2, 3, 4\}$; $E = [\![(1, 2), (1, 3), (2, 4), (3, 4)]\!]$

□

Exercise 12.45 Prove that, given any tree T, removing an edge from T—without removing any vertices—will produce a graph that is not connected. □

Exercise 12.46 Prove that, given any tree T, adding a new edge to T—without adding any new vertices—will produce a graph with a cycle. □

Exercise 12.47 Consider a graph, $G = (V, E)$, such that G is a tree. Given any two vertices, $v_1, v_2 \in V$ how many paths between v_1 and v_2 are there? □

12.8 Weighted graphs

At the start of this chapter, we motivated the study of graph theory by stating that it is central to many areas of computer science, with the means by which computers are networked together being a prime example of such an application. If we were to connect a series of computers, then we would expect each computer to be linked to every other computer either directly or indirectly; we would expect the means in which the machines are connected to have the properties of a connected graph.

In many practical situations, such as the networking of computers, it is not sufficient simply to denote the ways in which the machines may be linked; in some situations it is necessary to consider the cost of different potential configurations and select the most appropriate one (often the cheapest one) accordingly.

To reason about such problems, i.e., those problems in which values such as cost and distance are regarded as important, we need to consider *weighted graphs*. We define a weighted graph to be a simple graph in which each edge has an associated *weight*.

Example 12.19 Consider five computers—COM_1, COM_2, COM_3, COM_4, and COM_5— each of which is housed in a different building. The cost of linking each pair of machines is illustrated by the following graph.

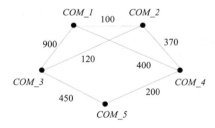

□

Example 12.20 A travel company offers flights between five different cities, namely Bangkok, Sydney, London, Los Angeles, and New York. The cost of flying between each pair of cities is illustrated by the following graph.

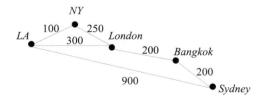

□

In Example 12.19, we showed the possible ways in which the computers could be linked and the cost of each link. Of course, one would not install all of these links in reality; one would only need to install sufficient links so that all of the machines are connected. As an example, the following installation would be sufficient (and the cheapest available).

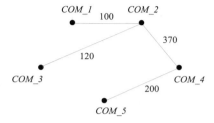

This graph satisfies the properties necessary for it to be a tree—it is a connected graph, and contains no cycles. Furthermore, it has identical vertices to the original graph and its edges are a subset of those of the original graph. We refer to such a tree as a *spanning tree*.

Another spanning tree of our network of computer is given below.

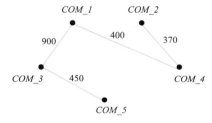

This graph also satisfies the properties necessary of a spanning tree. However, it does have one disadvantage when compared to our first spanning tree—the first spanning tree had an associated cost of 790, whereas this second spanning tree has an associated cost of 2120. As such, our first spanning tree is better in one important regard—it is cheaper! What is more, there are no spanning trees of our network which are cheaper than our original spanning tree. As such, it is referred to as a *minimal spanning tree*.

Example 12.21 The minimal spanning tree of the graph of Example 12.20 is given below.

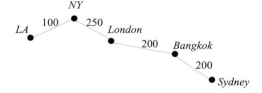

□

One way of determining a minimal spanning tree is as follows. Start at an arbitrary initial vertex and choose the edge with the lowest weight of all the edges that are incident to that vertex. Mark that edge and the two vertices that are connected by the edge. Next, consider the edge with the least weight that links one of the vertices that have been marked and one of the vertices that have not been marked. Mark that edge and the two vertices that are connected by the edge. This process continues until all of the vertices have been marked and all of the edges that have been marked form a spanning tree. This process is known as *Prim's algorithm*.

Example 12.22 We can apply Prim's algorithm to the weighted graph of Example 12.19 in the following way.

Initially, we have considered no edges. Starting with COM_1, we find that the cheapest edge is that connecting COM_1 and COM_2. As such, our tree becomes

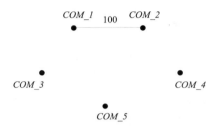

The cheapest edge linking one of COM_1 and COM_2, and one of COM_3, COM_4 and COM_5 is that connecting COM_2 and COM_3. As such, we add this edge.

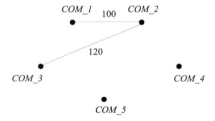

The cheapest edge linking one of COM_1, COM_2 and COM_3, and one of COM_4 and COM_5 is that connecting COM_2 and COM_4. As such, we add this edge.

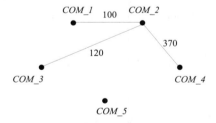

Finally, the cheapest edge linking one of COM_1, COM_2, COM_3 and COM_4, and COM_5 is that connecting COM_4 and COM_5. As such, we add this edge.

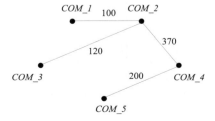

This is our minimal spanning tree. □

Exercise 12.48 Use Prim's algorithm to determine the minimal spanning tree of the graph of Example 12.20. □

Another algorithm for calculating minimal spanning trees is *Kruskal's algorithm*. This algorithm works as follows. Start by selecting the edge with the lowest cost. Mark this edge. Then select the edge with the next lowest cost. Mark this second edge. Next, select the edge with the next lowest cost. If a graph consisting of this edge and the two that have been marked previously would include a cycle then ignore it; otherwise mark this edge. This process continues until the marked edges form a spanning tree.

Example 12.23 We can apply Kruskal's algorithm to the graph of Example 12.19 in the following way.

Initially, the cheapest edge is that connecting COM_1 and COM_2. As such, our tree becomes

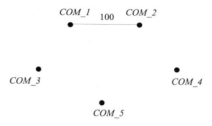

The next cheapest edge is that connecting COM_2 and COM_3. Furthermore, this edge does not create a cycle. As such, we add this edge.

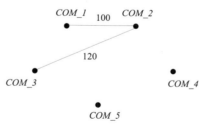

The next cheapest edge is that connecting COM_4 and COM_5. Furthermore, this edge does not create a cycle. As such, we add this edge.

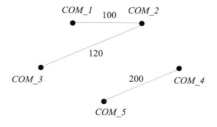

The next cheapest edge is that connecting COM_2 and COM_4. Furthermore, this edge does not create a cycle. As such, we add this edge.

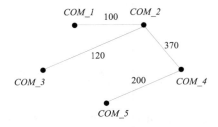

This is our minimal spanning tree. □

Exercise 12.49 Use Kruskal's algorithm to determine the minimal spanning tree of the graph of Example 12.20. □

12.9 Directed graphs

In many applications, undirected graphs—the type of graph that we have considered thus far—are a perfectly satisfactory representation of the problem at hand. However, there are some applications in which connections between vertices exist in one direction only. Consider, as an example, a flight network between cities A, B, and C, such that it is possible to fly from A to B directly, from B to C directly and from C to A directly, but it is not possible to fly in the opposite directions directly. A representation of this network defined in terms of undirected graphs would not provide the information which we desire: it would merely denote the fact that A and B are connected, B and C are connected, and C and A are connected. *Directed graphs*, on the other hand, do allow us to represent such a network appropriately.[5] The interested reader is referred to [Jen00] for an introduction to applications of directed graphs.

Formally, we may represent the components of directed graphs in exactly the same way as we represent undirected graphs: in terms of a set of vertices and a bag of edges, with the only substantial difference between directed graphs and undirected graphs being that, in a directed graph, the direction of each edge is made explicit by arrowheads.

Example 12.24 Take, as an example, the following graph representing the air network described above.

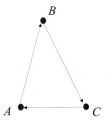

Here, the fact that there is an edge *from A to B* is denoted by the fact that there is an arrowhead on the B end of that edge. Similarly, it can be seen that there are edges from B to C and from C to A. □

Of course, just because a graph is directed, it does not mean that edges cannot exist between two nodes in opposite directions. For example, if the airline has expanded its portfolio of flights then there may be an edge from A to C, as well as one from C to A. This is illustrated below.

---

[5]In some texts directed graphs are referred to as *digraphs*.

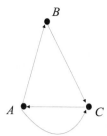

Furthermore, a directed graph can also have loops.

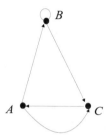

We may, of course, represent this directed graph in terms of both relations and matrices. Its representation in terms of relations is given by

$$V = \{A, B, C\}$$
$$E = [\![(A, C), (A, B), (B, B), (B, C), (C, A)]\!]$$

In addition, this graph's representation in terms of matrices (assuming A is vertex v_1, B is vertex v_2, and C is vertex v_3) is given by

$$\begin{pmatrix} 0 & 0 & 1 \\ 1 & 1 & 0 \\ 1 & 1 & 0 \end{pmatrix}$$

Of course, matrix representations of directed graphs are not symmetric, unlike matrix representations of undirected graphs. As such, the 'bottom left-hand half' matrix representations that were possible for representations of undirected graphs are not suitable for representing directed graphs.

Furthermore, with regards to our sets and bags representations of graphs, the ordering of the components of the elements of E is now important: in the context of undirected graphs the fact that a is connected to b can either be written as $(a, b) \in E$ or $(b, a) \in E$. In the context of directed graphs, $(a, b) \in E$ denotes the fact that there is an edge from a to b whereas $(b, a) \in E$ denotes the fact that there is an edge from b to a—these are two very different things.

Exercise 12.50 Draw the following directed graphs.

1. $V = \{a, b, c, d, e\}$; $E = [\![(a, b), (b, d), (d, e), (a, e)]\!]$

2. $V = \{a, b, c, d, e\}$; $E = [\![(b, c), (b, b), (c, b)]\!]$

3. $V = \{a, b, c, d, e\}$; $E = [\![(a, b), (b, c), (c, d), (d, e), (e, a)]\!]$

☐

Exercise 12.51 Construct matrix representations for the directed graphs of the previous exercise, assuming that $v_1 = a$, $v_2 = b$, $v_3 = c$, $v_4 = d$, and $v_5 = e$. ☐

12.10   Binary trees

Binary trees are structures which are fundamental to both discrete mathematics and computing. Essentially, binary trees are trees which satisfy certain properties.[6]

A binary tree has a special element, called a *root node*, and—traditionally (and somewhat counter-intuitively)—binary trees are drawn with the root node at the 'top' of the tree. Furthermore, each node (vertices are commonly referred to as *nodes* in the context of binary trees)—including the root node—in the tree can have up to two *descendants* or *children*.

Consider the binary tree given below.

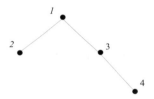

Here, the root node is node 1. Furthermore, this node has two children: node 2 and node 3, with node 2 being referred to as the *left child* of node 1 and node 3 being referred to as the *right child* of node 1. Conversely, node 1 may be referred to as the *parent* of nodes 2 and 3. In turn node 3 has one child: node 4. Nodes 2 and 4, on the other hand, have no children. Such nodes are referred to as *leaf* nodes.

Note how this binary tree is drawn: the root node is placed at the top level, with its children placed on the next level down. The descendants of the root node's children are placed one level further down, and so on. Note also that the collection of nodes descended from the left 'branch' of the root node also form a tree, as do the collection of nodes descended from the right branch. We may refer to the collection of nodes to the left of the root node as the *left subtree* of the root node. Similarly, we may refer to the collection of nodes to the right of the root node as the *right subtree* of the root node. In turn, node 3 has a right subtree, but no left subtree. Furthermore, as nodes 2 and 4 are both leaf nodes, neither has a subtree.

Recalling the notion of a partial ordering introduced in Chapter 8, it is clear that binary trees give a graphical representation of such an ordering—the properties of reflexivity, antisymmetry and transitivity all hold.

First, as an illustration, let $x \leq y$ denote the fact that node x is descended from or is equal to node y. It is clear that reflexivity holds as, given any node x that appears in the tree, it is the case that $x \leq x$ holds. Second, the property of transitivity holds as if node x is descended from node y and node y is descended from node z then it follows that node x is descended from node

---

[6]Note that binary trees are referred to as *rooted trees* in some texts.

z. Finally, antisymmetry holds as if it is the case that, for any nodes x and y, $x \leq y$ and $y \leq x$ both hold then it must be the case that $x = y$ holds.

Exercise 12.52 Draw the following binary tree.

$$V = \{a, b, c, d, e, f\}$$
$$E = [[(d, e), (d, f), (b, a), (a, d), (b, c)]]$$

☐

The *height* of a binary tree is given by the length of the path from the root node to one of the leaves at the lowest level of the tree. In the special case of the empty tree, the height is 0.

Example 12.25 Consider again the following binary tree.

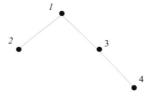

The height of this tree is 2. ☐

Exercise 12.53 What is the height of the binary tree of Exercise 12.52? ☐

Exercise 12.54 What is the maximum height of a binary tree with n nodes? ☐

Exercise 12.55 What is the minimum height of a binary tree with n nodes? ☐

A binary tree may be *traversed* (i.e., every node is visited, starting with the root node) in one of three ways: in *pre-order* fashion, in *in-order* fashion, or in *post-order* fashion. Essentially, traversing a binary tree allows us to list all of the nodes appearing in that tree. The different methods of traversing lead to the nodes being listed in different orders. We consider each approach in turn.

To traverse a binary tree in pre-order fashion involves starting at the root node, then travers-ing the left subtree, and then traversing the right subtree. This pattern follows through the tree: list the root node, then the nodes of the left subtree, then the nodes of the right subtree.

Example 12.26 Consider again the following binary tree.

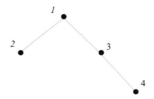

Traversing this tree in a *pre-order* fashion will lead to the nodes of this tree being listed in the order 1, 2, 3, 4. ☐

To traverse a binary tree in in-order fashion involves traversing the left subtree, listing the root node, and then traversing the right subtree. This pattern follows through the tree: list the nodes of the left subtree, then the root node, then the nodes of the right subtree.

Example 12.27 Traversing the above tree in an *in-order* fashion will lead to the nodes of this tree being listed in the order 2, 1, 3, 4. □

Finally, to traverse a binary tree in post-order fashion involves traversing the left subtree, then traversing the right subtree, and then listing the root node. This pattern follows through the tree: list the nodes of the left subtree, then the nodes of the right subtree, then the root node.

Example 12.28 Traversing the above tree in a *post-order* fashion will lead to the nodes of this tree being listed in the order 2, 4, 3, 1. □

Exercise 12.56 Consider the following binary tree.

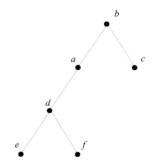

List the nodes of this tree in the following fashions.

1. Pre-order.

2. In-order.

3. Post-order.

□

12.11   Further exercises

Exercise 12.57 Draw the following (undirected) graphs.

1. $V = \{a, b, c, d, e\}$; $E = [\![(a, b), (b, e), (c, e), (d, e)]\!]$
2. $V = \{a, b, c, d, e\}$; $E = [\![(a, c), (a, e), (b, c), (b, d), (b, e)]\!]$

□

Exercise 12.58 Give a matrix representation of the graphs of the previous question, assuming that $v_0 = a$, $v_1 = b$, etc. □

Exercise 12.59 Give the adjacency matrices of the previous exercise in terms of matrices with values contained in only the bottom left-hand half. □

Exercise 12.60 Prove that none of the following graphs are isomorphic.

$$V_1 = \{a, b, c, d\} \; ; \; E_1 = [\![(a, b), (c, d), (a, d)]\!]$$
$$V_2 = \{x, y, z\} \quad ; \; E_2 = [\![(x, y), (y, z), (x, z)]\!]$$
$$V_3 = \{1, 2, 3, 4\} \; ; \; E_3 = [\![(1, 3), (2, 3), (3, 4)]\!]$$
$$V_4 = \{a, b, c, d\} \; ; \; E_4 = [\![(a, b), (c, d)]\!]$$

☐

Exercise 12.61 List all of the Eulerian paths contained in the following graph.

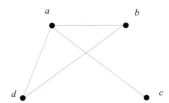

☐

Exercise 12.62 List all of the Hamiltonian paths contained in the graph of the previous exercise.
☐

Exercise 12.63 How many cycles does the graph of Exercise 12.61 have? ☐

Exercise 12.64 Consider the following binary tree.

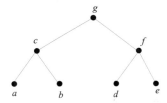

Traverse this binary tree in the following ways.

1. Pre-order.

2. In-order.

3. Post-order.

☐

Exercise 12.65 What is the height of the tree of the previous exercise? ☐

Exercise 12.66 Consider the following set of vertices.

$$V = \{a, b, c\}$$

1. How many possible binary trees featuring *one* element of V are there?

2. How many possible binary trees featuring *two* elements of V are there?

3. How many possible binary trees featuring *three* elements of V are there?

☐

12.12   Solutions

Solution 12.1 The following pairs of vertices are adjacent: 1 and 3, 1 and 4, 2 and 3, and 3 and 4. ☐

Solution 12.2 Edge a is incident to vertices 2 and 3; edge b is incident to vertices 1 and 3; edge c is incident to vertices 3 and 4; and edge d is incident to vertices 1 and 4. ☐

Solution 12.3 1 is not a simple graph, as it has a loop. 2 is a simple graph, as it has no loops or parallel edges. 3 is not a simple graph, as it has parallel edges. ☐

Solution 12.4 Yes, as if a graph has no edges then it can have no loops or parallel edges. As such, it must be a simple graph. ☐

Solution 12.5 This is given by

$$(n - 1) + (n - 2) + (n - 3) + .. + 1$$

which is equivalent to

$$\frac{n \times (n - 1)}{2}$$

☐

Solution 12.6

1.

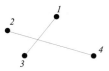

2.

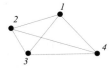

☐

Solution 12.7

$$deg\,(1) = 2$$
$$deg\,(2) = 3$$
$$deg\,(3) = 4$$
$$deg\,(4) = 2$$
$$deg\,(5) = 3$$
$$deg\,(6) = 0$$

□

Solution 12.8 Only vertex 6 is an isolated vertex; all others have a degree of at least 1. □

Solution 12.9 We may establish that this equation holds in the following way.

$$2 + 3 + 4 + 2 + 3 + 0 = 14$$
$$= 2 \times 7$$

□

Solution 12.10 The lemma holds as two vertices (2 and 5) have degrees which are odd. □

Solution 12.11 1 and 3 are connected graphs. □

Solution 12.12 It must have a minimum of $n - 1$ edges. □

Solution 12.13 Graph 2 is complete; the other two graphs are not. □

Solution 12.14 The graph must have

$$(n - 1) + (n - 2) + .. + 1$$

edges, which is equivalent to

$$\frac{n \times (n - 1)}{2}$$

□

Solution 12.15 Assume some graph G, which is complete. By definition, every vertex of this graph is adjacent to every other. It follows that we can move from every vertex to every other. As such, G is connected. □

Solution 12.16 E is defined in terms of a bag and not a set so that we may reason about parallel edges. Consider the following graph.

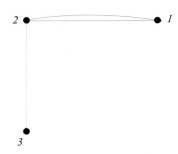

Here, there are parallel edges. If E was a set and not a bag, then E would be given by

$$E = \{(1,2), (2,3)\}$$

On the other hand, by defining E in terms of bags, we can see that there are two edges between vertices 1 and 2.

$$E = [\![(1,2), (1,2), (2,3)]\!]$$

Of course, if we were only to consider simple graphs, then representing E as a set would be perfectly satisfactory. □

Solution 12.17

1.

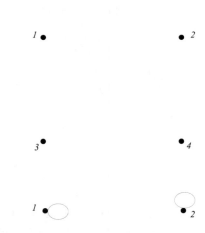

2.

3.

□

Solution 12.18

1. $V = \{1, 2, 3, 4\}$, $E = [\![(1,2), (1,3), (2,4), (3,4)]\!]$
2. $V = \{1, 2, 3, 4\}$, $E = [\![(1,1), (1,2), (2,2), (2,4), (4,4)]\!]$

□

Solution 12.19

1. $\begin{pmatrix} 0 & 1 & 0 \\ 1 & 0 & 0 \\ 0 & 0 & 0 \end{pmatrix}$

2. $\begin{pmatrix} 0 & 1 & 0 & 0 \\ 1 & 0 & 1 & 0 \\ 0 & 1 & 0 & 1 \\ 0 & 0 & 1 & 0 \end{pmatrix}$

□

Solution 12.20

1.

2.

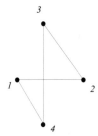

□

Solution 12.21 1 represents a simple graph: it consists of no edges (and, hence, no loops or parallel edges). 2 does not represent a simple graph as it involves one loop (at vertex v_1). Finally, 3 does not represent a simple graph as it involves parallel edges (between vertices v_1 and v_2). □

Solution 12.22

1. $\begin{pmatrix} 0 \\ 1\ 0 \\ 0\ 0\ 0 \end{pmatrix}$

2. $\begin{pmatrix} 0 \\ 1\ 0 \\ 0\ 1\ 0 \\ 0\ 0\ 1\ 0 \end{pmatrix}$

□

Solution 12.23

1.

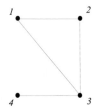

2.

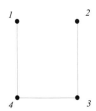

□

Solution 12.24 There will be $(n^2 + n)/2$ entries. □

Solution 12.25 No. For G_1 and G_2 to be isomorphic, we require that a total bijection $f : V_1 \rightarrowtail$ V_2 exists. However, if the cardinalities of V_1 and V_2 differ, then no such function can exist. As such, G_1 and G_2 cannot be isomorphic. □

Solution 12.26 No. For G_1 and G_2 to be isomorphic, we require that u and v are adjacent in G_1 if, and only if, $f(u)$ and $f(v)$ are adjacent in G_2. However, if the cardinalities of E_1 and E_2 differ, then this property cannot hold. As such, G_1 and G_2 cannot be isomorphic. □

Solution 12.27

1. $E_2 = [\![(1,1),(1,2),(3,4)]\!]$
2. $E_2 = [\![(1,4),(3,4),(2,3),(2,4)]\!]$

□

Solution 12.28 (V_1, E_1) and (V_3, E_3) are isomorphic, as evidenced by the total bijection f:

$$f = \{1 \mapsto 4, 3 \mapsto 2, 4 \mapsto 3, 2 \mapsto 1\}$$

None of the other graphs are isomorphic. □

Solution 12.29 The paths are given below.

$$\langle a, b, d \rangle, \langle a, b, a \rangle, \langle a, d, a \rangle, \langle a, d, b \rangle, \langle a, d, c \rangle, \langle b, a, b \rangle,$$
$$\langle b, a, d \rangle, \langle b, d, a \rangle, \langle b, d, b \rangle, \langle b, d, c \rangle, \langle c, d, c \rangle, \langle c, d, a \rangle,$$
$$\langle c, d, b \rangle, \langle d, a, b \rangle, \langle d, a, d \rangle, \langle d, b, a \rangle, \langle d, b, d \rangle, \text{ and } \langle d, c, d \rangle.$$

□

Solution 12.30 The paths are given below.

$$\langle a, b, d \rangle, \langle a, d, b \rangle, \langle a, d, c \rangle, \langle b, a, d \rangle, \langle b, d, a \rangle,$$
$$\langle b, d, c \rangle, \langle c, d, a \rangle, \langle c, d, b \rangle, \langle d, a, b \rangle, \text{ and } \langle d, b, a \rangle.$$

□

Solution 12.31

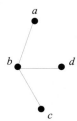

□

Solution 12.32 The circuits are

$$\langle a, b, d, a \rangle, \langle a, d, b, a \rangle, \langle b, a, d, b \rangle, \langle b, d, a, b \rangle, \langle d, a, b, d \rangle, \text{ and } \langle d, b, a, d \rangle.$$

□

Solution 12.33 They all are. □

Solution 12.34

1. 1
2. 3
3. 6
4. 10

□

Solution 12.35 There are eight Eulerian paths in this graph, and are listed below.

$$\langle 1, 2, 3, 1, 4, 3 \rangle, \langle 1, 2, 3, 4, 1, 3 \rangle, \langle 1, 3, 2, 1, 4, 3 \rangle, \langle 1, 3, 4, 1, 2, 3 \rangle,$$
$$\langle 3, 2, 1, 3, 4, 1 \rangle, \langle 3, 2, 1, 4, 3, 1 \rangle, \langle 3, 4, 1, 3, 2, 1 \rangle, \text{ and } \langle 3, 4, 1, 2, 3, 1 \rangle.$$

□

Solution 12.36 3 is semi-Eulerian and 4 is Eulerian. □

Solution 12.37 There are 12 Hamiltonian paths in this graph, and are listed below.

$$\langle 1, 2, 3, 4 \rangle, \langle 1, 4, 3, 2 \rangle, \langle 2, 1, 4, 3 \rangle, \langle 2, 1, 3, 4 \rangle, \langle 2, 3, 1, 4 \rangle, \langle 2, 3, 4, 1 \rangle,$$
$$\langle 3, 2, 1, 4 \rangle, \langle 3, 4, 1, 2 \rangle, \langle 4, 1, 2, 3 \rangle, \langle 4, 1, 3, 2 \rangle, \langle 4, 3, 1, 2 \rangle, \text{ and } \langle 4, 3, 2, 1 \rangle$$

□

Solution 12.38 2, 3 and 4 are Hamiltonian, whereas 1 is not. □

Solution 12.39 Yes, if the two edges are parallel edges between two vertices. □

Solution 12.40 Yes, if the edge is a loop. □

Solution 12.41 Three. □

Solution 12.42

1. There is one cycle: $\langle 2, 4, 5, 2 \rangle$.

2. There are three cycles. The first is $\langle 1, 2, 5, 1 \rangle$, the second is $\langle 1, 2, 3, 4, 1 \rangle$, and the third is $\langle 1, 5, 2, 3, 4, 1 \rangle$.

3. There are three cycles. The first is $\langle 1, 4, 5, 1 \rangle$, the second is $\langle 1, 4, 2, 3, 5, 1 \rangle$, and the third is $\langle 2, 3, 5, 4, 2 \rangle$.

□

Solution 12.43

$$\# E = 3$$

□

Solution 12.44 1 and 3 are trees. 2 and 4 are not trees, as they both contain cycles. □

Solution 12.45 Assume that T has n vertices. As T is a tree, T has $n - 1$ edges. Removing one edge will mean that a graph with n vertices and $n - 2$ edges is produced. This number of edges is insufficient for the new graph to be connected. □

Solution 12.46 Assume that T has n vertices. As T is a tree, T has $n - 1$ edges. Adding one edge will mean that the graph with n vertices and n edges is produced. As there are as many edges as vertices, the new graph must have a cycle. □

Solution 12.47 There is always exactly one path between any two vertices in a tree. □

Solution 12.48 Initially, we have considered no edges. Starting with Los Angeles, we find that the cheapest edge is that connecting Los Angeles and New York. As such, our tree becomes

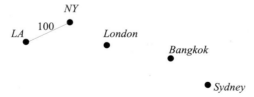

The cheapest edge linking one of New York and Los Angeles, and one of London, Sydney and Bangkok is that connecting New York and London. As such, we add this edge.

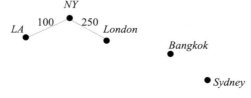

The cheapest edge linking one of New York, Los Angeles and London, and one of Sydney and Bangkok is that connecting London and Bangkok. As such, we add this edge.

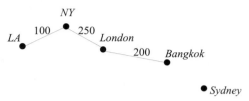

Finally, the cheapest edge linking one of New York, Los Angeles, London and Bangkok, and Sydney is that connecting Bangkok and Sydney. As such, we add this edge.

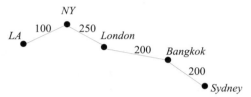

This is our minimal spanning tree. □

Solution 12.49 Initially, the cheapest edge is that connecting Los Angeles and New York. As such, our tree becomes

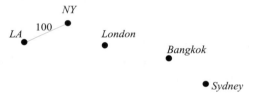

The next cheapest edge is either that between London and Bangkok, or that between Bangkok and Sydney. The addition of either of these edges would not create a cycle. We choose the former. As such, we add this edge.

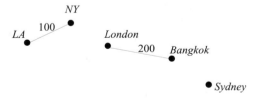

The next cheapest edge is that between Bangkok and Sydney. The addition of this edge would not create a cycle. As such, we add this edge.

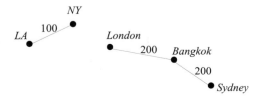

Finally, the next cheapest edge is that between London and New York. The addition of this edge would not create a cycle. As such, we add this edge.

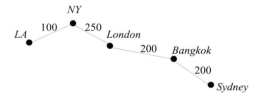

This is our minimal spanning tree. □

Solution 12.50

1.

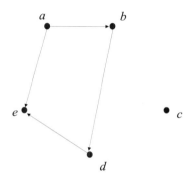

2.

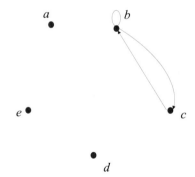

3.

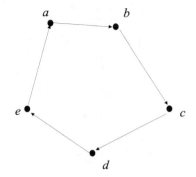

□

Solution 12.51

1. $\begin{pmatrix} 0\,0\,0\,0\,0 \\ 1\,0\,0\,0\,0 \\ 0\,0\,0\,0\,0 \\ 0\,1\,0\,0\,0 \\ 1\,0\,0\,1\,0 \end{pmatrix}$

2. $\begin{pmatrix} 0\,0\,0\,0\,0 \\ 0\,1\,1\,0\,0 \\ 0\,1\,0\,0\,0 \\ 0\,0\,0\,0\,0 \\ 0\,0\,0\,0\,0 \end{pmatrix}$

3. $\begin{pmatrix} 0\,0\,0\,0\,1 \\ 1\,0\,0\,0\,0 \\ 0\,1\,0\,0\,0 \\ 0\,0\,1\,0\,0 \\ 0\,0\,0\,1\,0 \end{pmatrix}$

□

Solution 12.52 There are several possible solutions, one of which is given below.

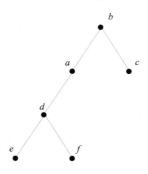

□

Solution 12.53 The height of the tree is 3. □

Solution 12.54 The maximum height of a binary tree with n nodes is n. □

Solution 12.55 The minimum height of a binary tree with n nodes is $\lceil log_2 (n + 1) \rceil$. □

Solution 12.56

1. b, a, d, e, f, c

2. e, d, f, a, b, c

3. e, f, d, a, c, b

□

Solution 12.57

1.

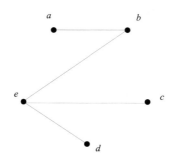

2.

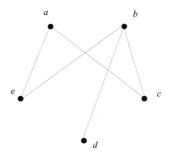

□

Solution 12.58

1. $\begin{pmatrix} 0\,1\,0\,0\,0 \\ 1\,0\,0\,0\,1 \\ 0\,0\,0\,0\,1 \\ 0\,0\,0\,0\,1 \\ 0\,1\,1\,1\,0 \end{pmatrix}$

2. $\begin{pmatrix} 0\,0\,1\,0\,1 \\ 0\,0\,1\,1\,1 \\ 1\,1\,0\,0\,0 \\ 0\,1\,0\,0\,0 \\ 1\,1\,0\,0\,0 \end{pmatrix}$

□

Solution 12.59

1. $\begin{pmatrix} 0 \\ 1\,0 \\ 0\,0\,0 \\ 0\,0\,0\,0 \\ 0\,1\,1\,1\,0 \end{pmatrix}$

2. $\begin{pmatrix} 0 \\ 0\,0 \\ 1\,1\,0 \\ 0\,1\,0\,0 \\ 1\,1\,0\,0\,0 \end{pmatrix}$

□

Solution 12.60 The graph represented by (V_2, E_2) has three vertices whereas the other graphs have four, and, as such, cannot be isomorphic to any of the others. The graph represented by (V_4, E_4) has two edges whereas the other graphs have three, and, as such, cannot be isomorphic to any of the others. Finally, it is impossible to exhibit a total bijection, f, from V_1 to V_3 such that $(u, v) \in E_1$ if, and only if, $(f(u), f(v)) \in E_3$. □

Solution 12.61 There are four, which are

$$\langle c, a, b, d, a \rangle, \langle c, a, d, b, a \rangle, \langle a, b, d, a, c \rangle, \text{ and } \langle a, d, b, a, c \rangle.$$

□

Solution 12.62 There are four, which are

$$\langle c, a, b, d \rangle, \langle c, a, d, b \rangle, \langle b, d, a, c \rangle, \text{ and } \langle d, b, a, c \rangle.$$

□

Solution 12.63 One: $\langle a, b, d, a \rangle$. □

Solution 12.64

1. g, c, a, b, f, d, e
2. a, c, b, g, d, f, e
3. a, b, c, d, e, f, g

□

Solution 12.65 The height of the tree is 2. □

Solution 12.66

1. There are three: one for each possible root node.
2. There are three possibilities for the root node: a, b, or c. If a is the root node, then it can have either a left or a right descendant (but not both). Both of these may be either b or c. As such, there are four possible binary trees if a is the root node. Similarly, there are four possible binary trees if b is the root node and four possible binary trees if c is the root node. As such, there are 12 possibilities.
3. There are three possibilities for the root node: a, b, or c. If a is the root node, then either it has two descendants, or it has a left descendant, or it has a right descendant. In turn, if a has one descendant, then a's descendant can have either a left descendant or a right descendant. As such, if a is the root node, then there are five possible shapes for the binary tree. As b or c can take either place in each possible binary tree, there are ten possible binary trees if a is the root node. Similarly, there are ten possible binary trees if b is the root node and ten possible binary trees if c is the root node. As such, there are 30 possibilities.

□

Chapter 13

Combinatorics

In this chapter we consider the topic of combinatorics. Essentially, the material described in this chapter is concerned with determining how many possible outcomes there are for certain scenarios. Such situations are common in computing. For example, we might wish to determine how many possible ways a network of computers can be connected, or we might wish to know how many possible results there are for a particular sequence of operations, or we might wish to determine the efficiency of a particular algorithm in terms of the number of operations it requires.

For reasons of space, we provide only a brief introduction to the topic of combinatorics in this chapter; for further exercises and examples on this subject the reader is referred to [LL92], and for a more in-depth introduction to the subject the reader is referred to [Bol86], [Eri96], [Bru99], or [Gri99].

We start by considering a mathematical function which is fundamental to the study of combinatorics—the factorial function.

13.1   The factorial function

The factorial function—the understanding of which is essential for this chapter—is defined as follows. For any natural number n, the factorial of n, denoted $n!$, is given by

$$n! = 1 \times 2 \times 3 \times .. \times n$$

That is $n!$ is calculated by multiplying all natural numbers between 1 and n. Furthermore, in the special case of $n = 0$, we have $0! = 1$.

We can, of course, provide a recursive definition of the factorial function in the following way.

$$0! = 1$$
$$1! = 1$$
$$n! = n \times (n - 1)! \text{ if } n > 1$$

Example 13.1 3! can be calculated as follows.

$$3! = 3 \times 2!$$
$$= 3 \times 2 \times 1!$$
$$= 3 \times 2 \times 1$$
$$= 6$$

□

The factorial function has a number of pleasing properties. One of these properties, which we shall make great use of in the following, is given below.

Law 13.1 For any natural number n,

$$\frac{(n+1)!}{n!} = n + 1$$

□

Example 13.2

$$\frac{4!}{3!} = 4$$

□

A generalised form of this law is given below.

Law 13.2 For any natural numbers m and n,

$$\frac{(n+m)!}{n!} = \prod_{i=n+1}^{n+m} i$$

□

Example 13.3 If $n = 7$ and $m = 3$, then we have the following.

$$\frac{10!}{7!} = \prod_{i=8}^{10} i$$

□

Exercise 13.1 Calculate each of the following.

1. $1!$
2. $3!$
3. $5!$
4. $-1!$

5. $\dfrac{1001!}{1000!}$

6. $\dfrac{20!}{18!}$

7. $\dfrac{\left(\dfrac{20!}{19!}\right)!}{18!}$

$\square$

Exercise 13.2 Refer to Exercise 12.34. Write a formula, in terms of !, for determining the maximum number of edges (including loops) a graph with n (where $n > 2$) vertices can have without it being connected. You may assume that there are no parallel edges. $\square$

13.2 Binomial coefficients

The *binomial coefficient*, $\begin{pmatrix} n \\ r \end{pmatrix}$, which is read as "*n choose r*", is defined, for natural numbers n and r, such that $r \leq n$, as follows.

$$\begin{pmatrix} n \\ r \end{pmatrix} = \frac{n!}{(n-r)! \times r!}$$

Essentially, the binomial coefficient $\begin{pmatrix} n \\ r \end{pmatrix}$ states how many times r objects can be chosen from a total of n.

Consider, as an example, a competition involving three prizes: a car, a house, and a donkey. If we were required to choose *two* prizes from the three available, then there are three possible combinations. These are listed below.

1. car + donkey

2. house + donkey

3. car + house

As such, for our example, the binomial coefficient is calculated in the following way.

$$\begin{pmatrix} 3 \\ 2 \end{pmatrix} = \frac{3!}{(3-2)! \times 2!}$$
$$= \frac{6}{2}$$
$$= 3$$

If, on the other hand, we were required to choose *one* prize from the three available, then this binomial coefficient would be calculated as follows.

$$\binom{3}{1} = \frac{3!}{(3-1)! \times 1!}$$
$$= \frac{6}{2}$$
$$= 3$$

As a further example, the number of ways in which we may choose two prizes from a total of four can be calculated as follows.

$$\binom{4}{2} = \frac{(4-2)!}{(4-2)! \times 2!}$$
$$= \frac{24}{4}$$
$$= 6$$

A pleasing property of binomial coefficients is given below.

Law 13.3 For any natural numbers n and r,

$$\binom{n}{r} = \binom{n}{n-r}$$

□

Example 13.4 Consider

$$\binom{20}{15}$$

By our definition of binomial coefficients, this is equal to

$$\frac{20!}{(20-15)! \times 15!}$$

By arithmetic, this is equal to

$$\frac{20!}{(20-5)! \times 5!}$$

Finally, by our definition of binomial coefficients, this is equal to

$$\binom{20}{5}$$

□

Our next law states that n elements can be chosen from a total of n elements in exactly one way.

Law 13.4 For any natural number n,

$$\binom{n}{n} = 1$$

□

Example 13.5 If we could choose any three prizes from an available three prizes, then there is only one possible outcome: to receive all three prizes. □

Next, 0 elements can also be chosen from a total of n elements in exactly one way.

Law 13.5 For any natural number n,

$$\binom{n}{0} = 1$$

□

Example 13.6 If we could choose zero prizes from an available three prizes, then there is only one possible outcome: to receive no prizes. □

Finally, the following property holds of binomial coefficients.

Law 13.6 For any natural numbers n and r, such that $n \geq 1$,

$$\binom{n}{r} = \binom{n-1}{r-1} + \binom{n-1}{r}$$

□

Example 13.7 Consider $n = 5$ and $r = 3$. Here, we wish to prove that

$$\binom{5}{3} = \binom{4}{2} + \binom{4}{3}$$

By our definition of binomial coefficients, the left-hand side is equal to

$$\frac{5!}{2! \times 3!}$$

In turn, this is equal to

$$\frac{120}{12}$$

which, of course, gives the value 10.

We now consider the right-hand side of the equation. By our definition of binomial coefficients, this is equal to

$$\frac{4!}{2! \times 2!} + \frac{4!}{1! \times 3!}$$

In turn, this is equal to

$$\frac{24}{4} + \frac{24}{6}$$

which, of course, gives the value $6 + 4$.

As such, the two sides of the equation are equal. $\square$

Exercise 13.3 Calculate each of the following.

1. $\binom{0}{0}$

2. $\binom{4}{4}$

3. $\binom{4}{3}$

4. $\binom{4}{1}$

5. $\binom{4}{2}$

6. $\binom{1001}{1000}$

$\square$

Exercise 13.4 Prove Law 13.3. $\square$

Exercise 13.5 Prove Law 13.4. $\square$

Exercise 13.6 Prove Law 13.5. $\square$

Exercise 13.7 *Pascal's triangle*[1] is a way of illustrating binomial coefficients, and is based on Law 13.6. The first three rows of Pascal's triangle are illustrated below.

```
    1
  1   1
1   2   1
```

---
[1] Named after Blaise Pascal (1623 - 1662).

The first row represents $n = 0$, the second row represents $n = 1$, and the third row represents $n = 2$. In addition, the first element of each row represents $r = 0$, the second element of each row represents $r = 1$, the third element of each row represents $r = 2$, and so on. As such, the first entry in the table represents "0 from 0", while the entries in the second row represent "0 from 1" and "1 from 1" respectively. In addition, we can see that there are two ways of choosing one element from a total of two. This entry can be calculated by adding the two elements which appear above the one that we are concerned with (recall that the construction of this triangle is based on Law 13.6). This rule is valid for all entries except those on the 'slope' of the triangle, which are always equal to one (these entries represent $r = 0$ and $r = n$). So, for example, adding the next row will give

$$
\begin{array}{ccccccc}
 & & & 1 & & & \\
 & & 1 & & 1 & & \\
 & 1 & & 2 & & 1 & \\
1 & & 3 & & 3 & & 1 \\
\end{array}
$$

Continue Pascal's triangle for the next three rows. $\square$

Exercise 13.8 Referring to the triangle developed in the previous question, find the following entries.

1. $n = 3, r = 2$
2. $n = 4, r = 2$
3. $n = 5, r = 3$
4. $n = 6, r = 3$

$\square$

Exercise 13.9 Referring to the triangle of Exercise 13.7, calculate the following.

1. $\dbinom{7}{2}$

2. $\dbinom{7}{3}$

3. $\dbinom{7}{4}$

4. $\dbinom{7}{5}$

$\square$

13.3   Counting

The fundamental principle of counting assumes that a series of events can occur in different numbers of ways. As an example, assume that we are concerned with three events—e_1, e_2, and e_3—and that e_1 can occur in n_1 different ways, e_2 can occur in n_2 different ways, and e_3 can occur in n_3 different ways. It follows that the series e_1, e_2, e_3 can occur in $n_1 \times n_2 \times n_3$ different ways.

Example 13.8 Assume that we are to toss a coin, then roll a die, then draw a card at random from a pack of playing cards. The first event has two possible outcomes, the second event has six possible outcomes, and the third event has 52 possible outcomes. As such, the series of events has

$$2 \times 6 \times 52 = 624$$

possible outcomes. □

It is important to note that each event is independent of each other—the fact that a head is tossed does not make a 6 being rolled more or less likely. Similarly, a 3 being rolled does not increase the likelihood of the Queen of Hearts being drawn from the pack of cards.

The counting principle is closely related to that of the Cartesian product. Recall that, given two sets, X and Y, the Cartesian product $X \times Y$ is defined by

$$X \times Y = \{x : X; \ y : Y \bullet (x, y)\}$$

Recall also that the cardinality of such a Cartesian product is given by

$$\# (X \times Y) = (\# X) \times (\# Y)$$

So, for example, given the following sets

$$Bread = \{brown, white\}$$
$$Filling = \{ham, cheese, ham\_and\_cheese\}$$

we find that

$$\# (Bread \times Filling) = 2 \times 3$$
$$= 6$$

That is, there are six possible outcomes.

In a more general sense, given sets $X_1, X_2, \ldots, X_n$,

$$\# (X_1 \times X_2 \times \ldots \times X_n) = \prod_{i=1}^{n} (\# X_i)$$

Exercise 13.10 Assume that in a particular computer system, the default for a user's electronic mail password is their date of birth, given in the form DDMMYY. What is the maximum number of unique passwords that can be generated by this system? □

Exercise 13.11 Consider again the system of the previous exercise. How many unique passwords can be generated by the system if it is known that all of the system's users were born in a particular four-year period? □

Exercise 13.12 Assume that in a different computer system the default for a person's electronic mail password is a random sequence of eight characters, which may be one of A..Z,a..z,0..9, apart from the first character, which cannot be a digit. What is the maximum number of unique passwords that can be generated by this system? □

Exercise 13.13 A sandwich shop offers three different types of filling: egg, chicken and tuna, and three different types of bread: wholemeal, rye, and granary. How many different types of sandwich does this shop sell? □

Exercise 13.14 Referring to the sandwich shop of the previous question, how many types of sandwiches are available for people who do not eat any sort of animal produce? (We shall assume, for the sake of argument, that a tuna fish is an animal.) □

Exercise 13.15 A tennis tournament involves 64 competitors in the men's competition and 64 competitors in the ladies' competition. How many possible combination of winners are there? □

13.4   Permutations

Counting was concerned with determining the number of different ways in which a series of events may occur. There, a notion of independence was assumed. In this section, we consider permutations, in which an assumption of independence is not appropriate. Here, we are concerned mainly with the number of ways in which certain related objects or events can be ordered.

As an example, assume the set *Colour*, which is defined as

$$Colour = \{red, yellow, green\}$$

If we were to consider how these elements may be ordered—that is, if we were to concern ourselves with the *permutations* of the elements of *Colour*—then the following possibilities arise.

> *red, yellow, green*
> *red, green, yellow*
> *yellow, red, green*
> *yellow, green, red*
> *green, red, yellow*
> *green, yellow, red*

Here, of course, the notion of independence that under-pinned combinations does not exist: if *red* is chosen first, then this influences what can be chosen next in the sense that it cannot be *red*. Similarly, if *yellow* is chosen first, then that cannot appear in positions 2 or 3. Thus, the number of possible orderings is given by

$$3 \times 2 \times 1 = 6$$

This, of course, is the same as 3!.

Given a set of n objects, any arrangement of these objects is called a *permutation*. The first element chosen can be one of n elements; the second element chosen can be one of $n - 1$ elements; the third element chosen can be one of $n - 2$ elements; and so on. Thus, the total number of possibilities is given by

$$n \times (n - 1) \times (n - 2) \times \ldots \times 1$$

Therefore, the number of permutations of the n objects is given by $n!$.

Given a set of n objects, which we assume is denoted X, we may define the set of permutations of X formally, using our sequence notation.

$$\{s : \text{iseq } X \mid \# s = \# X\}$$

Example 13.9 Given the set *Friends*, which is defined by

$$Friends = \{richard, george, john\}$$

then the set of permutations of *Friends* is given by

$$\{\langle richard, george, john\rangle, \langle richard, john, george\rangle, \langle george, richard, john\rangle,$$
$$\langle george, john, richard\rangle, \langle john, richard, george\rangle, \langle john, george, richard\rangle\}$$

Here, there are six possible permutations (recall that $3! = 6$). □

Any arrangement of any $r \leq n$ objects is called an *r-permutation*. The number of r-permutations of n objects is denoted $P(n, r)$ and defined as follows.

$$P(n, r) = \frac{n!}{(n-r)!}$$

Again, we may define the set of r-permutations of a set X formally in terms of our sequence notation.

$$\{s : \text{iseq } X \mid \# s = r\}$$

Example 13.10 Consider again the set *Colour*.

$$S = \{red, yellow, green\}$$

If we wished to consider how any two elements of *Colour* may be arranged, then we would have the following possibilities.

$$\{\langle red, yellow\rangle, \langle red, green\rangle, \langle yellow, red\rangle,$$
$$\langle yellow, green\rangle, \langle green, red\rangle, \langle green, yellow\rangle\}$$

Here,

$$P(3, 2) = \frac{3!}{(3-2)!}$$
$$= \frac{6}{1}$$
$$= 6$$

□

Example 13.11 The number of 3-permutations of a set containing ten elements can be calculated in the following way.

$$P(10,3) = \frac{10!}{(10-3)!}$$
$$= \frac{10!}{7!}$$
$$= 10 \times 9 \times 8$$
$$= 720$$

□

Exercise 13.16 In how many ways can the letters A, B, C, D, and E be arranged? □

Exercise 13.17

1. In how many ways can *two* of the letters A, B, C, D, and E be arranged?
2. In how many ways can *three* of the letters A, B, C, D, and E be arranged?
3. In how many ways can *four* of the letters A, B, C, D, and E be arranged?

□

Exercise 13.18 There are 12 bottles of wine available for a party. Eight are to be drunk. How many possible permutations are there? □

Exercise 13.19 There are ten horses competing in a race. Prize money is available for the first three horses. How many possible permutations of prize winners are there? □

Exercise 13.20 There are ten chairs available at a party. Given that five people have attended the party, how many different permutations of seating arrangements are there? □

Exercise 13.21 Eight people have entered a competition, of which only three different people can win different prizes. How many different permutations of prize winners are there? □

13.5   Combinations

In Section 13.4, we saw that the set $\{red, yellow, green\}$ has six *permutations*: there, order was of importance, as, for example, the permutation $\langle red, yellow, green \rangle$ is a different permutation to $\langle green, yellow, red \rangle$. On the other hand, a *combination* of the elements of a given set is a selection of objects in which order does not matter: a combination is simply a collection of objects for which order is not important. For example, *red, yellow* and *yellow, red* are equivalent combinations drawn from the set $\{red, yellow, green\}$. Combinations are the subject of this section.

An r-combination of a set of n objects, denoted $C(n,r)$, is any subset of r elements of a set of n objects. $C(n,r)$ is defined in the following way.

$$C(n,r) = \binom{n}{r}$$

Whereas we defined the permutations of a set X in terms of sequences, we define r-combinations in terms of sets.

$$\{s : \mathbb{P}\ X \mid \#\,s = r\}$$

Example 13.12 The set of 2-combinations of $\{red, yellow, green\}$ is given by

$$\{\{red, yellow\}, \{red, green\}, \{yellow, green\}\}$$

In addition,

$$\begin{aligned} C\,(3, 2) &= \binom{3}{2} \\ &= \frac{3!}{1! \times 2!} \\ &= 3 \end{aligned}$$

□

Example 13.13 The number of 3-combinations of a set containing ten elements can be calculated in the following way.

$$\begin{aligned} C\,(10, 3) &= \binom{10}{3} \\ &= \frac{10!}{7! \times 3!} \\ &= \frac{10 \times 9 \times 8}{3!} \\ &= 120 \end{aligned}$$

□

Exercise 13.22 Prove that

$$C\,(n, r) = \frac{P\,(n, r)}{r!}$$

□

Exercise 13.23 A standard pack of playing cards contains 52 cards. In how many ways might the following number of cards be chosen from such a pack?

1. 51

2. 50

3. 49

□

Exercise 13.24 A bag contains six differently coloured balls. If three balls are to be drawn from the bag, how many different combinations are there? □

Exercise 13.25 Consider the set of letters $\{A, B, C, D, E\}$.

1. How many combinations of *two* letters from this set are there?

2. How many combinations of *three* letters from this set are there?

3. How many combinations of *four* letters from this set are there?

□

Exercise 13.26 There are 12 bottles of wine available for a party. Eight are to be drunk. How many possible combinations are there? □

Exercise 13.27 There are ten horses competing in a race. Prize money is available for the first three horses. How many possible combinations of prize winners are there? □

Exercise 13.28 There are ten chairs available at a party. Given that five people have attended the party how many possible combinations of seated people are there? □

Exercise 13.29 Eight people have entered a competition, of which only three different people can win different prizes. How many different combinations of prize winners are there? □

13.6 Tree diagrams

Tree diagrams allow us to enumerate visually all of the different possibilities of a sequence of events (assuming that each of these events may occur in a finite number of different ways).

As an example, consider a game which involves tossing a coin and then choosing a non-zero natural number, such that we are concerned with whether this number is odd or even. The possible outcomes of the game are illustrated below.

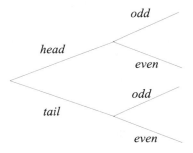

Here, there are four possible outcomes: a head is tossed and an odd number is chosen; a head is tossed and an even number is chosen; a tail is tossed and an odd number is chosen; and a tail is tossed and an even number is chosen. These distinct outcomes are illustrated by the tree diagram.

Note that we are not concerned with the likelihood of the different outcomes, but simply with enumerating the different possibilities. Note also that we may only construct tree diagrams for finite sequences of events, such that each such event has a finite number of possible outcomes.

Example 13.14 The sets *Bread* and *Filling* are given as follows.

$Bread = \{brown, white\}$

$Filling = \{ham, cheese, ham\_and\_cheese\}$

The different combinations of sandwiches that can be made from these types of bread and filling are given by the following.

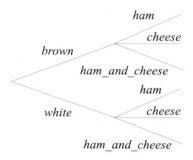

☐

While counting enabled us to determine the number of possible outcomes of a series of independent events, tree diagrams allow us to represent such outcomes visually.

Exercise 13.30 Consider a game in which a card is drawn from a standard pack of playing cards, and then a coin is tossed. If the card is a heart or a spade and a head is tossed, then the player wins; otherwise the player loses. Show the possible outcomes in terms of a tree diagram. ☐

Exercise 13.31 A particular two-player game consists of up to four legs. If one player achieves a two-leg lead, then he or she wins. Alternatively, if the score reaches 2 - 2, then the match is a draw. Use a tree diagram to enumerate the possible outcomes of such a game. ☐

Exercise 13.32 Dave can either go to a party on Friday evening, or he can stay in. In addition, he can either go to the cinema, go for a meal, or stay in on Saturday evening. Draw a tree diagram to represent the possibilities for Dave's weekend. ☐

13.7 Sampling

When we considered tree diagrams, in each case we tended to be concerned with a series of independent events. For example, the choice of filling in our sandwich example was not influenced by the choice of bread. In addition, the outcome of tossing a coin cannot influence (nor can be influenced by) which card is drawn from a pack of playing cards. These examples are not untypical of the sort of events that one might wish to model in computing. On other occasions, however, we might wish to consider a series of events, such that a single event is repeated over, and over again. Our game of Exercise 13.31 was such an example. As a further example, consider the following tree diagram.

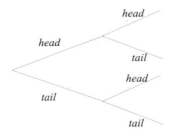

Here, the same event—the tossing of a coin—occurs twice. Of course, the result of the tossing of the second coin is independent of the result of the tossing of the first coin.

Now consider a game in which a player chooses two coins from a possible three, with the three coins on offer being a one pound coin, a five pence piece, and a two pence piece. We may represent the possible outcomes of this game in the following fashion.

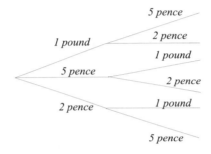

Here, although the second event—the choosing of a coin at random—is the same as the first event, the set of possible outcomes of the second occurrence of the event are directly influenced by the outcome of the first occurrence. For example, if a one pound coin is chosen initially, then a one pound coin cannot also be chosen subsequently.

The set of possible outcomes of a series of events is known as the *sample space*. There are two forms of sampling: *sampling with replacement* and *sampling without replacement*. We consider the difference between these forms of sampling by means of an example.

Consider a bag containing three balls: one red ball (denoted r), one yellow ball (denoted y), and one green ball (denoted g). If we were to select a first ball from the bag, then a second ball *without having replaced the first ball*, then there are six possible outcomes: rg, ry, yr, yg, gr, and gy. This is an example of sampling without replacement, with the number of possible outcomes being given by $P(n, r)$. An alternative means of calculating the number of such possibilities for this example is to reason that there are three possibilities for the first ball and—given the choice of the first ball—two possibilities for the second ball. Hence, there are

$$3 \times 2 = 6$$

possibilities. This, of course, is the same as $P(3, 2)$.

Now consider the same scenario, but with the first ball being replaced in the bag after it has been chosen. Here, there are nine possible outcomes: rr, rg, ry, gr, gg, gy, yr, yg, and yy. This is an example of sampling with replacement, with the number of possible outcomes being given by n^r. Of course, this is equivalent to reasoning that there are three possibilities for the first ball and three possibilities for the second ball. Hence, there are

$$3 \times 3 = 9$$

possibilities.

Thus, given a sample space of n, if the sample space is sampled r times *without* replacement then the number of possible outcomes is given by $P(n, r)$. On the other hand, if the sample space is sampled r times *with* replacement then the number of possible outcomes is given by n^r.

Example 13.15 If three cards are chosen at random from a standard pack of playing cards with replacement, then there are

$$52 \times 52 \times 52 = 140,608$$

possible outcomes. On the other hand, if three cards are chosen at random from a standard pack of playing cards without replacement, then there are

$$52 \times 51 \times 50 = 132,600$$

possible outcomes. □

Exercise 13.33 A committee consists of a president, a vice-president, a treasurer, and a secretary. The posts are to be drawn from 30 people, and no person can hold more than one post. In how many ways can this committee be organised? □

Exercise 13.34 Assume the committee described in the previous question now allows a person to hold more than one post, provided that one of them is not president. In how many ways can this committee be organised? □

Exercise 13.35 Consider the committee of the previous two exercises. Assume that any person may hold any number of different posts. In how many ways can this committee be organised? □

Exercise 13.36 Consider the set of letters $\{A, B, C, D, E\}$.

1. In how many ways can *two* letters be drawn from this set without replacement?

2. In how many ways can *three* letters be drawn from this set without replacement?

3. In how many ways can *four* letters be drawn from this set without replacement?

□

Exercise 13.37 Consider the set of letters $\{A, B, C, D, E\}$.

1. In how many ways can *two* letters be drawn from this set with replacement?

2. In how many ways can *three* letters be drawn from this set with replacement?

3. In how many ways can *four* letters be drawn from this set with replacement?

□

13.8   Further exercises

Exercise 13.38 Calculate the following.

1. $\binom{6}{2}$

2. $\binom{10}{8}$

3. $\binom{17}{14}$

□

Exercise 13.39 In Exercise 13.7, we considered rows 0 to 6 of Pascal's triangle. Write down the values of rows 7, 8, and 9. □

Exercise 13.40 Consider the sets X, Y and Z, such that $\# X = 4$, $\# Y = 5$, and $\# Z = 10$. Calculate the cardinality of $X \times Y \times Z$. □

Exercise 13.41 The combination lock for a safe consists of four digits, each in the range $0 .. 9$. How many possible values are there for the combination that opens the safe? □

Exercise 13.42 Consider a country in which car number plates consist of a letter, followed by three digits, followed by three letters. How many possible number plates of this type are there? □

Exercise 13.43 A country's postcodes consist of two letters, followed by two digits, followed by two letters. How many possible postcodes of this type are there? □

Exercise 13.44 Calculate the following.

1. $P(8, 4)$

2. $P(10, 4)$

3. $P(12, 4)$

□

Exercise 13.45 In how many ways can the letters of the following words be organised?

1. BASE

2. FOOT

3. BALL

4. FOOTBALL

5. BASEBALL

□

Exercise 13.46 Calculate the following.

1. $C(8,4)$

2. $C(10,4)$

3. $C(12,4)$

□

Exercise 13.47 A sports competition involves six teams. The teams are to be split into two groups of three teams. How many different combinations are there? □

Exercise 13.48 The number of questions which are asked in a test is 35. The total number of questions available from a 'question bank' is 50. How many different combinations of questions are there? □

Exercise 13.49 The number of questions which are asked in a test is 35. The total number of questions available from a 'question bank' is 50. How many different permutations of questions are there? □

Exercise 13.50 A set menu consists of: two starters (soup and bread); two main courses (pasta and fish); and two desserts (ice cream and cheesecake). Draw a tree diagram to illustrate all the possible combinations of courses. □

Exercise 13.51 A personality quiz asks people to list which three different personality traits out of a total of ten that most suit them. How many different combinations of answers are there? □

Exercise 13.52 A personality quiz asks people to list which three different personality traits out of a total of ten that most suit them. How many different permutations of answers are there? □

Exercise 13.53 Consider the digits 4, 7, and 9.

1. How many two-digit numbers can be formed from these digits if replacement is allowed?

2. How many two-digit numbers can be formed from these digits if replacement is not allowed?

□

13.9   Solutions

Solution 13.1

1. 1

2. 6

3. 120

4. undefined

5. 1001

6. 380

7. 380

☐

Solution 13.2 First, consider the number of possible loops. If a graph has n vertices, then it can have a maximum of n loops.

Next, consider the maximum number of edges that a graph can have without it being connected. To establish this, we may reason that the most connected unconnected graph is one in which *all* vertices are connected, with the exception of exactly one. So, for example, if $n = 4$, then three vertices are connected to each other (a total of three edges). As another example, if $n = 5$, then four vertices are connected to each other (a total of six edges). As such, our formula is given by

$$\frac{(n-1)!}{2} + n$$

☐

Solution 13.3

1. $\begin{pmatrix} 0 \\ 0 \end{pmatrix} = 1$

2. $\begin{pmatrix} 4 \\ 4 \end{pmatrix} = 1$

3. $\begin{pmatrix} 4 \\ 3 \end{pmatrix} = \dfrac{4!}{1! \times 3!}$
   $= 4$

4. $\begin{pmatrix} 4 \\ 1 \end{pmatrix} = \dfrac{4!}{3! \times 1!}$
   $= 4$

5. $\begin{pmatrix} 4 \\ 2 \end{pmatrix} = \dfrac{4!}{2! \times 2!}$
   $= 6$

6. $\begin{pmatrix} 1001 \\ 1000 \end{pmatrix} = \dfrac{1001!}{1! \times 1000!}$
   $= 1001$

☐

Solution 13.4 Consider $\begin{pmatrix} n \\ r \end{pmatrix}$. This is equal to

$$\frac{n!}{(n-r)! \times r!}$$

As multiplication is commutative, this is the same as

$$\frac{n!}{r! \times (n-r)!}$$

In turn, this is equal to

$$\frac{n!}{(n-(n-r))! \times (n-r)!}$$

which is, of course, the same as

$$\binom{n}{n-r}$$

$\square$

Solution 13.5 Consider some natural number n. Here, we have the following.

$$\begin{aligned}\binom{n}{n} &= \frac{n!}{(n-n)! \times n!} \\ &= \frac{n!}{0! \times n!} \\ &= \frac{n!}{1 \times n!} \\ &= 1\end{aligned}$$

As such,

$$\binom{n}{n} = 1$$

$\square$

Solution 13.6 Consider some natural number n. Here, we have the following.

$$\begin{aligned}\binom{n}{0} &= \frac{n!}{(n-0)! \times 0!} \\ &= \frac{n!}{n! \times 1} \\ &= 1\end{aligned}$$

As such,

$$\binom{n}{0} = 1$$

$\square$

Solution 13.7

```
              1
           1     1
        1     2     1
     1     3     3     1
  1     4     6     4     1
1     5    10    10     5    1
1  6   15    20    15   6   1
```

□

Solution 13.8

1. 3
2. 6
3. 10
4. 20

□

Solution 13.9

1. 21
2. 35
3. 35
4. 21

□

Solution 13.10 Assuming a maximum of 31 days per month, 12 months per year and years in the range 00–99, there are

$$31 \times 12 \times 100 = 37,200$$

possible combinations. □

Solution 13.11 Assuming a choice of four years, there are

$$31 \times 12 \times 4 = 1,488$$

possible combinations. □

Solution 13.12 There are

$$52 \times 62^7 = 183,123,959,522,816$$

possible combinations. □

Solution 13.13 There are

$$3 \times 3 = 9$$

possible combinations. □

Solution 13.14 There are

$$0 \times 3 = 0$$

possible combinations. □

Solution 13.15 There are

$$64 \times 64 = 4,096$$

possible combinations. □

Solution 13.16 There are

$$
\begin{aligned}
P\left(5,5\right) &= \frac{5!}{(5-5)!} \\
&= \frac{120}{1} \\
&= 120
\end{aligned}
$$

possible permutations. □

Solution 13.17

1. $$
\begin{aligned}
P\left(5,2\right) &= \frac{5!}{(5-2)!} \\
&= \frac{120}{6} \\
&= 20
\end{aligned}
$$

2. $$
\begin{aligned}
P\left(5,3\right) &= \frac{5!}{(5-3)!} \\
&= \frac{120}{2} \\
&= 60
\end{aligned}
$$

3. $P(5,4) = \dfrac{5!}{(5-4)!}$

$\qquad\qquad = \dfrac{120}{1}$

$\qquad\qquad = 120$

$\square$

Solution 13.18

$$P(12,8) = \dfrac{12!}{(12-8)!}$$

$$\qquad = \dfrac{12!}{4!}$$

$$\qquad = 19,958,400$$

$\square$

Solution 13.19

$$P(10,3) = \dfrac{10!}{(10-3)!}$$

$$\qquad = \dfrac{10!}{7!}$$

$$\qquad = 720$$

$\square$

Solution 13.20

$$P(10,5) = \dfrac{10!}{(10-5)!}$$

$$\qquad = \dfrac{10!}{5!}$$

$$\qquad = 30,240$$

$\square$

Solution 13.21

$$P(8,3) = \dfrac{8!}{(8-3)!}$$

$$\qquad = \dfrac{8!}{5!}$$

$$\qquad = 336$$

$\square$

Solution 13.22

$$C\left(n, r\right) = \binom{n}{r}$$

$$= \frac{n!}{\left(n - r\right)! \times r!}$$

$$= \frac{n!}{\left(n - r\right)!} \times \frac{1}{r!}$$

$$= P\left(n, r\right) \times \frac{1}{r!}$$

$$= \frac{P\left(n, r\right)}{r!}$$

$\square$

Solution 13.23

1. $C\left(52, 51\right) = \binom{52}{51}$

$$= \frac{52!}{1! \times 51!}$$

$$= 52$$

2. $C\left(52, 50\right) = \binom{52}{50}$

$$= \frac{52!}{2! \times 50!}$$

$$= \frac{52 \times 51}{2}$$

$$= 1,326$$

3. $C\left(52, 49\right) = \binom{52}{49}$

$$= \frac{52!}{3! \times 49!}$$

$$= \frac{52 \times 51 \times 50}{6}$$

$$= 22,100$$

$\square$

Solution 13.24

$$C(6,3) = \binom{6}{3}$$
$$= \frac{6!}{3! \times 3!}$$
$$= \frac{720}{36}$$
$$= 20$$

☐

Solution 13.25

1. $C(5,2) = \binom{5}{2}$
$$= \frac{5!}{3! \times 2!}$$
$$= 10$$

2. $C(5,3) = \binom{5}{3}$
$$= \frac{5!}{2! \times 3!}$$
$$= 10$$

3. $C(5,4) = \binom{5}{4}$
$$= \frac{5!}{1! \times 4!}$$
$$= 5$$

☐

Solution 13.26

$$C(12,8) = \binom{12}{8}$$
$$= \frac{12!}{4! \times 8!}$$
$$= 495$$

☐

Solution 13.27

$$C(10, 3) = \binom{10}{3}$$
$$= \frac{10!}{7! \times 3!}$$
$$= 120$$

□

Solution 13.28

$$C(10, 5) = \binom{10}{5}$$
$$= \frac{10!}{5! \times 5!}$$
$$= 252$$

□

Solution 13.29

$$C(8, 5) = \binom{8}{5}$$
$$= \frac{8!}{3! \times 5!}$$
$$= 56$$

□

Solution 13.30

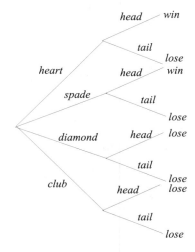

□

Solution 13.31

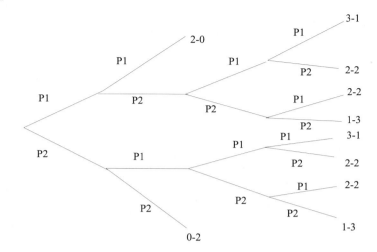

□

Solution 13.32

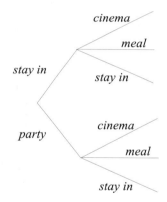

□

Solution 13.33

$30 \times 29 \times 28 \times 27 = 657,720$

□

Solution 13.34

$30 \times 29 \times 29 \times 29 = 731,670$

□

Solution 13.35

$$30 \times 30 \times 30 \times 30 = 810,000$$

□

Solution 13.36

1. $5 \times 4 = 20$
2. $5 \times 4 \times 3 = 60$
3. $5 \times 4 \times 3 \times 2 = 120$

□

Solution 13.37

1. $5 \times 5 = 25$
2. $5 \times 5 \times 5 = 125$
3. $5 \times 5 \times 5 \times 5 = 625$

□

Solution 13.38

1. $\binom{6}{2} = \dfrac{6!}{4! \times 2!}$
   $= 15$

2. $\binom{10}{8} = \dfrac{10!}{2! \times 8!}$
   $= 45$

3. $\binom{17}{14} = \dfrac{17!}{3! \times 14!}$
   $= 680$

□

Solution 13.39 Row 7 is given by

$$1 \ 7 \ 21 \ 35 \ 35 \ 21 \ 7 \ 1$$

Row 8 is given by

$$1 \ 8 \ 28 \ 56 \ 70 \ 56 \ 28 \ 8 \ 1$$

Finally, row 9 is given by

$$1 \ 9 \ 36 \ 84 \ 126 \ 126 \ 84 \ 36 \ 9 \ 1$$

□

Solution 13.40 The cardinality of $X \times Y \times Z$ is given by

$$4 \times 5 \times 10$$

which is 200. □

Solution 13.41 There are

$$10^4 = 10,000$$

possibilities. □

Solution 13.42 There are

$$26 \times 10^3 \times 26^3 = 456,976,000$$

possible number plates. □

Solution 13.43 There are

$$26^2 \times 10^2 \times 26^2 = 45,697,600$$

possible postcodes. □

Solution 13.44

1. $P(8,4) = \dfrac{8!}{(8-4)!}$

   $\qquad\quad = \dfrac{8!}{4!}$

   $\qquad\quad = 8 \times 7 \times 6 \times 5$

   $\qquad\quad = 1,680$

2. $P(10,4) = \dfrac{10!}{(10-4)!}$

   $\qquad\qquad = \dfrac{10!}{6!}$

   $\qquad\qquad = 10 \times 9 \times 8 \times 7$

   $\qquad\qquad = 5,040$

3. $P(12,4) = \dfrac{12!}{(12-4)!}$

   $\qquad\qquad = \dfrac{12!}{8!}$

   $\qquad\qquad = 12 \times 11 \times 10 \times 9$

   $\qquad\qquad = 11,880$

□

Solution 13.45

1. There are 4! ways of organising these letters, which is equivalent to 24.

2. There are 4! ways of organising these letters. However, two of them are the same (swapping the positions of the two 0s makes no difference to the sequence of letters generated). As such, there are

$$\frac{4!}{2!} = 12$$

different ways of organising these letters.

3. Again, there are 12 different ways of organising these letters.

4. There are 8! ways of organising these letters. However, there are two pairs of letters, and swapping the two Ls or the two Os makes no difference to the sequence of letters generated. As such, there are

$$\frac{8!}{2! \times 2!} = 10,080$$

different ways of organising these letters.

5. There are 8! ways of organising these letters. However, there are three pairs of letters, and swapping the two Bs, the two As or the two Ls makes no difference to the sequence of letters generated. As such, there are

$$\frac{8!}{2! \times 2! \times 2!} = 5,040$$

different ways of organising these letters.

□

Solution 13.46

1. $C(8,4) = \begin{pmatrix} 8 \\ 4 \end{pmatrix}$

$$= \frac{8!}{4! \times 4!}$$
$$= 70$$

2. $C(10,4) = \begin{pmatrix} 10 \\ 4 \end{pmatrix}$

$$= \frac{10!}{6! \times 4!}$$
$$= 210$$

3. $C(12,4) = \begin{pmatrix} 12 \\ 4 \end{pmatrix}$

$$= \frac{12!}{8! \times 4!}$$

$$= 495$$

□

Solution 13.47 The total number of combinations is

$$\begin{pmatrix} 6 \\ 3 \end{pmatrix} = \frac{6!}{3! \times 3!}$$

$$= \frac{720}{36}$$

$$= 20$$

□

Solution 13.48 The total number of combinations is

$$\begin{pmatrix} 50 \\ 35 \end{pmatrix} = \frac{50!}{15! \times 35!}$$

□

Solution 13.49 The total number of permutations is

$$\frac{50!}{(50-35)!} = \frac{50!}{15!}$$

□

Solution 13.50

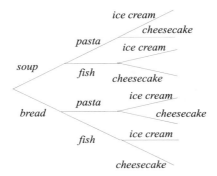

□

Solution 13.51 There are

$$C(10,3) = \binom{10}{3}$$
$$= \frac{10!}{7! \times 3!}$$
$$= \frac{10 \times 9 \times 8}{3!}$$
$$= 120$$

possible combinations. □

Solution 13.52 There are

$$P(10,3) = \frac{10!}{7!}$$
$$= 10 \times 9 \times 8$$
$$= 720$$

possible permutations. □

Solution 13.53

1. If replacement is allowed, then there are $3 \times 3 = 9$ possibilities.

2. If replacement is not allowed, then there are $3 \times 2 = 6$ possibilities.

□

Chapter 14

Examples

In this chapter we provide a number of examples to illustrate how the techniques described in the previous chapters can be used to model and reason about different types of systems. The examples vary both in size and complexity, but what is consistent throughout is that they all use the power of *abstraction* to concentrate only on those aspects of the problem at hand that are of concern to us in our modelling.

14.1 Modelling program variables

Variables in high-level programming languages allow us to capture the values of entities that may change over time. For example, the value of *current_temperature* for a particular application may be 21 initially, but this could change over time; certainly, if a program deemed current temperature to be important enough, one would expect the value of the variable to change to reflect reality. Furthermore, if the safe execution of this program depended on an accurate representation of the current temperature, then one would expect this value to be updated regularly.

We might consider a type, *VARIABLE*, which contains all possible variable names (of type $\mathbb{N}$, for the sake of simplicity), and $\mathbb{N}$, which contains all possible values which such a variable might take. The function *value* might then be defined in the following way.

$$value : VARIABLE \nrightarrow \mathbb{N}$$

We shall assume that an element of *VARIABLE* only appears in the domain of *value* if it has been declared within the program with which we are concerned.

In addition, we assume that all such variables have the value 0 initially.

14.1.1 A programming language

In this case study, we shall deal with a small programming language consisting of the following instructions.

- `declare`: A variable may be declared using this statement—no variables may be used without first being declared. For example, `declare(x)` declares a variable, x.

- `write`: The contents of a particular variable are updated with a specific value. For example, `write(x,4)` indicates that the variable x has been updated with the value 4.

- **read**: The contents of a particular variable are read. For example, **read(x)** returns the value associated with **x**.

- **swap**: The contents of two variables are swapped. For example, **swap(x,y)** swaps the contents of **x** and **y**.

In addition, the instructions **begin** and **end** initiate and terminate programs.

Exercise 14.1 What is the value of the function *value* initially? □

Exercise 14.2 Describe formally the effect of the statement **declare(x)** on *value* (assuming that *x* hasn't been previously been declared). □

Exercise 14.3 How might one denote the effect of the following statements on *value*?

1. write(x,4)

2. write(x,read(y))

3. swap(x,y)

□

Exercise 14.4 Give the value of the function *value* after each step in the following program.

1. begin

2. declare(x)

3. declare(y)

4. declare(z)

5. write(x,1)

6. write(y,2)

7. write(z,3)

8. swap(x,y)

9. swap(y,z)

10. end

□

Exercise 14.5 Represent the following statements in terms of propositional logic.

1. The value of x is the same as that of y.

2. If the value of x is 3, then the value of y is 2.

3. If the value of x and y are both 0, then z must not have been declared.

4. If x and y are the only variables that have been declared, then the value of y cannot be greater than that of x.

5. The values of x and y are equal if, and only if, the values of both are 0.

□

Exercise 14.6 Represent the following statements in terms of predicate logic.

1. All declared variables with a non-zero value have a value of at least 10.

2. All declared variables have a distinct value.

3. There is at least one declared variable with the value 0.

4. There is exactly one declared variable with a value greater than that of x.

5. If there is one declared variable with a non-zero value, then all declared variables have non-zero values.

□

Exercise 14.7 Define a sequence, *ordered*, via an axiomatic definition, which lists all of the variables used in a program in descending order of the value of their contents. □

Exercise 14.8 Define a recursive function, *add*, which sums all of the values contained in *ordered*.
□

Exercise 14.9 Define a recursive function, *double*, which produces a sequence containing, in order, all of the doubles of the elements of *ordered*. So, for example, if

$$value = \{x \mapsto 3, y \mapsto 2, z \mapsto 1\}$$

then

$$double \langle x, y, z \rangle = \langle 6, 4, 2 \rangle$$

□

Exercise 14.10 Prove, via structural induction, that

$$add\,(double\,ordered) = 2 \times (add\,ordered)$$

holds. □

14.2 Meta search engines

The concept of a search engine is familiar to all who use the World Wide Web: current favourites, such as Yahoo, Lycos and Alta Vista, take a user's query, process that query, and then return a list of up to hundreds, thousands, or even millions of links to World Wide Web pages that are somehow related to that query. The concept of a meta search engine goes a little further than this: a meta search engine receives a query, submits that query to a collection of search engines, collates the results, ranks them according to some criteria, and presents these meta search results to the user. Such a system is ideal for formal modelling.

In the following, we shall assume that *URL* represents the set of all URLs (Uniform Resource Locators, or Web Site addresses) and *ENGINE* represents the set of all search engines.

14.2.1 Modelling using sets

Consider the following scenario. Our meta search engine has submitted queries to three different search engines—$S1$, $S2$ and $S3$—and the following results have been received.

S1	S2	S3
www.xyz.com	www.abc.com	www.xyz.com
www.lmn.com	www.fgh.com	www.lmn.com
www.abc.com	www.xyz.com	
www.fgh.com		

We can, of course, represent each of the findings in terms of sets. The search engine S1 returned four results, while S2 and S3 returned three and two respectively. We may represent this information in terms of sets as follows.

$S1 = \{www.xyz.com, www.lmn.com, www.abc.com, www.fgh.com\}$
$S2 = \{www.abc.com, www.fgh.com, www.xyz.com\}$
$S3 = \{www.xyz.com, www.lmn.com\}$

We are now concerned with what to present to the user. We may decide to present only those URLs listed by *each* of $S1$, $S2$ and $S3$ (the generalised intersection of our three sets). Alternatively, we may decide to present those URLs listed by *at least one* of $S1$, $S2$ and $S3$ (the generalised union of our three sets).

Exercise 14.11 Calculate the generalised intersection of $S1$, $S2$, and $S3$. □

Exercise 14.12 Calculate the generalised union of $S1$, $S2$, and $S3$. □

14.2.2 Modelling using sequences

We can assume that the order in which the results were presented by $S1$, $S2$ and $S3$ are important: the first result matches the user's enquiry the closest, the second result is the next closest match, and so on. As such, it might be more natural to represent the information with which we are concerned in terms of sequences. Thus, a more concrete representation of the above information might be given as follows.

$S1 = \langle www.xyz.com, www.lmn.com, www.abc.com, www.fgh.com \rangle$
$S2 = \langle www.abc.com, www.fgh.com, www.xyz.com \rangle$
$S3 = \langle www.xyz.com, www.lmn.com \rangle$

Exercise 14.13 Is S1 an injective sequence? What about S2 and S3? □

Exercise 14.14 Calculate the following.

1. *reverse S2*

2. *# tail S3*

3. *head tail S1*

4. $(S1 \frown S2) \restriction (\mathrm{ran}\ S3)$

5. $\mathrm{ran}\,((S1 \frown S2) \restriction (\mathrm{ran}\ S3))$

□

Exercise 14.15 Write the following statements in terms of predicate logic, assuming the existence of sequences $S1$, $S2$ and $S3$.

1. If a URL appears in $S1$ then it appears in $S2$.

2. No URL appears in both $S1$ and $S2$.

3. If a URL u appears before another URL v in $S1$ and they both appear in $S2$ then u also appears before v in $S2$.

4. If a URL appears in position 1 in $S1$, then it also appears in that position in $S2$ and $S3$.

□

14.2.3   Modelling using relations

Thus far, we have considered $S1$, $S2$, and $S3$—be they in their set or sequence form—to be independent entities. By using a relation, we can combine the information associated with each search engine in order to determine our overall meta search engine's result.

Assuming the types *ENGINE* such that, for example, $S1 \in ENGINE$ and *URL* such that, for example, $www.xyz.com \in URL$, we may define the relation $result \in ENGINE \leftrightarrow \mathrm{seq}\ URL$ as follows.

$$\begin{aligned} result = \{ S1 &\mapsto \langle www.xyz.com, www.lmn.com, www.abc.com, www.fgh.com \rangle, \\ S2 &\mapsto \langle www.abc.com, www.fgh.com, www.xyz.com \rangle, \\ S3 &\mapsto \langle www.xyz.com, www.lmn.com \rangle \} \end{aligned}$$

Exercise 14.16 Calculate the following, assuming the relation *result* given above.

1. dom *result*

2. ran *result*

3. $\bigcup \{s : \mathrm{ran}\ result \bullet \mathrm{ran}\ s\}$

4. $result^{\sim}$

5. $result (\!| \{S2\} |\!)$

6. $\{S1\} \lhd result$

7. $result \rhd \{s : \mathrm{seq}\ URL \mid www.fgh.com \in \mathrm{ran}\ s\}$

□

Exercise 14.17 Is *result* a homogeneous or heterogeneous relation?   □

Exercise 14.18 Show how the following sets can be constructed, via set comprehension, from *result*.

1. $\{S2, S3\}$

2. $\{S1 \mapsto \langle www.xyz.com, www.lmn.com, www.abc.com, www.fgh.com \rangle\}$

3. $\{S1 \mapsto www.xyz.com, S2 \mapsto www.abc.com, S3 \mapsto www.xyz.com\}$

4. $\{www.xyz.com \mapsto S1, www.abc.com \mapsto S2, www.xyz.com \mapsto S3\}$

$\square$

Exercise 14.19 Is *result* a function? $\square$

Exercise 14.20 Calculate the following.

1. *result* $S1$

2. $result \oplus \{e : \mathrm{dom}\ result \bullet e \mapsto tail\ result\ e\}$

3. $result \oplus \{e : \mathrm{dom}\ result \bullet e \mapsto tail\ (result\ e \restriction \{www.xyz.com, www.abc.com\})\}$

$\square$

Exercise 14.21 Is *result* injective? Is it surjective? $\square$

14.2.4 Moving on

Having retrieved the results from different search engines, the next task for our meta search engine is to determine which—if any—results are to be presented to the user, and in which order (if, indeed, we are concerned with order). There are two possible approaches to consider: one that is defined in terms of sets, and one that is defined in terms of sequences.

Approach 1

The first approach that we shall consider doesn't concern itself with the order in which results are presented to the user—it simply considers which results have been presented to it by the search engines, and then determines which should be passed on to the user, according to some suitably defined criteria. The results to be passed to the user can be captured in terms of a number of rules. In addition, these rules can be represented formally in terms of set comprehensions.

For example, we may wish to consider those URLs which have featured in the results of all chosen search engines. Such a rule might be described as follows.

$$meta = \{u : URL \mid (\forall e : \mathrm{dom}\ result \bullet u \in \mathrm{ran}\ e)\}$$

Assuming the function *result* is given by

$$result = \{S1 \mapsto \langle www.xyz.com, www.lmn.com, www.abc.com, www.fgh.com \rangle,$$
$$S2 \mapsto \langle www.abc.com, www.fgh.com, www.xyz.com \rangle,$$
$$S3 \mapsto \langle www.xyz.com, www.lmn.com \rangle\}$$

this would result in the set *meta* taking the value $\{www.xyz.com\}$.

Exercise 14.22 Describe the following rules in terms of set comprehensions.

1. Only those URLS which have been returned by both S1 and S2 should be considered.

2. Only those URLS which have been returned by at least one of S1, S2, or S3 should be considered.

3. Only those URLS which have been returned at the top of the list by at least one search engine should be considered.

4. Only those URLS which have been returned at the top of the list by at least two search engines should be considered.

5. Only those URLS which appear in the top three results of S1 should be considered.

6. Only those URLS which appear in the top three results of S1, S2, or S3 should be considered.

□

Exercise 14.23 Assume the following value of *result*.

$$result = \{S1 \mapsto \langle www.xyz.com, www.lmn.com, www.abc.com, www.fgh.com\rangle,$$
$$S2 \mapsto \langle www.abc.com, www.fgh.com, www.xyz.com\rangle,$$
$$S3 \mapsto \langle www.xyz.com, www.lmn.com\rangle\}$$

Show the result of combining *result* with each occurrence of *meta* from the previous exercise. □

Approach 2

The first approach only considered which URLs to present to the user; a more refined approach would require that we consider the order in which these results should be presented to the user. Thus, rather than the overall result being a set of URLS, we require it to be a sequence.

We might, for example, give a score to each URL depending on the position it was returned by each search engine. Recall our original example, which was given as follows.

S1	S2	S3
www.xyz.com	www.abc.com	www.xyz.com
www.lmn.com	www.fgh.com	www.lmn.com
www.abc.com	www.xyz.com	
www.fgh.com		

Here, with regards to S1, `www.xyz.com` might be awarded 1 point, `www.lmn.com` might be awarded 2 points, and so on. Any URL not returned by that search engine may then be awarded some default value; let us assume for the moment that this value is 100. We may define a function, *position_scores*, which is of type $URL \times ENGINE \nrightarrow \mathbb{N}$ to perform this task.

$$\forall u : URL; \ e : ENGINE \bullet$$
$$u \in \text{ran } result(e) \Rightarrow position\_scores(u, e) =$$
$$(\mu n : \mathbb{N} \mid (result \ e) \ n = u)$$
$$\land$$
$$u \notin \text{ran } result(e) \Rightarrow position\_scores(u, e) = 100$$

For example,

$$position\_scores \, (www.abc.com, S1) = 3$$

and

$$position\_scores \, (www.xyz.com, S2) = 3$$

whereas

$$position\_scores \, (www.fgh.com, S3) = 100$$

Exercise 14.24 Calculate the following.

1. $position\_scores \, (www.abc.com, S1)$

2. $position\_scores \, (www.abc.com, S2)$

3. $position\_scores \, (www.abc.com, S3)$

□

Exercise 14.25 There is an implicit assumption in our definition of *position_scores* that each search engine returns an injective sequence of results. Produce a definition of *position_scores* which does not rely on this assumption, but associates the position of a given URL with its *first* appearance in a given sequence. □

The function *position_scores* represents the allocation of scores to URLs on a search engine by search engine basis. To be able to rank the URLs returned by our meta search engine, we require a function of type $URL \nrightarrow \mathbb{N}$ which adds these scores together. We may define such a function in the following way.

$$scores : URL \nrightarrow \mathbb{N}$$
$$\forall u : URL \bullet scores \, (u) = \sum_{e \in ENGINE} position\_scores \, (u, e)$$

Given the above, together with our earlier definition of *result*, which was given by

$$result = \{S1 \mapsto \langle www.xyz.com, www.lmn.com, www.abc.com, www.fgh.com \rangle,$$
$$S2 \mapsto \langle www.abc.com, www.fgh.com, www.xyz.com \rangle,$$
$$S3 \mapsto \langle www.xyz.com, www.lmn.com \rangle\}$$

we obtain the following result.

$$scores = \{www.xyz.com \mapsto 5, www.lmn.com \mapsto 104,$$
$$www.abc.com \mapsto 104, www.fgh.com \mapsto 106\}$$

Exercise 14.26 Assume the following value for *result*.

$$result = \{S1 \mapsto \langle www.abc.com, www.fgh.com, www.lmn.com, www.xyz.com\rangle,$$
$$S2 \mapsto \langle www.fgh.com, www.xyz.com, www.rst.com, www.lmn.com\rangle,$$
$$S3 \mapsto \langle www.xyz.com, www.rst.com, www.abc.com, www.lmn.com\rangle,$$
$$S4 \mapsto \langle www.fgh.com, www.lmn.com, www.rst.com\rangle\}$$

Given this value for *result*, calculate *scores*. □

The function *scores*, together with a suitably-defined set comprehension, *meta*, allows us to present the desired results to the user in terms of a final sequence. We can be sure that only those URLs satisfying the criteria defined by *meta* will appear in the sequence. Furthermore, the order in which the URLs appear is determined by the value of *scores*. We define such a function, via an axiomatic definition, as follows.

$$final : \text{iseq } URL$$

$$\text{ran } final = meta$$
$$\wedge$$
$$\forall m, n : \text{dom } final \bullet m < n \Rightarrow scores\ final\ m \leq scores\ final\ n$$

Exercise 14.27 Consider again the function *result*, with the following value.

$$result = \{S1 \mapsto \langle www.xyz.com, www.lmn.com, www.abc.com, www.fgh.com\rangle,$$
$$S2 \mapsto \langle www.abc.com, www.fgh.com, www.xyz.com\rangle,$$
$$S3 \mapsto \langle www.xyz.com, www.lmn.com\rangle\}$$

Consider also *scores*, which is given by

$$scores = \{www.xyz.com \mapsto 5, www.lmn.com \mapsto 104,$$
$$www.abc.com \mapsto 104, www.fgh.com \mapsto 106\}$$

Assuming the following definitions of *meta*, determine the value of *final* in each case.

1. Only those URLs that have been returned by both S1 and S2 should be considered.

2. Only those URLs that have been returned by at least one of S1, S2, or S3 should be considered.

3. Only those URLs that have been returned at the top of the list by at least one search engine should be considered.

4. Only those URLs that have been returned at the top of the list by at least two search engines should be considered.

5. Only those URLs that appear in the top three results of S1 should be considered.

6. Only those URLs that appear in the top three results of S1, S2, or S3 should be considered.

□

Exercise 14.28 Consider again the value of *result* from Exercise 14.26.

$$result = \{S1 \mapsto \langle www.abc.com, www.fgh.com, www.lmn.com, www.xyz.com \rangle,$$
$$S2 \mapsto \langle www.fgh.com, www.xyz.com, www.rst.com, www.lmn.com \rangle,$$
$$S3 \mapsto \langle www.xyz.com, www.rst.com, www.abc.com, www.lmn.com \rangle$$
$$S4 \mapsto \langle www.fgh.com, www.lmn.com, www.rst.com \rangle\}$$

Consider also the associated value of *scores*.

$$scores = \{www.abc.com \mapsto 204, www.fgh.com \mapsto 104,$$
$$www.lmn.com \mapsto 13, www.rst.com \mapsto 108, www.xyz.com \mapsto 107\}$$

Assuming the definitions of *meta* given in the previous exercise, determine the value of *final* in each case. □

14.3 Sequences for stacks and queues

Two data structures which should be familiar to all computer science students are *stacks* and *queues*. Both of these structures may be modelled in terms of sequences. We provide a brief recap of the two data structures before discussing how they may be modelled in terms of sequences.

A stack is a data structure which maintains data in a last in, first out order: the last item to join the data structure is the first one to leave it. This data structure is so-called because it is evocative of a stack of plates in a restaurant. There, plates are stacked one on top of another: when a plate is added it is placed at the top of the stack; when a plate is removed, it is removed from the top of the stack.

Essentially, a stack has four basic operators: `pop`, `push`, `is_empty`, and `top`. The operator `pop` removes the top element from a stack; the operator `push` pushes an element on to the top of the stack; the operator `is_empty` returns a Boolean value related to whether a stack is empty or not; and the operator `top` returns the value of the top element of a stack (without removing it).

As an example, assume the following stack of natural numbers.

```
7
9
11
```

Applying `pop` to this stack takes the top element away and leaves the following stack.

```
9
11
```

Note that the operation `pop` is only defined on non-empty stacks.

Pushing 5 on to the stack achieves the following result.

```
5
9
11
```

Applying `is_empty` to this stack will produce the value `false`. Furthermore, applying `top` to this stack will produce the value 5. As was the case with `pop`, the operation `top` is defined only on non-empty stacks.

Exercise 14.29 Consider the following stack.

```
10
12
14
```

Give the results of the following operations on *this* stack.

1. `pop`
2. `push(8)`
3. `top`
4. `is_empty`

□

A queue is a data structure which maintains data in a first in, first out order: the first item to join the data structure is the first one to leave it. This data structure is so-called because it is evocative of a queue at a bus stop. There, people queue one behind another: when a new person comes along, they join the queue at the rear; when a bus arrives, people board the bus (and, hence, leave the queue) from the front.

As was the case with a stack, a queue has four basic operators: `dequeue`, `enqueue`, `is_empty`, and `first`. The operator `dequeue` removes the first element from a queue; the operator `enqueue` adds an element to the rear of the queue; the operator `is_empty` returns a Boolean value related to whether a queue is empty or not; and the operator `first` returns the value of the first element of a queue (without removing it).

As an example, assume the following queue of natural numbers.

```
7, 9, 11
```

Applying `dequeue` to this queue takes the first element away and leaves the following queue.

```
9, 11
```

Adding 5 to the queue achieves the following result.

```
9, 11, 5
```

Applying `is_empty` to this queue will produce the value `false`. Furthermore, applying `first` to this queue will produce the value 9.

Exercise 14.30 Consider the following queue.

10, 12, 14

Give the results of the following operations on *this* queue.

1. dequeue

2. queue(8)

3. first

4. is_empty

□

We may provide a formal representation of both stacks and queues in terms of sequences. We consider our formal representation of stacks first.

Consider the set of stacks of some type X, which we shall denote $Stack[X]$. Here, we assume that

$$Stack[X] = \text{seq } X$$

So, for example, the sequence $\langle 7, 9, 11 \rangle$ represents the following stack.

7
9
11

The operation *pop* is defined for our formal representation in the following way.

$$\forall s : Stack[X] \mid s \neq \langle \rangle \bullet pop\ s = tail\ s$$

So, for example,

$$pop\ \langle 7, 9, 11 \rangle = \langle 9, 11 \rangle$$

We define *push* as follows.

$$\forall s : Stack[X];\ x : X \bullet push\,(s, x) = \langle x \rangle \frown s$$

Referring to the above example, this definition gives us the following.

$$push\,(\langle 9, 11 \rangle, 5) = \langle 5, 9, 11 \rangle$$

Next, we define *is_empty* in terms of a predicate.

$$\forall s : Stack[X] \bullet is\_empty\,(s) \Leftrightarrow s = \langle \rangle$$

Finally, the function *top* is defined as follows.

$$\forall s : Stack[X] \mid s \neq \langle \rangle \bullet top\ s = head\ s$$

So, for example,

$top \langle 5, 9, 11 \rangle = 5$

Exercise 14.31 Prove, via structural induction, that

$\forall s : Stack\,[X];\ x : X \bullet pop\,(push\,(s, x)) = s$

☐

Exercise 14.32 Prove the theorem of the previous exercise via equational reasoning. ☐

Exercise 14.33 Prove the following theorem via equational reasoning.

$is\_empty\,(push\,(pop\,(s), top\,(s))) \Leftrightarrow is\_empty\,(s)$

☐

We now consider the modelling of queues.

We shall denote the set of queues of type X by $Queue\,[X]$, such that

$Queue\,[X] = \text{seq}\ X$

So, for example, the sequence $\langle 7, 9, 11 \rangle$ represents the following queue.

7, 9, 11

The operation *dequeue* is given by

$\forall q : Queue\,[X] \mid q \neq \langle\rangle \bullet dequeue\ q = tail\ q$

So, for example,

$dequeue \langle 7, 9, 11 \rangle = \langle 9, 11 \rangle$

We define *enqueue* as follows.

$\forall q : Queue\,[X];\ x : X \bullet enqueue\,(q, x) = q \frown \langle x \rangle$

As an example,

$enqueue\,(\langle 9, 11 \rangle, 5) = \langle 9, 11, 5 \rangle$

Again, we define *is_empty* in terms of a predicate.

$\forall q : Queue\,[X] \bullet is\_empty\,(q) \Leftrightarrow q = \langle\rangle$

Finally, the function *first* is defined as follows.

$\forall q : Queue\,[X] \mid q \neq \langle\rangle \bullet first\ q = head\ q$

So, for example,

$first \langle 9, 11, 5 \rangle = 9$

Exercise 14.34 Is the following statement a theorem?

$$\forall\, q : Queue\,[X];\ x : X \bullet dequeue\,(enqueue\,(q, x)) = q$$

□

Exercise 14.35 Prove the following theorem via equational reasoning.

$$\forall\, x : X \bullet dequeue\,(enqueue\,(\langle\rangle, x)) = \langle\rangle$$

□

Exercise 14.36 Prove the following theorem via equational reasoning.

$$\forall\, q : Queue\,[X];\ x, y : X \bullet is\_empty\,(q) \Rightarrow \#\,enqueue\,(enqueue\,(q, x), y) = 2$$

□

14.4   Digital circuits

The classic application of Boolean algebra is, of course, in the context of digital circuits. We consider this application of Boolean algebra in this section. The interested reader is referred to [Men70] for a collection of problems and solutions for this topic.

The *binary* number system—the number system used by digital computers—consists of two symbols: 0 and 1. Digital computers operate by transmitting signals that can take only one of two values; the two symbols of the binary number system are usually used to represent these values.

A *logic gate circuit* performs digital computations. Logic gate circuits consist of a number of inputs—each of which may be in one of two states (which we denote by 0 and 1)—and one or more outputs—again, each of which may be in one of two states (denoted 0 and 1). Logic gate circuits are constructed from *logic gates*. As we shall see, the three logic gates that we consider, together with the values 0 and 1, form a Boolean algebra. It should be borne in mind that although these gates correspond to physical devices, we only concern ourselves with their abstract, logical behaviour. Technical issues such as power supplies and timings are of no consequence to this description.

The first logic gate that we are concerned with is the *NOT* gate (sometimes referred to as an inverter). This digital circuit accepts one input (denoted x) and produces one output (denoted y). A *NOT* gate is shown below.

Here, if 0 is input, then 1 is output. Alternatively, if 1 is input, then 0 is output. This gate corresponds to the complement operator of Boolean algebra, i.e., $y = x'$.

An *AND* gate is a digital circuit that accepts two inputs (x and y) and produces one output (z). An *AND* gate is shown below.

Here, if two 1s are input, then 1 is output. Alternatively, if at least one 0 is input, then 0 is output. This gate corresponds to the product operator of Boolean algebra, i.e., $z = x * y$.

An *OR* gate is a digital circuit that accepts two inputs (x and y) and produces one output (z). An *OR* gate is shown below.

Here, if at least one 1 is input, then 1 is output. Alternatively, if two 0s are input, then 0 is output. This gate corresponds to the sum operator of Boolean algebra, i.e., $z = x + y$.

Exercise 14.37 Show that the values 0 and 1, together with *NOT*, *AND* and *OR* form a Boolean algebra. □

Of course, as this forms a Boolean algebra, all of the derived laws that we met in Chapter 5 are applicable to our reasoning about logic gates.

Complex digital circuits can be made by combining the three gates described above. For example, the output from two AND gates might subsequently act as input to an OR gate. This state of affairs is illustrated below.

This circuit has four inputs—w, x, y and z—and one output: v. Furthermore, this circuit corresponds to the Boolean equation $v = wx + yz$.

Just as when evaluating a truth table we considered each contributing proposition in turn, so, when evaluating the likely outputs from digital circuits we evaluate each contributing gate in turn. To evaluate the output from the above circuit, first w *AND* x is evaluated, then y *AND* z is evaluated, and finally

$$(w \; AND \; x) \; OR \; (y \; AND \; z)$$

is evaluated.

Exercise 14.38 Calculate the output, v, from the above circuit for the following combinations of inputs.

1. $w = 0, x = 1, y = 1, z = 0$
2. $w = 1, x = 1, y = 0, z = 0$
3. $w = 0, x = 1, y = 0, z = 1$

□

Exercise 14.39 The proposition $x \lor y$ can be represented in terms of digital circuits as x *OR* y. Furthermore, the propositions $\neg \; x$ and $x \land y$ can be represented in terms of digital circuits as

NOT x and *x AND y*. Show how $x \Rightarrow y$ and $x \Leftrightarrow y$ might also be represented in terms of digital circuits. $\square$

Consider an alteration to the above circuit, such that it represents

$(x\ AND\ y)\ OR\ (x\ AND\ z)$

This is given below.

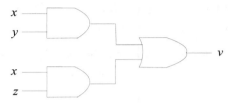

This circuit consists of three gates: two *AND* gates, and an *OR* gate. Recall that one of our distribution laws states that

$x\ AND\ (y\ OR\ z) = (x\ AND\ y)\ OR\ (x\ AND\ z)$

As such, the following circuit is equivalent to the one given above.

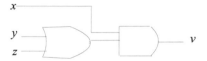

As this circuit consists of only two gates—one *AND* gate and an *OR* gate—it is considered to be preferable—in a real, physical sense—to the original. This is because as it has only two gates it requires less physical space to implement it. Furthermore, as only two computations are involved instead of three, output occurs quicker. Thus, the rules of Boolean algebra that we saw in Chapter 5 and the technique of equational reasoning that we met in Chapter 3 are applicable—and very important—in the context of digital circuits.

Exercise 14.40 Show how the circuit represented by

$((x\ AND\ 0)\ OR\ (x\ AND\ 1))\ OR\ ((NOT\ x)\ AND\ (NOT\ y))$

which consists of seven gates can be replaced by an equivalent circuit that has two gates. $\square$

Exercise 14.41 Prove the equivalence of the following pairs of circuits.

1. $(x\ AND\ y)\ OR\ (x\ AND\ (NOT\ y))$
   and
   x

2. $x\ OR\ (((NOT\ x)\ AND\ y)\ OR\ ((NOT\ z)\ AND\ y))$
   and
   $x\ OR\ y$

$\square$

14.5   A school database

In this section, we consider how functions may be used to model a database for a school. The interested reader is referred to [Hay93], which contains a collection of similar case studies in Z—a formal description technique which builds on the notation used in this case study.

First, we introduce the sets associated with our formal description. The first such set, *Teacher*, is the set of teachers. In a similar vein, we introduce the set *Student*. We assume a set *Subject*, which contains elements such as *history* and *mathematics*. We also assume a set *Classroom*. Next, we define by extension the set *Day*.

$$Day = \{monday, tuesday, wednesday, thursday, friday\}$$

We assume a relation *before*, of type $Day \times Day$, such that

$$before = \{monday \mapsto tuesday, monday \mapsto wednesday, monday \mapsto thursday,$$
$$monday \mapsto friday, tuesday \mapsto wednesday, tuesday \mapsto thursday,$$
$$tuesday \mapsto friday, wednesday \mapsto thursday,$$
$$wednesday \mapsto friday, thursday \mapsto friday\}$$

Finally, the set *Period* represents the different periods of each day in which lessons may take place; we assume that there are eight such periods per day.

$$Period = 1 \mathbin{..} 8$$

The first function that we consider is that which maps subjects to the set of teachers that teach it.

$$teacher : Subject \nrightarrow \mathbb{P}\ Teacher$$

Exercise 14.42 Would you expect the function *teacher* to be injective? □

Exercise 14.43 Define the following sets in terms of *teacher*.

1. The set of Geography teachers.

2. The set of teachers who teach more than one subject.

□

Next, we consider the function *chosen_subjects*. This maps students to the set of subjects that they have chosen.

$$chosen\_subjects : Student \nrightarrow \mathbb{P}\ Subject$$

Exercise 14.44 Show how, via axiomatic definitions, the following constraints may be put on *chosen_subjects*.

1. All students must have chosen at least two subjects, but no more than five.

2. A student may choose Humanities if, and only if, he or she has chosen neither History nor Geography.

3. Any student who has chosen at least four subjects must have chosen mathematics.

4. The conjunction of the above constraints.

☐

Exercise 14.45 Define a function *taught_by*, in terms of the existing functions *chosen_subjects* and *teacher*, which maps each student to the set of all teachers that teach that student's chosen subjects. ☐

Exercise 14.46 Define the following sets in terms of *chosen_subjects*.

1. The set of all students studying mathematics.

2. The set of all students who have chosen exactly three subjects.

3. The set of all subjects chosen by students who have chosen exactly three subjects.

☐

We now consider the first of two relationships that might hold between teachers and students. Every teacher is associated with a tutor group, which consists of a number of students. This information is represented by the function *tutees*.

$$tutees : Teacher \nrightarrow \mathbb{P} \; Student$$

Exercise 14.47 Show how, via axiomatic definitions, the following constraints may be imposed on *tutees*.

1. Every tutor group consists of between 15 and 30 (inclusive) students.

2. No student can appear in more than one tutor group.

3. The conjunction of the above constraints.

☐

Exercise 14.48 Define a function *tutor_teaches* in terms of *teacher* and *tutees*, which maps each student to the subjects that their tutor teaches. ☐

Exercise 14.49 Define the following in terms of *teacher*, *chosen_subjects*, and *tutees*.

1. The set of all tutors who have exactly 30 students in their tutor group.

2. The set of all tutors who have at least one tutee who has chosen five subjects.

3. The set of all pairs (s, t), such that t is the tutor of s and t teaches one of the subjects chosen by s.

☐

We now consider the information associated with students' marks. We represent this information as a function, *marks*, which maps pairs of type *Student* × *Subject* to natural numbers.

$$marks : Student \times Subject \nrightarrow \mathbb{N}$$

Exercise 14.50 Show how, via axiomatic definitions, the following constraints can be imposed on *marks*.

1. The maximum mark for any subject is 100.

2. No student may have a mark for more than five subjects.

3. The conjunction of the above.

□

Exercise 14.51 Define formally the relationship that should hold between the functions *marks* and *chosen_subjects*. □

Exercise 14.52 Define a relation *best_subject* in terms of *marks* which maps each student to the subject (or subjects) in which they have achieved their highest mark. □

Exercise 14.53 Define the following in terms of set comprehension.

1. The set of students who have scored over 50 in at least one of their chosen subjects.

2. The set of students who have scored over 50 in both History and Geography.

3. The set of students who have scored over 50 in a subject that their tutor teaches.

4. The set of subjects in which every student has scored over 50.

□

We now consider a second relationship between teachers and students. The *class* function maps tuples of type *Subject* × *Teacher* × *Day* × *Period* to a set of students. So, for example, *class* (s, t, d, p) denotes the set of students taught subject s by teacher t on day d at period p. This function is introduced in the following way.

$$class : Subject \times Teacher \times Day \times Period \rightarrow \mathbb{P} \; Student$$

Exercise 14.54 Show how, via axiomatic definitions, the following constraints may be imposed on *class*.

1. No student can be in more than one place at once.

2. No teacher can be in more than one place at once.

3. The conjunction of the above.

□

Exercise 14.55 State the relationships that should hold between the following pairs of functions.

1. *class* and *chosen_subjects*

2. *class* and *teacher*

3. *class* and *marks*

□

Exercise 14.56 Define a function *taught_tutees* that maps a given teacher *t* to the set of students that *t* both teaches and is tutor for. □

Exercise 14.57 Define the following in terms of set comprehension.

1. The set of all subjects taught on Monday, period 1.

2. The set of all students taught history on Monday, period 1.

3. The set of all pairs $(d, p) \in Day \times Period$ when History is taught.

□

Next, we consider the location of each class. This information is represented by the function *classroom*.

$$classroom : Subject \times Teacher \times Period \times Day \nrightarrow Classroom$$

Exercise 14.58 Show how, via axiomatic definitions, the following constraints may be imposed on *classroom*.

1. A teacher can be teaching in at most one classroom at any given time.

2. At most one teacher may be teaching in a given classroom at any one time.

3. The conjunction of the above.

□

Exercise 14.59 State the relationships that should hold between the following pairs of functions.

1. *classroom* and *class*

2. *classroom* and *teacher*

□

Exercise 14.60 Define the following functions in terms of *class*.

1. *location*, which, when given a teacher, a period and a day, returns a classroom.

2. *teaching_periods*, which maps each teacher to the number of periods that they teach.

□

Exercise 14.61 Define the following in terms of set comprehension.

1. The set of classrooms that are in use on Monday, period 1.

2. The set of classrooms in which History is taught.

3. The set of all pairs $(d, p) \in Day \times Period$ when all classrooms are in use.

□

Finally, we consider the function *timetable*, which associates a timetable (i.e., a sequence of tuples) with each student.

$$timetable : Student \nrightarrow \text{seq} \, (Period \times Day \times Subject \times Teacher \times Classroom)$$

Exercise 14.62 Show how, via axiomatic definitions, the following constraints may be imposed on *timetable*.

1. No student can be timetabled to be in two places at once.

2. Every student's timetable is ordered with respect to time, i.e., the tuple for Monday, period 1 appears before the tuple for Monday, period 2, and so on.

3. Every student has at most five subjects on their timetable.

□

Exercise 14.63 State the relationships that should hold between the following pairs of functions.

1. *timetable* and *classroom*

2. *timetable* and *class*

3. *timetable* and *marks*

4. *timetable* and *chosen_subjects*

□

Exercise 14.64 Define the following functions in terms of *timetable*.

1. *student_location*, which, when given a student, a day and a period, returns a classroom.

2. *teachers*, which, when given a student returns the set of all teachers that teach that student.

3. *taught_periods*, which maps each student to the number of periods in which they are taught.

□

Exercise 14.65 Define the following in terms of set comprehension.

1. The set of students who have a full timetable (recall that there are 40 periods in a week).

2. The set of students who have lessons on Monday, period 1.

3. The set of students who are taught History on Monday, period 1.

□

14.6 Knowledge-based systems

In computing, the field of expert systems is concerned with developing computer systems that represent the knowledge and experience of experts in specific fields. The two fundamental components of any expert system are the *knowledge base* and a set of *inference rules*. The former is a collection of known facts (essentially, a collection of true predicates), while the latter is a set of rules which allow us to derive new facts from existing ones (akin to our rules of natural deduction). It follows that the system can be queried, with responses being produced according to the information stored in the knowledge base.

We present a simple example of a knowledge-based system. First, we consider the following facts.

capital (*France, Paris*)	*currency* (*France, franc*)
capital (*Italy, Rome*)	*currency* (*Italy, lira*)
capital (*Spain, Madrid*)	*currency* (*Spain, peseta*)
capital (*Australia, Canberra*)	*currency* (*Australia, Aus dollar*)
capital (*USA, Washington*)	*currency* (*USA, US dollar*)
lives_in (*Pierre, Paris*)	*origin* (*Pierre, France*)
lives_in (*Igor, Venice*)	*origin* (*Igor, Italy*)
lives_in (*Anna, Madrid*)	*origin* (*Anna, Canada*)
lives_in (*Andrew, Brisbane*)	*origin* (*Andrew, UK*)
lives_in (*Cyril, Baltimore*)	*origin* (*Cyril, Zimbabwe*)

Here, *capital* (x, y) means that y is the capital city of country x. In addition, *currency* (x, y) means that y is the currency of country x. Next, *lives_in* (x, y) means that person x lives in country y. Finally, *origin* (x, y) means that person x is originally from country y.

A query is written

$$? - capital\,(a, b)$$

This asks the system if a is the capital of b. The system returns *true* if it can determine that this is a fact, and returns *false* otherwise. As an example, the query

$$? - capital\,(France, Paris)$$

returns the result *true*.

Exercise 14.66 Give the results of the following queries.

1. $? - capital\,(Australia, Sydney)$

2. $? - capital\,(USA, Washington)$

□

Having constructed the knowledge base of our system, we are now in a position to consider its inference rules. These rules will allow us to construct new predicates out of existing ones.

As an example, a particular relationship between the predicates *in* and *capital* can be stated as follows.

$$capital\,(x, y) \Rightarrow in\,(y, x)$$

Here, $in\,(y, x)$ means that city y is in country x. This will obviously hold if y is the capital of x. So, for example, $in\,(Paris, France)$ returns true, as we know that $capital\,(France, Paris)$ holds and, by the above implication, it follows that $in\,(Paris, France)$ is true. On the other hand, $in\,(Lille, France)$ cannot be concluded as there is no information to establish the truth of this predicate. As such, the system concludes that $in\,(Lille, France)$ is false.

Exercise 14.67 Given the facts and inference rules introduced thus far, give the results of the following queries.

1. $? - in\,(Madrid, Spain)$

2. $? - in\,(Barcelona, Spain)$

3. $? - in\,(Rome, Spain)$

□

Exercise 14.68 Define inference rules in terms of *capital*, *lives_in*, *currency* and *origin* to represent the following predicates.

1. *lives_in_capital*, such that *lives_in_capital* (x) represents the fact that x lives in a capital city.

2. *lives_in_home_country*, such that *lives_in_home_country* (x) denotes the fact that x lives in his or her home country.

3. *lives_in_foreign_country*, such that *lives_in_foreign_country* (x) denotes the fact that x lives in a foreign country.

4. *spends*, such that *spends* (x, y) denotes the fact that person x spends currency y in the country in which he or she lives.

□

Exercise 14.69 Calculate the results of the following queries.

1. $? - lives\_in\_capital\,(Igor)$

2. $? - lives\_in\_home\_country\,(Igor)$

3. $? - lives\_in\_foreign\_country\,(Igor)$

4. $? - spends\,(Igor, lira)$

□

Exercise 14.70 Some erroneous results were produced in response to the queries of the previous question. How might we adapt our system to correct this? □

In some circumstances, it is appropriate not to ask questions such as "does Igor live in a capital city?" but rather to enquire of the system exactly *which* people in our system live in a capital city. To do this, we can pass a *variable* (such as, for example, x) as a parameter, rather than a *value* (such as, for example, *Igor*). Under such circumstances, the query $p(x)$ will return the values of x for which the predicate p is true.

As an example, the query

$$? - lives\_in\_capital\,(x)$$

returns the results

$$x = Anna$$

and

$$x = Pierre$$

That is, it returns the names of *all* elements that satisfy the criteria specified of x.

Exercise 14.71 Calculate the following.

1. $spends\,(Anna, x)$
2. $spends\,(x, peseta)$
3. $spends\,(x, y)$
4. $lives\_in\_capital\,(x) \land lives\_in\_home\_country\,(x)$
5. $lives\_in\_capital\,(x) \land lives\_in\_home\_country\,(y)$

□

14.7   Solutions

Solution 14.1 Initially,

$$value = \emptyset$$

□

Solution 14.2 *value* is updated to become

$$value \cup \{x \mapsto 0\}$$

□

Solution 14.3

1. *value* becomes $value \oplus \{x \mapsto 4\}$
2. *value* becomes $value \oplus \{x \mapsto value\ y\}$

    3. *value* becomes *value* $\oplus \{x \mapsto value\ y, y \mapsto value\ x\}$

$\square$

Solution 14.4

    1. $\emptyset$

    2. $\{x \mapsto 0\}$

    3. $\{x \mapsto 0, y \mapsto 0\}$

    4. $\{x \mapsto 0, y \mapsto 0, z \mapsto 0\}$

    5. $\{x \mapsto 1, y \mapsto 0, z \mapsto 0\}$

    6. $\{x \mapsto 1, y \mapsto 2, z \mapsto 0\}$

    7. $\{x \mapsto 1, y \mapsto 2, z \mapsto 3\}$

    8. $\{x \mapsto 2, y \mapsto 1, z \mapsto 3\}$

    9. $\{x \mapsto 2, y \mapsto 3, z \mapsto 1\}$

    10. $\{x \mapsto 2, y \mapsto 3, z \mapsto 1\}$

$\square$

Solution 14.5

    1. *value* $x = $ *value* y

    2. *value* $x = 3 \Rightarrow$ *value* $y = 2$

    3. *value* $x = 0 \land$ *value* $y = 0 \Rightarrow z \notin$ dom *value*

    4. dom *value* $= \{x, y\} \Rightarrow$ *value* $x \geq$ *value* y

    5. *value* $x = $ *value* $y \Leftrightarrow$ *value* $x = 0 \land$ *value* $y = 0$

$\square$

Solution 14.6

    1. $\forall v :$ dom *value* $\mid$ *value* $v \neq 0 \bullet$ *value* $v \geq 10$

    2. $\forall v, w :$ dom *value* $\mid v \neq w \bullet$ *value* $v \neq$ *value* w

    3. $\exists v :$ dom *value* $\bullet$ *value* $v = 0$

    4. $\exists_1 v :$ dom *value* $\bullet$ *value* $v > $ *value* x

    5. $\exists v :$ dom *value* $\bullet$ *value* $v \neq 0 \Rightarrow \forall v :$ dom *value* $\bullet$ *value* $v \neq 0$

$\square$

Solution 14.7

$$
\begin{array}{|l}
\hline
ordered : \text{seq } VARIABLE \\
\hline
\# \ ordered = \# \ value \\
\wedge \\
\text{ran } ordered = dom \ value \\
\wedge \\
\forall \, m, n : \text{dom } ordered \bullet m < n \Leftrightarrow value \ ordered \ m \geq value \ ordered \ n \\
\end{array}
$$

□

Solution 14.8 First, we define the function add as follows.

$$
add \, (\langle\rangle) = 0
$$
$$
add \, (\langle x \rangle \frown s) = value \ x + add \, (s)
$$

Thus, $add \, (ordered)$ gives us the desired result. □

Solution 14.9 First, we define the function $double$ as follows.

$$
double \, (\langle\rangle) = \langle\rangle
$$
$$
double \, (\langle x \rangle \frown s) = \langle 2 \times value \ x \rangle \frown double \ s
$$

Thus, $double \, (ordered)$ gives us the desired result. □

Solution 14.10 We consider the base case first. This is established in the following way.

$$
\begin{aligned}
add \, (double \, \langle\rangle) &= add \, \langle\rangle \\
&= 0 \\
&= 2 \times 0 \\
&= 2 \times (add \, \langle\rangle)
\end{aligned}
$$

We now consider the inductive step. The inductive hypothesis is given by

$$
add \, (double \ s) = 2 \times (add \ s)
$$

We establish the inductive step as follows.

$$
\begin{aligned}
add \, (double \, \langle x \rangle \frown s) &= add \, (\langle 2 \times value \ x \rangle + double \ s) \\
&= (2 \times value \ x) + add \, (double \ s) \\
&= (2 \times value \ x) + (2 \times (add \ s)) \\
&= 2 \times (value \ x + add \ s) \\
&= 2 \times (add \, \langle x \rangle \frown s)
\end{aligned}
$$

□

Solution 14.11

$$\bigcap \{S1, S2, S3\} = \{www.xyz.com\}$$

□

Solution 14.12

$$\bigcup \{S1, S2, S3\} = \{www.xyz.com, www.lmn.com, www.abc.com, www.fgh.com\}$$

□

Solution 14.13 They are all injective sequences. □

Solution 14.14

1. $\langle www.xyz.com, www.fgh.com, www.abc.com \rangle$

2. 1

3. $www.lmn.com$

4. $\langle www.xyz.com, www.lmn.com, www.xyz.com \rangle$

5. $\{www.xyz.com, www.lmn.com\}$

□

Solution 14.15

1. $\forall\, u : URL \bullet u \in \operatorname{ran} S1 \Rightarrow u \in \operatorname{ran} S2$

2. $\forall\, u : URL \bullet u \notin \operatorname{ran} S1 \cap \operatorname{ran} S2$

3. $\forall\, u, v : URL \mid u, v \in \operatorname{ran} S1 \bullet$
   $\quad S1^\sim u < S1^\sim v \wedge u \in \operatorname{ran} S2 \wedge v \in \operatorname{ran} S2 \Rightarrow$
   $\quad\quad S2^\sim u < S2^\sim v$

4. $\forall\, u : URL \bullet S1^\sim u = 1 \Rightarrow S2^\sim u = 1 \wedge S3^\sim u = 1$

□

Solution 14.16

1. $\{S1, S2, S3\}$

2. $\{\langle www.xyz.com, www.lmn.com, www.abc.com, www.fgh.com \rangle,$
   $\quad \langle www.abc.com, www.fgh.com, www.xyz.com \rangle,$
   $\quad \langle www.xyz.com, www.lmn.com \rangle \}$

3. $\{www.xyz.com, www.lmn.com, www.abc.com, www.fgh.com\}$

4. $\{\langle www.xyz.com, www.lmn.com, www.abc.com, www.fgh.com \rangle \mapsto S1,$
   $\quad \langle www.abc.com, www.fgh.com, www.xyz.com \rangle \mapsto S2,$
   $\quad \langle www.xyz.com, www.lmn.com \rangle \mapsto S3\}$

5. $\{\langle www.abc.com, www.fgh.com, www.xyz.com\rangle\}$

6. $\{S1 \mapsto \langle www.xyz.com, www.lmn.com, www.abc.com, www.fgh.com\rangle\}$

7. $\{S3 \mapsto \langle www.xyz.com, www.lmn.com\rangle\}$

□

Solution 14.17 It is a heterogeneous relation. □

Solution 14.18

1. $\{e : \mathrm{dom}\ result \mid e \neq S1\}$

2. $\{e : \mathrm{dom}\ result \mid e = S1 \bullet e \mapsto result\ e\}$

3. $\{e : \mathrm{dom}\ result \bullet e \mapsto head\ result\ e\}$

4. $\{e : \mathrm{dom}\ result \bullet head\ result\ e \mapsto e\}$

□

Solution 14.19 Yes, it is a function. □

Solution 14.20

1. $\langle www.xyz.com, www.lmn.com, www.abc.com, www.fgh.com\rangle$

2. $\{S1 \mapsto \langle www.lmn.com, www.abc.com, www.fgh.com\rangle,$
   $\quad S2 \mapsto \langle www.fgh.com, www.xyz.com\rangle, S3 \mapsto \langle www.lmn.com\rangle\}$

3. $\{S1 \mapsto \langle www.abc.com\rangle, S2 \mapsto \langle www.xyz.com\rangle, S3 \mapsto \langle\rangle\}$

□

Solution 14.21 It is injective, but it is not surjective. □

Solution 14.22

1. $meta = \{u : URL \mid u \in (\mathrm{ran}\ result\ S1 \cap \mathrm{ran}\ result\ S2)\}$

2. $meta = \{u : URL \mid u \in (\mathrm{ran}\ result\ S1 \cup \mathrm{ran}\ result\ S2 \cup \mathrm{ran}\ result\ S3)\}$

3. $meta = \{e : \mathrm{dom}\ result \bullet head\ result\ e\}$

4. $meta = \{u : URL \mid (\exists\, e_1, e_2 : ENGINE \mid e_1 \neq e_2 \bullet$
   $\qquad\qquad\qquad head\ result\ e_1 = u \wedge head\ result\ e_2 = u)\}$

5. $meta = \{u : URL \mid (\exists\, i : 1 .. 3 \bullet (result\ S1)\ i = u)\}$

6. $meta = \{u : URL;\ e : \{S1, S2, S3\};\ i : 1 .. 3 \mid (result\ e)\ i = u \bullet u\}$

□

Solution 14.23

1. $meta = \{www.xyz.com, www.abc.com, www.fgh.com\}$

2. $meta = \{www.xyz.com, www.lmn.com, www.abc.com, www.fgh.com\}$

3. $meta = \{www.xyz.com, www.abc.com\}$

4. $meta = \{www.xyz.com\}$

5. $meta = \{www.xyz.com, www.lmn.com, www.abc.com\}$

6. $meta = \{www.xyz.com, www.lmn.com, www.abc.com, www.fgh.com\}$

□

Solution 14.24

1. 3

2. 1

3. 100

□

Solution 14.25

$$\forall\, u : URL;\ e : ENGINE \bullet$$
$$u \in \mathrm{ran}\ result\,(e) \Rightarrow position\_scores\,(u, e) =$$
$$(\mu\, n : \mathbb{N} \mid (result\ e)\, n = u \wedge$$
$$\forall\, m : \mathbb{N} \mid (result\ e)\, m = u \bullet m \geq n)$$
$$\wedge$$
$$u \notin \mathrm{ran}\ result\,(e) \Rightarrow position\_scores\,(u, e) = 100$$

□

Solution 14.26

$$scores = \{www.abc.com \mapsto 204, www.fgh.com \mapsto 104,$$
$$www.lmn.com \mapsto 13, www.rst.com \mapsto 108, www.xyz.com \mapsto 107\}$$

□

Solution 14.27

1. $\langle www.xyz.com, www.abc.com, www.fgh.com \rangle$

2. $\langle www.xyz.com, www.lmn.com, www.abc.com, www.fgh.com \rangle$ or
   $\langle www.xyz.com, www.abc.com, www.lmn.com, www.fgh.com \rangle$

3. $\langle www.xyz.com, www.abc.com \rangle$

4. $\langle www.xyz.com \rangle$

5. $\langle www.xyz.com, www.lmn.com, www.abc.com \rangle$ or
   $\langle www.xyz.com, www.abc.com, www.lmn.com \rangle$

6. ⟨*www.xyz.com, www.lmn.com, www.abc.com, www.fgh.com*⟩ or
   ⟨*www.xyz.com, www.abc.com, www.lmn.com, www.fgh.com*⟩

☐

Solution 14.28

1. ⟨*www.lmn.com, www.fgh.com, www.xyz.com*⟩

2. ⟨*www.lmn.com, www.fgh.com, www.xyz.com, www.rst.com, www.abc.com*⟩

3. ⟨*www.fgh.com, www.xyz.com, www.abc.com*⟩

4. ⟨*www.fgh.com*⟩

5. ⟨*www.lmn.com, www.fgh.com, www.abc.com*⟩

6. ⟨*www.lmn.com, www.fgh.com, www.xyz.com, www.rst.com, www.abc.com*⟩

☐

Solution 14.29

1. 12
   14

2. 8
   10
   12
   14

3. 10

4. false

☐

Solution 14.30

1. 12, 14

2. 10, 12, 14, 8

3. 10

4. false

☐

Solution 14.31 We consider the base case first. This can be established in the following way.

$$pop\,(push\,(\langle\rangle, x)) = pop\,(\langle x\rangle)$$
$$= \langle\rangle$$

We now consider the inductive step. The inductive hypothesis is given by

$$pop\,(push\,(s, x)) = s$$

We may use this to establish the inductive step. However, we can prove the inductive step without using the inductive hypothesis, as shown below.

$$pop\,(push\,((\langle y\rangle \frown s, x)) = pop\,((\langle x\rangle \frown \langle y\rangle \frown s)$$
$$= \langle y\rangle \frown s$$

□

Solution 14.32 As this is a theorem concerning sequences, which are a special kind of relation, we need to establish that

$$(a, b) \in pop\,(push\,(s, x)) \Leftrightarrow (a, b) \in s$$

We may accomplish this as follows.

$$(a, b) \in pop\,push\,(s, x)$$
$$\Leftrightarrow (a, b) \in tail\,push\,(s, x) \quad [\text{Definition of } pop]$$
$$\Leftrightarrow (a, b) \in tail\,\langle x\rangle \frown s \quad [\text{Definition of } push]$$
$$\Leftrightarrow (a, b) \in s \quad\quad\quad\quad [\text{Definition of } tail]$$

□

Solution 14.33

$$is\_empty\,(push\,(pop\,(s), top\,(s)))$$
$$\Leftrightarrow push\,(pop\,(s), top\,(s)) = \langle\rangle \quad [\text{Definition of } is\_empty]$$
$$\Leftrightarrow push\,(tail\,s, top\,(s)) = \langle\rangle \quad [\text{Definition of } pop]$$
$$\Leftrightarrow push\,(tail\,s, head\,s) = \langle\rangle \quad [\text{Definition of } top]$$
$$\Leftrightarrow \langle head\,s\rangle \frown tail\,s = \langle\rangle \quad [\text{Definition of } push]$$
$$\Leftrightarrow s = \langle\rangle \quad\quad\quad\quad\quad [\text{Law } 10.5]$$
$$\Leftrightarrow is\_empty\,(s) \quad\quad\quad [\text{Definition of } is\_empty]$$

□

Solution 14.34 No. Consider, as a counter-example $q = \langle 1, 2\rangle$ and $x = 3$. Here,

$$dequeue\,(enqueue\,(q, x)) = \langle 2, 3\rangle$$

which is not the same as q. □

Solution 14.35

$$(a, b) \in dequeue\,(enqueue\,(\langle\rangle, x))$$
$$\Leftrightarrow (a, b) \in dequeue\,\langle x\rangle \quad [\text{Definition of } enqueue]$$
$$\Leftrightarrow (a, b) \in \langle\rangle \quad\quad\quad [\text{Definition of } dequeue]$$

□

Solution 14.36

$$is\_empty\,(q) \Rightarrow \# \,enqueue\,(enqueue\,(q, x), y) = 2$$
$$\Leftrightarrow \neg \,is\_empty\,(q) \vee \# \,enqueue\,(enqueue\,(q, x), y) = 2 \quad [\text{Law } 3.18]$$
$$\Leftrightarrow \neg \,is\_empty\,(q) \vee \# \,enqueue\,(q \frown \langle x \rangle, y) = 2 \qquad [\text{Defn of } enqueue]$$
$$\Leftrightarrow \neg \,is\_empty\,(q) \vee \# \,q \frown \langle x \rangle \frown \langle y \rangle = 2 \qquad [\text{Defn of } enqueue]$$
$$\Leftrightarrow \neg \,is\_empty\,(q) \vee \# \,q = 0 \qquad\qquad\qquad [\text{Defn of } \#]$$
$$\Leftrightarrow \neg \,is\_empty\,(q) \vee q = \langle \rangle \qquad\qquad\qquad [\text{Defn of } \#]$$
$$\Leftrightarrow \neg \,is\_empty\,(q) \vee is\_empty\,(q) \qquad\qquad [\text{Defn of } is\_empty]$$
$$\Leftrightarrow true \qquad\qquad\qquad\qquad\qquad\qquad [\text{Law } 3.15]$$

□

Solution 14.37 The values 0 and 1, together with NOT, AND and OR form a Boolean algebra, as each of the following is true.

$$x \; OR \; y = y \; OR \; x$$
$$x \; AND \; y = y \; AND \; x$$
$$x \; OR \; (y \; AND \; z) = (x \; OR \; y) \; AND \; (x \; OR \; z)$$
$$x \; AND \; (y \; OR \; z) = (x \; AND \; y) \; OR \; (x \; AND \; z)$$
$$x \; OR \; 0 = x$$
$$x \; AND \; 1 = x$$
$$x \; OR \; (NOT \; x) = 1$$
$$x \; AND \; (NOT \; x) = 0$$

□

Solution 14.38

1. $v = 0$
2. $v = 1$
3. $v = 0$

□

Solution 14.39 The proposition $x \Rightarrow y$ can be represented as

$$(NOT \; x) \; OR \; y$$

while the proposition $x \Leftrightarrow y$ can be represented as

$$((NOT \; x) \; OR \; y) \; AND \; ((NOT \; y) \; OR \; x)$$

□

Solution 14.40 First,

$$((x \; AND \; 0) \; OR \; (x \; AND \; 1)) \; OR \; ((NOT \; x) \; AND \; (NOT \; y))$$

is equivalent to

$$(0 \; OR \; (x \; AND \; 1)) \; OR \; ((NOT \; x) \; AND \; (NOT \; y))$$

which, in turn, is equivalent to

$$(0 \; OR \; x) \; OR \; ((NOT \; x) \; AND \; (NOT \; y))$$

In turn, this is equivalent to

$$x \; OR \; ((NOT \; x) \; AND \; (NOT \; y))$$

Via distribution, this is equivalent to

$$(x \; OR \; (NOT \; x)) \; AND \; (x \; OR \; (NOT \; y))$$

As $x \; OR \; NOT \; x$ is equivalent to 1, this circuit is equivalent to

$$x \; OR \; (NOT \; y)$$

□

Solution 14.41

1. $(x \; AND \; y) \; OR \; (x \; AND \; (NOT \; y)) = x \; AND \; (y \; OR \; (NOT \; y))$
$$= x \; AND \; 1$$
$$= x$$

2. $x \; OR \; (((NOT \; x) \; AND \; y) \; OR \; ((NOT \; z) \; AND \; y))$
$$= x \; OR \; (y \; AND \; ((NOT \; x) \; OR \; (NOT \; z)))$$
$$= (x \; OR \; y) \; AND \; (x \; OR \; ((NOT \; x) \; OR \; (NOT \; z)))$$
$$= (x \; OR \; y) \; AND \; ((x \; OR \; (NOT \; x)) \; OR \; (NOT \; z))$$
$$= (x \; OR \; y) \; AND \; (1 \; OR \; (NOT \; z))$$
$$= (x \; OR \; y) \; AND \; 1$$
$$= (x \; OR \; y)$$

□

Solution 14.42 Not necessarily. Consider the case in which History and Geography are taught by the same subset of *Teacher*. In these circumstances, *teacher* is not injective. □

Solution 14.43

1. $\{t : Teacher \mid t \in teacher \; geography\}$
2. $\{t : Teacher \mid (\exists \, s_1, s_2 : Subject \bullet s_1 \neq s_2 \wedge t \in teacher \; s_1 \wedge t \in teacher \; s_2)\}$

□

Solution 14.44

1.

$$
\begin{array}{|l}
chosen\_subjects : Student \nrightarrow \mathbb{P}\ Subject \\
\hline
\forall\, s : \mathrm{dom}\ chosen\_subjects \bullet \\
\quad \#\,(chosen\_subjects\ s) \geq 2 \wedge \#\,(chosen\_subjects\ s) \leq 5
\end{array}
$$

2.

$$
\begin{array}{|l}
chosen\_subjects : Student \nrightarrow \mathbb{P}\ Subject \\
\hline
\forall\, s : \mathrm{dom}\ chosen\_subjects \bullet \\
\quad humanities \in chosen\_subjects\ s \Leftrightarrow \\
\qquad \{history, geography\} \cap chosen\_subjects\ s = \emptyset
\end{array}
$$

3.

$$
\begin{array}{|l}
chosen\_subjects : Student \nrightarrow \mathbb{P}\ Subject \\
\hline
\forall\, s : \mathrm{dom}\ chosen\_subjects \bullet \\
\quad \#\,(chosen\_subjects\ s) \geq 4 \Rightarrow mathematics \in chosen\_subjects\ s
\end{array}
$$

4.

$$
\begin{array}{|l}
chosen\_subjects : Student \nrightarrow \mathbb{P}\ Subject \\
\hline
\forall\, s : \mathrm{dom}\ chosen\_subjects \bullet \\
\quad \#\,(chosen\_subjects\ s) \geq 2 \wedge \#\,(chosen\_subjects\ s) \leq 5 \\
\wedge \\
\forall\, s : \mathrm{dom}\ chosen\_subjects \bullet \\
\quad humanities \in chosen\_subjects\ s \Leftrightarrow \\
\qquad \{history, geography\} \cap chosen\_subjects\ s = \emptyset \\
\wedge \\
\forall\, s : \mathrm{dom}\ chosen\_subjects \bullet \\
\quad \#\,(chosen\_subjects\ s) \geq 4 \Rightarrow mathematics \in chosen\_subjects\ s
\end{array}
$$

□

Solution 14.45

$$
taught\_by = \{\, s : Student \bullet s \mapsto \bigcup \mathrm{ran}\,(\{chosen\_subjects\ s\} \lhd teacher)\,\}
$$

□

Solution 14.46

1. $\{\, s : Student \mid mathematics \in chosen\_subjects\ s \,\}$
2. $\{\, s : Student \mid \#\,(chosen\_subjects\ s) = 3 \,\}$

3. $\{s : Subject \mid (\exists\, x : Student \bullet \#\, (chosen\_subjects\ x) = 3 \wedge s \in chosen\_subjects\ x)\}$

$\square$

Solution 14.47

1.

$$
\begin{array}{|l}
tutees : Teacher \nrightarrow \mathbb{P}\ Student \\
\hline
\forall\, s : \mathrm{ran}\ tutees \bullet \#\, s \geq 15 \wedge \#\, s \leq 30
\end{array}
$$

2.

$$
\begin{array}{|l}
tutees : Teacher \nrightarrow \mathbb{P}\ Student \\
\hline
\forall\, s, t : \mathrm{dom}\ tutees \mid s \neq t \bullet tutees\ s \cap tutees\ t = \emptyset
\end{array}
$$

3.

$$
\begin{array}{|l}
tutees : Teacher \nrightarrow \mathbb{P}\ Student \\
\hline
\forall\, s : \mathrm{ran}\ tutees \bullet \#\, s \geq 15 \wedge \#\, s \leq 30 \\
\wedge \\
\forall\, s, t : \mathrm{dom}\ tutees \mid s \neq t \bullet tutees\ s \cap tutees\ t = \emptyset
\end{array}
$$

$\square$

Solution 14.48

$$
tutor\_teaches = \{s : Student;\ x : Subject;\ t : Teacher \mid \\
s \in tutees\ t \wedge t \in teacher\ x \bullet (s, x)\}
$$

$\square$

Solution 14.49

1. $\{t : Teacher \mid \#\, (tutees\ t) = 30\}$
2. $\{t : Teacher \mid (\exists\, s : Student \bullet s \in tutees\ t \wedge \#\, (chosen\_subjects\ s) = 5)\}$
3. $\{s : Student;\ t : Teacher \mid$
   $(\exists\, x : Subject \bullet x \in chosen\_subjects\ s \wedge t \in teacher\ x \wedge s \in tutees\ t)\}$

$\square$

Solution 14.50

1.

$$
\begin{array}{|l}
marks : Student \times Subject \nrightarrow \mathbb{N} \\
\hline
\forall\, n : \mathrm{ran}\ marks \bullet n \leq 100
\end{array}
$$

2.

$$\begin{array}{|l}
marks : Student \times Subject \nrightarrow \mathbb{N} \\
\hline
\forall s : Student \bullet \# \{x : Subject \mid (s, x) \in \mathrm{dom} \ marks\} \leq 5
\end{array}$$

3.

$$\begin{array}{|l}
marks : Student \times Subject \nrightarrow \mathbb{N} \\
\hline
\forall n : \mathrm{ran} \ marks \bullet n \leq 100 \\
\wedge \\
\forall s : Student \bullet \# \{x : Subject \mid (s, x) \in \mathrm{dom} \ marks\} \leq 5
\end{array}$$

□

Solution 14.51

$$\forall s : Student; \ t : Subject \bullet t \in chosen\_subjects \ s \Leftrightarrow (s, t) \in \mathrm{dom} \ marks$$

□

Solution 14.52

$$best\_subject = \{s : Student; \ x : Subject \mid$$
$$(s, x) \in \mathrm{dom} \ marks \wedge$$
$$(\forall y : Subject \mid (s, y) \in \mathrm{dom} \ marks \bullet$$
$$marks \ (s, x) \geq marks \ (s, y))\}$$

□

Solution 14.53

1. $\{s : Student \mid (\exists x : Subject \bullet marks \ (s, x) \geq 50)\}$

2. $\{s : Student \mid marks \ (s, history) \geq 50 \wedge marks \ (s, geography) \geq 50\}$

3. $\{s : Student \mid$
   $(\exists x : Subject; \ t : Teacher \bullet$
   $s \in tutees \ t \wedge t \in teacher \ x \wedge marks \ (s, x) \geq 50)\}$

4. $\{s : Subject \mid$
   $(\forall x : Student \bullet (x, s) \in \mathrm{dom} \ marks \Rightarrow marks \ (x, s) \geq 50)\}$

□

Solution 14.54

1.

> $class : Subject \times Teacher \times Day \times Period \nrightarrow \mathbb{P}\ Student$
> ___
> $\forall\, s : Student;\ d : Day;\ p : Period\ \bullet$
> $\exists_1\, x : Subject;\ t : Teacher\ \bullet\ s \in class\,(x, t, d, p)$
> $\vee$
> $\neg\ \exists\, x : Subject;\ t : Teacher\ \bullet\ s \in class\,(x, t, d, p)$

2.

> $class : Subject \times Teacher \times Day \times Period \nrightarrow \mathbb{P}\ Student$
> ___
> $\forall\, t : Teacher;\ d : Day;\ p : Period\ \bullet$
> $\exists_1\, s : Subject\ \bullet\ (s, t, d, p) \in \mathrm{dom}\ class$
> $\vee$
> $\neg\ \exists\, s : Subject\ \bullet\ (s, t, d, p) \in \mathrm{dom}\ class$

3.

> $class : Subject \times Teacher \times Day \times Period \nrightarrow \mathbb{P}\ Student$
> ___
> $(\forall\, s : Student;\ d : Day;\ p : Period\ \bullet$
> $\exists_1\, x : Subject;\ t : Teacher\ \bullet\ s \in class\,(x, t, d, p)$
> $\vee$
> $\neg\ \exists\, x : Subject;\ t : Teacher\ \bullet\ s \in class\,(x, t, d, p))$
> $\wedge$
> $(\forall\, t : Teacher;\ d : Day;\ p : Period\ \bullet$
> $\exists_1\, s : Subject\ \bullet\ (s, t, d, p) \in \mathrm{dom}\ class$
> $\vee$
> $\neg\ \exists\, s : Subject\ \bullet\ (s, t, d, p) \in \mathrm{dom}\ class)$

$\square$

Solution 14.55

1. $\forall\, s : Student;\ x : Subject\ \bullet$
   $x \in chosen\_subjects\ s \Leftrightarrow$
      $\exists\, d : Day;\ t : Teacher;\ p : Period\ \bullet\ s \in class\,(x, t, d, p)$

2. $\forall\, t : Teacher;\ x : Subject\ \bullet$
   $t \in teacher\ x \Leftrightarrow \exists\, d : Day;\ p : Period\ \bullet\ (x, t, d, p) \in \mathrm{dom}\ class$

3. $\forall\, s : Student;\ x : Subject\ \bullet$
   $(s, x) \in \mathrm{dom}\ marks \Leftrightarrow$
      $\exists\, t : Teacher;\ d : Day;\ p : Period\ \bullet\ s \in class\,(x, t, d, p)$

$\square$

Solution 14.56

$$taught\_tutees =$$
$$\{t : Teacher \bullet t \mapsto (tutees\ t$$
$$\cap$$
$$\{s : Student \mid (\exists d : Day;\ p : Period;\ x : Subject \bullet$$
$$s \in class\ (x, d, t, p))\})\}$$

□

Solution 14.57

1. $\{s : Subject \mid (\exists t : Teacher \bullet (s, t, monday, 1) \in \mathrm{dom}\ class)\}$
2. $\{s : Student \mid (\exists t : Teacher \bullet s \in class\ (history, t, monday, 1))\}$
3. $\{d : Day;\ p : Period \mid (\exists t : Teacher \bullet (history, t, d, p) \in \mathrm{dom}\ class)\}$

□

Solution 14.58

1.

> $classroom : Subject \times Teacher \times Period \times Day \nrightarrow Classroom$
>
> $\forall x, y : \mathrm{dom}\ classroom \bullet$
> $x.2 = y.2 \wedge x.3 = y.3 \wedge x.4 = y.4 \Rightarrow$
> $x = y \wedge classroom\ x = classroom\ y$

2.

> $classroom : Subject \times Teacher \times Period \times Day \nrightarrow Classroom$
>
> $\forall x, y : \mathrm{dom}\ classroom \bullet$
> $x.3 = y.3 \wedge x.4 = y.4 \wedge classroom\ x = classroom\ y \Rightarrow$
> $x.2 = y.2$

3.

> $classroom : Subject \times Teacher \times Period \times Day \nrightarrow Classroom$
>
> $\forall x, y : \mathrm{dom}\ classroom \bullet$
> $x.2 = y.2 \wedge x.3 = y.3 \wedge x.4 = y.4 \Rightarrow$
> $x = y \wedge classroom\ x = classroom\ y$
> $\wedge$
> $\forall x, y : \mathrm{dom}\ classroom \bullet$
> $x.3 = y.3 \wedge x.4 = y.4 \wedge classroom\ x = classroom\ y \Rightarrow$
> $x.2 = y.2$

□

Solution 14.59

1. dom *classroom* = dom *class*

2. $\forall x : Subject \times Teacher \times Period \times Day \bullet$
   $\quad x \in \text{dom } classroom \Leftrightarrow x.2 \in teacher\ x.1$

□

Solution 14.60

1. $location = \{x : \text{dom } classroom \bullet (x.2, x.3, x.4) \mapsto classroom\ x\}$

2. $teaching\_periods = \{t : Teacher \bullet t \mapsto \# \{x : \text{dom } classroom \mid x.2 = t\}\}$

□

Solution 14.61

1. $\{x : \text{dom } classroom \mid x.3 = 1 \wedge x.4 = monday \bullet classroom\ x\}$

2. $\{x : \text{dom } classroom \mid x.1 = history \bullet classroom\ x\}$

3. $\{d : Day;\ p : Period \mid$
   $\quad (\forall c : Classroom \bullet \exists s : Subject;\ t : Teacher \bullet$
   $\quad\quad classroom\ (s, t, p, d) = c)\}$

□

Solution 14.62

1.

$$
\begin{array}{|l}
\hline
timetable : Student \nrightarrow \text{seq}\,(Period \times Day\times \\
\qquad\qquad\qquad\qquad\qquad Subject \times Teacher \times Classroom) \\
\hline
\forall s : Student \bullet \forall x, y : \text{ran } timetable\ s \bullet x.1 = y.1 \wedge x.2 = y.2 \Rightarrow x = y
\end{array}
$$

2.

$$
\begin{array}{|l}
\hline
timetable : Student \nrightarrow \text{seq}\,(Period \times Day\times \\
\qquad\qquad\qquad\qquad\qquad Subject \times Teacher \times Classroom) \\
\hline
\forall s : Student \bullet \forall x, y : \text{dom } timetable\ s \bullet \\
\quad x < y \Leftrightarrow \\
\qquad (((timetable\ s\ x).2, (timetable\ s\ y).2) \in before) \\
\qquad \vee \\
\qquad ((timetable\ s\ x).2 = (timetable\ s\ y).2 \wedge \\
\qquad\quad (timetable\ s\ x).1 < (timetable\ s\ y).1)
\end{array}
$$

3.

$$
\begin{array}{|l}
\hline
timetable : Student \nrightarrow \text{seq} \, (Period \times Day \times \\
\qquad\qquad\qquad\qquad\qquad Subject \times Teacher \times Classroom) \\
\hline
\forall \, s : Student \bullet \# \, \{x : \text{ran } timetable \, s \bullet x.3\} \leq 5 \\
\end{array}
$$

□

Solution 14.63

1. $\forall \, s : Student; \; x : Period \times Day \times Subject \times Teacher \times Classroom \bullet$
   $x \in \text{ran } timetable \, s \Leftrightarrow classroom \, (x.3, x.4, x.1, x.2) = x.5$

2. $\forall \, s : Student; \; x : Period \times Day \times Subject \times Teacher \times Classroom \bullet$
   $x \in \text{ran } timetable \, s \Leftrightarrow s \in class \, (x.3, x.4, x.2, x.1)$

3. $\forall \, s : Student; \; x : Period \times Day \times Subject \times Teacher \times Classroom \bullet$
   $x \in \text{ran } timetable \, s \Leftrightarrow (s, x.3) \in \text{dom } marks$

4. $\forall \, s : Student; \; x : Period \times Day \times Subject \times Teacher \times Classroom \bullet$
   $x \in \text{ran } timetable \, s \Leftrightarrow x.3 \in chosen\_subjects \, s$

□

Solution 14.64

1. $student\_location = \{s : Student; \; d : Day; \; p : Period; \; c : Classroom \; |$
   $\qquad\qquad\qquad \exists \, y : \text{ran } timetable \, s \bullet y.1 = p \land y.2 = d \land y.5 = c \bullet$
   $\qquad\qquad (s, d, p) \mapsto c\}$

2. $teachers = \{s : Student \bullet s \mapsto \{x : \text{ran } timetable \, s \bullet x.4\}\}$

3. $taught\_periods = \{s : Student \bullet s \mapsto \# \, (\text{ran } timetable \, s)\}$

□

Solution 14.65

1. $\{s : Student \; | \; \# \, (timetable \, s) = 40\}$

2. $\{s : Student \; | \; (\exists \, x : \text{ran } timetable \, s \bullet x.1 = 1 \land x.2 = monday)\}$

3. $\{s : Student \; |$
   $\qquad (\exists \, x : \text{ran } timetable \, s \bullet x.1 = 1 \land x.2 = monday \land x.3 = history)\}$

□

Solution 14.66 The first query returns the result *false*, while the second query returns the result *true*. □

Solution 14.67 The first query returns the result *true*. The second query returns *false*, as there is no knowledge in the system indicating that Barcelona is in Spain. The third query returns *false* for the same reason. □

Solution 14.68

1. $lives\_in\,(x, y) \wedge capital\,(z, y) \Rightarrow lives\_in\_capital\,(x)$

2. $lives\_in\,(x, y) \wedge origin\,(x, z) \wedge in\,(y, z) \Rightarrow lives\_in\_home\_country\,(x)$

3. $lives\_in\,(x, y) \wedge origin\,(x, z) \wedge \neg\,in\,(y, z) \Rightarrow lives\_in\_foreign\_country\,(x)$

4. $lives\_in\,(x, y) \wedge in\,(y, z) \wedge currency\,(z, w) \Rightarrow spends\,(x, w)$

$\square$

Solution 14.69 Query 1 returns the answer *false*, as one would expect. Query 2 returns the result *false* due to the fact that, because of our definition of *in*, the system cannot establish that Venice is in Italy. Query 3 returns the result *true* due to our definition of *in* (the system can't determine that Venice is in Italy, and, as such, concludes that it is not in Italy). Finally, query 4 returns the result *false*, again, due to our definition of *in*. $\square$

Solution 14.70 We need to add the following fact.

$$in\,(\,Venice, Italy\,)$$

It would, of course, also be sensible to add

$$in\,(\,Brisbane, Australia\,)$$

and

$$in\,(\,Baltimore, USA\,)$$

to the knowledge base. $\square$

Solution 14.71

1. $x = peseta$

2. $x = Anna$

3. $x = Pierre, y = franc$
   $x = Igor, y = lira$
   $x = Anna, y = peseta$
   $x = Andrew, y = Aus\ dollar$
   $x = Cyril, y = US\ dollar$

4. $x = Pierre$

5. $x = Pierre, y = Pierre$
   $x = Pierre, y = Igor$
   $x = Anna, y = Pierre$
   $x = Anna, y = Igor$

$\square$

Bibliography

[Bal97]   V. K. Balakrishnan. *Schaum's Outline of Graph Theory, Including Hundreds of Solved Problems*. McGraw-Hill, 1997.

[BMN97]   M. Bergmann, J. Moor, and J. Nelson. *The Logic Book with Student Solutions Manual*. McGraw-Hill, 1997.

[Bol86]   B. Bollobas. *Combinatorics*. Cambridge University Press, 1986.

[Bru99]   R. A. Brualdi. *Introductory Combinatorics*. Prentice-Hall, third edition, 1999.

[CLP00]   R. Cori, D. Lascar, and D. Pelletier. *Mathematical Logic: A Course with Exercises Part 1: Propositional Calculus, Boolean Algebras, Predicate Calculus, Completeness Theorems*. Oxford University Press, 2000.

[CLP01]   R. Cori, D. Lascar, and D. Pelletier. *Mathematical Logic: A Course with Exercises: Part 2: Recursion Theory, Gödel's Theorem, Set Theory and Model Theory*. Oxford University Press, 2001.

[Dau90]   J. W. Dauben. *Georg Cantor*. Princeton University Press, 1990.

[DG00]   I. Duntsch and G. Gediga. *Sets, Relations, Functions*. Methodos Publishers, 2000.

[Dun99]   W. Dunham, editor. *Euler: The Master of Us All*. The Mathematical Association of America, 1999.

[End77]   H. B. Enderton. *Elements of Set Theory*. Academic Press, 1977.

[Eri96]   M. J. Erickson. *Introduction to Combinatorics*. John Wiley & Sons, 1996.

[Gar92]   A. R. Garciadiego. *Bertrand Russell and the Origins of the Set-theoretic 'Paradoxes'*. Birkhauser Verlag AG, 1992.

[GGB97]   I. Grattan-Guinness and G. Bornet. *George Boole: Selected Manuscripts on Logic and its Philosophy*. Birkhauser Verlag AG, 1997.

[Gre98]   J. Gregg. *Ones and Zeros: Understanding Boolean Algebra, Digital Circuits, and the Logic of Sets*. Institute of Electrical & Electronic Engineers, 1998.

[Gri99]   R. P. Grimaldi. *Discrete and Combinatorial Mathematics: An Applied Introduction*. Addison-Wesley, fourth edition, 1999.

[Hay93]   I. J. Hayes. *Specification Case Studies*. Prentice-Hall, second edition, 1993.

[Hen86] J. M. Henle. *An Outline of Set Theory.* Springer-Verlag, 1986.

[Hoa85] C. A. R. Hoare. *Communicating Sequential Processes.* Prentice-Hall, 1985.

[Hun84] G. M. K. Hunt. *Tactics and Strategies in Natural Deduction.* Department of Philosophy, University of Warwick, 1984.

[Hun96] E. Hunt. *Venn Diagrams.* Hunt Publications, 1996.

[Jen00] B. Jensen. *Directed Graphs with Applications.* John Wiley & Sons, 2000.

[Ken80] H. C. Kennedy. *Peano: Life and Works of Giuseppe Peano.* Kluwer Academic Publishers, 1980.

[LL92] S. Lipschutz and M. L. Larson. *2000 Solved Problems in Discrete Mathematics.* McGraw-Hill, 1992.

[Men70] E. Mendelson. *Schaum's Outline of Theory and Problems of Boolean Algebra and Switching Circuits.* Schaum, 1970.

[Men87] E. Mendelson. *Introduction to Mathematical Logic.* Wadsworth and Brookes/Cole, third edition, 1987.

[Mer90] D. D. Merrill. *Augustus De Morgan and the Logic of Relations.* Kluwer Academic Publishers, 1990.

[Mer00] J. Merris. *Graph Theory.* John Wiley & Sons, 2000.

[Mon89] J. D. Monk. *Handbook of Boolean Algebras: Volume 1.* North-Holland, 1989.

[Mon97] R. Monk. *Bertrand Russell: The Spirit of Solitude 1872 - 1921.* Free Press, 1997.

[Ros97] A. W. Roscoe. *The Theory and Practice of Concurrency.* Prentice-Hall, 1997.

[Sol90] D. Solow. *How to Read and Do Proofs.* Wiley, second edition, 1990.

[Spi92] J. M. Spivey. *The Z Notation: A Reference Manual.* Prentice-Hall, second edition, 1992.

[Sup60] P. C. Suppes. *Axiomatic Set Theory.* Van Nostrand, 1960.

[Wan80] M. Wand. *Induction, Recursion, and Programming.* Elsevier North Holland, 1980.

[WD96] J. C. P. Woodcock and J. W. Davies. *Using Z: Specification, refinement, and proof.* Prentice-Hall, 1996.

[Wes95] D. B. West. *Introduction to Graph Theory.* Prentice-Hall, 1995.

[Wil96] R. J. Wilson. *Introduction to Graph Theory.* Longman Higher Education, 1996.

[You64] B. K. Youse. *Mathematical Induction.* Prentice-Hall, 1964.

Index

abbreviations (set theory), 130
adjacency matrices, 327
adjacent vertices, 318
AND gates, 410
antecedent, 29
antisymmetry, 217
associativity (Boolean algebra), 105
associativity (concatenation), 280
associativity (conjunction), 24
associativity (disjunction), 26
associativity (logical equivalence), 31
associativity (set intersection), 75
associativity (set union), 72
assumptions, 38
asymmetry, 216
atomic propositions, 17
axiomatic definitions, 137
axiomatic set theory, 122

bag brackets ($[\![\,]\!]$), 274
bags, 273
base 2 logarithms, 8
bijections, 258
bijective functions, 258
binary, 410
binary relations, 195
binary trees, 346
binding occurrences, 162
binomial coefficients, 367
Boolean algebra, 101, 410
Boolean algebra of binary arithmetic, 104
Boolean algebra of propositions, 108

Boolean algebra of sets, 107
Boolean algebra, definition of, 103
Boolean algebra, laws of, 104
bound variables, 161

Cantor's paradox, 122
cardinality, 81
cardinality symbol ($\#$), 81
Cartesian product symbol ($\times$), 131
Cartesian products, 131
characteristic tuples, 129
circuits, 333
closures, 220
combinations, 375
commutativity (Boolean algebra), 104
commutativity (conjunction), 23
commutativity (disjunction), 26
commutativity (equivalence), 31
commutativity (set intersection), 74
commutativity (set union), 72
complement (Boolean algebra), 104, 105
complement of a graph, 320
complete graphs, 324
component selection, 136
concatenation, 279
concatenation symbol ($\frown$), 279
conclusion, 37
conjunction, 21
conjunction elimination, 40
conjunction introduction, 40
conjunction symbol ($\wedge$), 21
connected graphs, 323, 334

consequent, 29
contingencies, 34
contradictions, 33
counting, 371
counting principle, 372
cycles, 337

de Morgan's laws (Boolean algebra), 105
de Morgan's laws (propositional logic), 25
deduction, 37
degree, 321
digital circuits, 410
digraphs, 344
directed graphs, 344
discharge of assumptions, 38
disjunction, 24
disjunction elimination, 41
disjunction introduction, 41
disjunction symbol ($\vee$), 24
distributivity (Boolean algebra), 104
distributivity (propositional logic), 27
distributivity (set theory), 75
domain, 197
domain co-restriction, 202
domain co-restriction symbol ($\vartriangleright$), 202
domain restriction, 202
domain restriction symbol ($\vartriangleleft$), 202
domain symbol (dom), 198
duality, 109

edges, 317
elements, 61
empty sequence, 277
empty sequence symbol ($\langle\rangle$), 277
empty set, 64
empty set symbol ($\emptyset$), 64
equal sets, 63
equational reasoning (predicate), 173
equational reasoning (propositional), 35
equational reasoning (relations), 197
equational reasoning(set theory), 80
equivalence elimination, 42
equivalence introduction, 42
equivalence relations, 220
equivalence symbol ($\Leftrightarrow$), 30
Eulerian circuits, 334

Eulerian graphs, 335
Eulerian paths, 334
existential quantification, 156
existential quantification elimination, 178
existential quantification introduction, 178
existential quantifier ($\exists$), 156

factorial, 365
false, 17
false introduction, 43
finite sets, 82
forests, 339
free variables, 161
function application, 251
function override symbol ($\oplus$), 253
function overriding, 253
functions, 247
fundamental products, 112

generalised intersection symbol ($\bigcap$), 86
generalised set intersection, 86
generalised set operations, 85
generalised set union, 85
generalised union symbol ($\bigcup$), 85
graph theory, 317
graphs, 317

Hamiltonian graphs, 336
Hamiltonian paths, 336
handshaking lemma, 322
head, 281
head (of a sequence), 281
height of binary trees, 347
heterogeneous relations, 211
homogeneous relations, 211

idempotence (Boolean algebra), 105
idempotence (conjunction), 23
idempotence (disjunction), 26
idempotence (set intersection), 74
idempotence (set union), 72
identity (Boolean algebra), 104
identity relation (Id), 221
implication, 28
implication elimination, 38
implication introduction, 38
implication symbol ($\Rightarrow$), 28

in-order traversal, 348
incident (graph theory), 318
induction, 303
infinite sets, 82
injections, 255
injective functions, 255
injective sequence symbol (iseq), 288
injective sequences, 288
integer division (div), 5
integers, 11
isolated vertices, 322
isomorphic Boolean algebras, 108
isomorphic graphs, 331

Kruskal's algorithm, 343

law of the excluded middle, 27
leaf nodes, 346
length (sequences), 278
lines, 317
literals, 112
logarithms (*log*), 8
logic gate circuits, 410
logic gates, 410
logical equivalence, 30
loops, 319

maplet, 196
maps to symbol ($\mapsto$), 196
mathematical induction, 303
members, 61
minimal spanning trees, 341
modelling sequences, 276
modulus (mod), 5
modus ponens, 38
mu operator (μ), 171
mu-expressions, 171

n-ary relations, 225
natural deduction, 37
natural deduction (predicate logic), 176
natural numbers, 5
negation, 19
negation elimination, 43
negation introduction, 42
negation of quantifiers, 159
negation symbol ($\neg$), 19

nodes, 317
NOT gates, 410

one-point rule, 180
OR gates, 411
ordered pairs, 131

paradoxes, 122
parallel edges, 319
partial bijection symbol ($\rightarrowtail\!\!\!\rightarrow$), 259
partial function symbol ($\nrightarrow$), 248
partial injection symbol($\rightarrowtail$), 255
partial orderings, 219
partial surjection symbol ($\twoheadrightarrow$), 257
Pascal's triangle, 370
paths, 332
Peano arithmetic, 9, 303
permutations, 373
points, 317
post-order traversal, 348
potato diagrams, 197
power set symbol ($\mathbb{P}$), 83
power sets, 83
pre-order traversal, 347
precedence (Boolean algebra), 107
precedence (propositional logic), 32
predicate logic, 153
premise, 37
Prim's algorithm, 342
product ($\prod$), 6
product operator (Boolean algebra), 105
proof trees, 37
proper subset symbol ($\subset$), 67
proper subsets, 67
proper superset symbol ($\supset$), 70
proper supersets, 70
properties of relations, 211
propositions, 17

quantification, 153
queues, 406

range, 197
range co-restriction, 204
range co-restriction symbol ($\rhd\!\!\!-$), 204
range restriction, 203
range restriction symbol ($\rhd$), 203

range symbol (ran), 198

real numbers, 11

reasoning about sets, 80

recursively defined functions, 259, 289

reflexive closure notation (R^R), 220

reflexive closures, 220

reflexive transitive closure notation (R^*), 223

reflexive transitive closures, 223

reflexivity, 211

relation symbol ($\leftrightarrow$), 196

relational composition, 206

relational composition symbol ($\stackrel{\circ}{\,}$), 207

relational image, 205

relational image brackets ($(\!|\ |\!)$), 205

relational inverse, 199

relational inverse symbol ($R^\sim$), 199

relational inverse symbol (R^{-1}), 199

relations, 195

representing graphs (matrices), 327

representing graphs (sets and bags), 325

restriction, 167

restriction (sequences), 283

reverse of a sequence (*reverse*), 286

reversing sequences, 286

root nodes, 346

rounding down brackets ($\lfloor\ \rfloor$), 9

rounding up brackets ($\lceil\ \rceil$), 9

Russell's paradox, 122

sample space, 379

sampling, 378

sampling with replacement, 379

sampling without replacement, 379

satisfaction, 159

scope, 162

semi-Eulerian graphs, 335

sequence brackets ($\langle\ \rangle$), 275

sequence filter symbol ($\upharpoonright$), 283

sequence symbol (seq), 276

sequences, 273

sequential substitutions, 165

set brackets ($\{\ \}$), 61

set comprehension, 125

set difference, 76

set difference symbol ($\backslash$), 76

set intersection, 73

set intersection symbol ($\cap$), 73

set membership, 64

set membership symbol ($\in$), 64

set of integers ($\mathbb{Z}$), 11

set of natural numbers ($\mathbb{N}$), 5

set of real numbers ($\mathbb{R}$), 11

set theory, 61

set union, 70

set union symbol ($\cup$), 70

sets, 61

simple graphs, 319

simple paths, 333

simultaneous substitutions, 165

singleton sets, 63

source, 197

spanning trees, 341

stacks, 406

structural induction, 307

subset symbol ($\subseteq$), 66

subsets, 65

substitution, 163

subtrees, 346

sum ($\sum$), 6

sum operator (Boolean algebra), 105

superset symbol ($\supseteq$), 70

supersets, 70

surjections, 257

surjective functions, 257

symmetric closure notation (R^S), 224

symmetric closures, 224

symmetry, 214

tail, 281

tail (of a sequence), 281

target, 197

tautology, 33

theorem, 37

three-valued logic, 18

total bijection symbol ($\rightarrowtail\!\!\!\rightarrow$), 259

total function symbol ($\rightarrow$), 249

total functions, 249

total injection symbol ($\rightarrowtail$), 255

total orderings, 220

total surjection symbol ($\twoheadrightarrow$), 257

totality, 218

transitive closure notation (R^+), 222

transitive closures, 222
transitivity, 213
traversing binary trees, 347
tree diagrams, 377
trees, 338
true, 17
true introduction, 40
truth tables, 20, 34
truth values, 18
tuples, 135
typed set theory, 121

undefined (truth value), 18
unique existential quantifier ($\exists_1$), 170
uniqueness, 170
unit element, 104
universal quantification, 153
universal quantification elimination, 176
universal quantification introduction, 177
universal quantifier ($\forall$), 153
unsatisfiability, 159

validity, 159
variable capture, 166
Venn diagrams, 65
vertices, 317

weighted graphs, 340
well-defined recursive functions, 261

zero element, 104